Praise for
The Age of Spiritual Machines

The Age of Spiritual Machines "ranges widely over such juicy topics as entropy, chaos, the big bang, quantum theory, DNA computers, quantum computers, Godel's theorem, neural nets, genetic algorithms, nanoengineering, the Turing test, brain scanning, the slowness of neurons, chess playing programs, the Internet— the whole world of information technology past, present, and future. This is a book for anyone who wonders where human technology is going next."
—*The New York Times Book Review*

"A mind-expanding account of the rise of intelligent machines. . . . . Nothing less than a blueprint for how to shove Homo sapiens off centre-stage in evolution's endless play. . . . If you buy into [Kurzweil's Law of Accelerating Returns]—and all empirical evidence currently available supports it completely—then the replacement of humans by machines as the primary intellectual force on Earth is indeed imminent."
—John Casti, *Nature*

"A welcome challenge to beliefs we hold dear . . . Kurzweil paints a tantalizing— and sometimes terrifying—portrait of a world where the line between humans and machines has become thoroughly blurred."
—Chet Raymo, *The Boston Globe*

"Brilliant . . . Kurzweil clearly takes his place as a leading futurist of our time. He links the relentless growth of our future technology to a universe in which Artificial Intelligence and Nanotechnology combine to bring unimaginable wealth and longevity, not merely to our descendants, but to some of those living today."
—Marvin Minsky, Professor of Media Arts and Sciences, MIT

"*The Age of Spiritual Machines* makes all other roads to the computer future look like goat paths in Patagonia."
—George Gilder, author of *Wealth and Poverty* and *Life After Television*

"A compelling vision of the future from one of our nation's leading innovators. Kurzweil brings serious science and a twinkling sense of humor to the question of where we are headed . . . With his pioneering inventions, and his penetrating ideas, Kurzweil convincingly takes us through what promises to be the most pivotal of centuries."
—Mike Brown, Chairman of the Nasdaq Stock Market

"An extremely provocative glimpse into what the next few decades may well hold . . . Kurzweil's broad outlook and fresh approach make his optimism hard to resist."
—*Kirkus Reviews*

Ray Kurzweil's inventions include reading machines for the blind, music synthesizers used by Stevie Wonder and many others, and marketing leading speech-recognition technology. He is the author of *The Age of Intelligent Machines*, which won the Association of American Publishers' Award for the Most Outstanding Computer Science Book of 1990, and *The 10% Solution for a Healthy Life*. He was awarded the Dickson Prize, Carnegie Mellon's top science prize, in 1994. The Massachusetts Institute of Technology named him Inventor of the Year in 1988. He is also the recipient of nine honorary degrees and honors from two U.S. presidents. Kurzweil lives in a suburb of Boston.

THE AGE OF SPIRITUAL MACHINES

WHEN COMPUTERS EXCEED HUMAN INTELLIGENCE

RAY KURZWEIL

PENGUIN BOOKS

PENGUIN BOOKS
Published by the Penguin Group
Penguin Putnam Inc., 375 Hudson Street,
New York, New York 10014, U.S.A.
Penguin Books Ltd, 27 Wrights Lane,
London W8 5TZ, England
Penguin Books Australia Ltd, Ringwood,
Victoria, Australia
Penguin Books Canada Ltd, 10 Alcorn Avenue,
Toronto, Ontario, Canada M4V 3B2
Penguin Books (N.Z.) Ltd, 182–190 Wairau Road,
Auckland 10, New Zealand

Penguin Books Ltd, Registered Offices:
Harmondsworth, Middlesex, England

First published in the United States of America by Viking Penguin,
a member of Penguin Putnam Inc. 1999
Published in Penguin Books 2000

9   10   8

Illustrations credits
Pages 24, 26–27, 104, 156: Concept and text by Ray Kurzweil.
Illustration by Rose Russo and Robert Brun.
Page 72: © 1977 by Sidney Harris.
Pages 167–168: Paintings by Aaron, a computerized robot built and programmed by Harold Cohen.
Photographed by Becky Cohen.
Page 188: Roz Chast © 1998. From The Cartoon Bank. All rights reserved.
Page 194: Danny Shananhan © 1994. From The New Yorker Collection. All rights reserved.
Page 219: Peter Steiner © 1997. From The New Yorker Collection. All rights reserved.

THE LIBRARY OF CONGRESS HAS CATALOGED THE HARDCOVER EDITION AS FOLLOWS:
Kurzweil, Ray.
The age of spiritual machines/ Ray Kurzweil.
p.   cm.
Includes bibliographical references and index.
ISBN 0-670-88217-8 (hc.)
ISBN 0 14 02.8202 5 (pbk.)
1. Artificial intelligence.   2. Computers.   I. Title.
Q335.K88   1999
006.3—dc21   98–388804

Printed in the United States of America
Set in Berkeley Oldstyle

A NOTE TO THE READER

As a photon wends its way through an arrangement of glass panes and mirrors, its path remains ambiguous. It essentially takes every possible path available to it (apparently these photons have not read Robert Frost's poem "The Road Not Taken"). This ambiguity remains until observation by a conscious observer forces the particle to decide which path it had taken. Then the uncertainty is resolved—retroactively—and it is as if the selected path had been taken all along.

Like these quantum particles, you—the reader—have choices to make in your path through this book. You can read the chapters as I intended them to be read, in sequential order. Or, after reading the Prologue, you may decide that the future can't wait, and you wish to immediately jump to the chapters in Part III on the twenty-first century (the table of contents on the next pages offers a description of each chapter). You may then make your way back to the earlier chapters that describe the nature and origin of the trends and forces that will manifest themselves in this coming century. Or, perhaps, your course will remain ambiguous until the end. But when you come to the Epilogue, any remaining ambiguity will be resolved, and it will be as if you had always intended to read the book in the order that you selected.

CONTENTS

Before the next century is over, human beings will no longer be the most intelligent or capable type of entity on the planet. Actually, let me take that back. The truth of that last statement depends on how we define *human*.

PART ONE: PROBING THE PAST

For the past forty years, in accordance with Moore's Law, the power of transistor-based computing has been growing exponentially. But by the year 2020, transistor features will be just a few atoms thick, and Moore's Law will have run its course. What then? To answer this critical question, we need to understand the exponential nature of time.

Can an intelligence create another intelligence more intelligent than itself? Are we more intelligent than the evolutionary process that created us? In turn, will the intelligence that we are creating come to exceed that of its creator?

"I am lonely and bored, please keep me company." If your computer displayed this message on its screen, would that convince you that it is conscious and has feelings? Before you say no too quickly, we need to consider how such a plaintive message originated.

Intelligence rapidly creates satisfying, sometimes surprising plans that meet an array of constraints. Clearly, no simple formula can emulate this most powerful of phenomena. Actually, that's wrong. All that is needed to solve a surprisingly wide range of intelligent problems is exactly this: simple methods combined with heavy doses of computation, itself a simple process.

It is sensible to remember today's insights for tomorrow's challenges. It is not fruitful to rethink every problem that comes along. This is particularly true for humans, due to the extremely slow speed of our computing circuitry.

PART TWO: PREPARING THE PRESENT

Evolution has found a way around the computational limitations of neural circuitry. Cleverly, it has created organisms who in turn invented a computational technology a million times faster than carbon-based neurons. Ultimately, the computing conducted on extremely slow mammalian neural circuits will be ported to a far more versatile and speedier electronic (and photonic) equivalent.

A disembodied mind will quickly get depressed. So what kind of bodies will we provide for our twenty-first-century machines? Later on, the question will become: What sort of bodies will they provide for themselves?

If all the computers in 1960 stopped functioning, few people would have noticed. Circa 1999 is another matter. Although computers still lack a sense of humor, a gift for small talk, and other endearing qualities of human thought, they are nonetheless mastering an increasingly diverse array of tasks that previously required human intelligence.

PART THREE: TO FACE THE FUTURE

CHAPTER NINE: 2009

It is now 2009. A $1,000 personal computer can perform about a trillion calculations per second. Computers are imbedded in clothing and jewelry. Most routine business transactions take place between a human and a virtual personality. Translating telephones are commonly used. Human musicians routinely jam with cybernetic musicians. The neo-Luddite movement is growing.

CHAPTER TEN: 2019

A $1,000 computing device is now approximately equal to the computational ability of the human brain. Computers are now largely invisible and are embedded everywhere. Three-dimensional virtual-reality displays, embedded in glasses and contact lenses, provide the primary interface for communication with other persons, the Web, and virtual reality. Most interaction with computing is through gestures and two-way natural-language spoken communication. Realistic all-encompassing visual, auditory, and tactile environments enable people to do virtually anything with anybody, regardless of physical proximity. People are beginning to have relationships with automated personalities as companions, teachers, caretakers, and lovers.

CHAPTER ELEVEN: 2029

A $1,000 unit of computation has the computing capacity of approximately one thousand human brains. Direct neural pathways have been perfected for high-bandwidth connection to the human brain. A range of neural implants is becoming available to enhance visual and auditory perception and interpretation, memory, and reasoning. Computers have read all available human- and machine-generated literature and multimedia material. There is growing discussion about the legal rights of computers and what constitutes being human. Machines claim to be conscious and these claims are largely accepted.

CHAPTER TWELVE: 2099

There is a strong trend toward a merger of human thinking with the world of machine intelligence that the human species initially created. There is no longer any clear distinction between humans and computers. Most conscious entities do not have a permanent physical presence. Machine-based intelligences derived from extended models of human intelligence claim to be human. Most of these intelligences are not tied to a specific computational processing unit. The number of software-based humans vastly exceeds those still using native neuron-cell-based computation. Even among those human intelligences still using carbon-based neurons, there is ubiquitous use of neural-implant technology that provides enormous augmentation of human perceptual and cognitive abilities. Humans who do not utilize such implants are unable to meaningfully participate in dialogues with those who do. Life expectancy is no longer a viable term in relation to intelligent beings.

EPILOGUE: THE REST OF THE UNIVERSE REVISITED

Intelligent beings consider the fate of the universe.

ACKNOWLEDGMENTS

I would like to express my gratitude to the many persons who have provided inspiration, patience, ideas, criticism, insight, and all manner of assistance for this project. In particular, I would like to thank:

- My wife, Sonya, for her loving patience through the twists and turns of the creative process
- My mother for long engaging walks with me when I was a child in the woods of Queens (yes, there were forests in Queens, New York, when I was growing up) and for her enthusiastic interest in and early support for my not-always-fully-baked ideas
- My Viking editors, Barbara Grossman and Dawn Drzal, for their insightful guidance and editorial expertise and the dedicated team at Viking Penguin, including Susan Petersen, publisher; Ivan Held and Paul Slovak, marketing executives; John Jusino, copy editor; Betty Lew, designer; Jariya Wanapun, editorial assistant, and Laura Ogar, indexer
- Jerry Bauer for his patient photography
- David High for actually devising a spiritual machine for the cover
- My literary agent, Loretta Barrett, for helping to shape this project
- My wonderfully capable researchers, Wendy Dennis and Nancy Mulford, for their dedicated and resourceful efforts, and Tom Garfield for his valuable assistance
- Rose Russo and Robert Brun for turning illustration ideas into beautiful visual presentations

- Aaron Kleiner for his encouragement and support
- George Gilder for his stimulating thoughts and insights
- Harry George, Don Gonson, Larry Janowitch, Hannah Kurzweil, Rob Pressman, and Mickey Singer for engaging and helpful discussions on these topics
- My readers: Peter Arnold, Melanie Baker-Futorian, Loretta Barrett, Stephen Baum, Bryan Bergeron, Mike Brown, Cheryl Cordima, Avi Coren, Wendy Dennis, Mark Dionne, Dawn Drzal, Nicholas Fabijanic, Gil Fischman, Ozzie Frankell, Vicky Frankell, Bob Frankston, Francis Ganong, Tom Garfield, Harry George, Audra Gerhardt, George Gilder, Don Gonson, Martin Greenberger, Barbara Grossman, Larry Janowitch, Aaron Kleiner, Jerry Kleiner, Allen Kurzweil, Amy Kurzweil, Arielle Kurzweil, Edith Kurzweil, Ethan Kurzweil, Hannah Kurzweil, Lenny Kurzweil, Missy Kurzweil, Nancy Kurzweil, Peter Kurzweil, Rachel Kurzweil, Sonya Kurzweil, Jo Lernout, Jon Lieff, Elliot Lobel, Cyrus Mehta, Nancy Mulford, Nicholas Mullendore, Rob Pressman, Vlad Sejnoha, Mickey Singer, Mike Sokol, Kim Storey, and Barbara Tyrell for their compliments and criticisms (the latter being the most helpful) and many invaluable suggestions
- Finally, all the scientists, engineers, entrepreneurs, and artists who are busy creating the age of spiritual machines.

PROLOGUE:
AN INEXORABLE EMERGENCE

The gambler had not expected to be here. But on reflection, he thought he had shown some kindness in his time. And this place was even more beautiful and satisfying than he had imagined. Everywhere there were magnificent crystal chandeliers, the finest handmade carpets, the most sumptuous foods, and, yes, the most beautiful women, who seemed intrigued with their new heaven mate. He tried his hand at roulette, and amazingly his number came up time after time. He tried the gaming tables, and his luck was nothing short of remarkable: He won game after game. Indeed his winnings were causing quite a stir, attracting much excitement from the attentive staff, and from the beautiful women.

This continued day after day, week after week, with the gambler winning every game, accumulating bigger and bigger earnings. Everything was going his way. He just kept on winning. And week after week, month after month, the gambler's streak of success remained unbreakable.

After a while, this started to get tedious. The gambler was getting restless; the winning was starting to lose its meaning. Yet nothing changed. He just kept on winning every game, until one day, the now anguished gambler turned to the angel who seemed to be in charge and said that he couldn't take it anymore. Heaven was not for him after all. He had figured he was destined for the "other place" nonetheless, and indeed that is where he wanted to be.

"But this is the other place," came the reply.

That is my recollection of an episode of *The Twilight Zone* that I saw as a young child. I don't recall the title, but I would call it "Be Careful What You Wish For."[1] As this engaging series was wont to do, it illustrated one of the paradoxes of human nature: We like to solve problems, but we don't want them all solved, not too quickly, anyway. We are more attached to the problems than to the solutions.

Take death, for example. A great deal of our effort goes into avoiding it. We make extraordinary efforts to delay it, and indeed often consider its intrusion a tragic event. Yet we would find it hard to live without it. Death gives meaning to our lives. It gives importance and value to time. Time would become meaningless if there were too much of it. If death were indefinitely put off, the human psyche would end up, well, like the gambler in *The Twilight Zone* episode.

We do not yet have this predicament. We have no shortage today of either death or human problems. Few observers feel that the twentieth century has left us with too much of a good thing. There is growing prosperity, fueled not incidentally by information technology, but the human species is still challenged by issues and difficulties not altogether different than those with which it has struggled from the beginning of its recorded history.

The twenty-first century will be different. The human species, along with the computational technology it created, will be able to solve age-old problems of need, if not desire, and will be in a position to change the nature of mortality in a postbiological future. Do we have the psychological capacity for all the good things that await us? Probably not. That, however, might change as well.

Before the next century is over, human beings will no longer be the most intelligent or capable type of entity on the planet. Actually, let me take that back. The truth of that last statement depends on how we define human. And here we see one profound difference between these two centuries: The primary political and philosophical issue of the next century will be the definition of who we are.[2]

But I am getting ahead of myself. This last century has seen enormous technological change and the social upheavals that go along with it, which few pundits circa 1899 foresaw. The pace of change is accelerating and has been since the inception of invention (as I will discuss in the first chapter, this acceleration is an inherent feature of technology). The result will be far greater transformations in the first two decades of the twenty-first century than we saw in the entire twentieth century. However, to appreciate the inexorable logic of where the twenty-first century will bring us, we have to go back and start with the present.

TRANSITION TO THE TWENTY-FIRST CENTURY

Computers today exceed human intelligence in a broad variety of intelligent yet narrow domains such as playing chess, diagnosing certain medical conditions, buying and selling stocks, and guiding cruise missiles. Yet human intelligence overall remains far more supple and flexible. Computers are still unable to describe the objects on a crowded kitchen table, write a summary of a movie, tie a pair of shoelaces, tell the difference between a dog and a cat (although this feat, I believe, is becoming feasible today with contemporary neural nets—computer simulations of human neurons),[3] recognize humor, or perform other subtle tasks in which their human creators excel.

One reason for this disparity in capabilities is that our most advanced computers are still simpler than the human brain—currently about a million times simpler (give or take one or two orders of magnitude depending on the assumptions used). But this disparity will not remain the case as we go through the early part of the next century. Computers doubled in speed every three years at the beginning of the twentieth century, every two years in the 1950s and 1960s, and are now doubling in speed every twelve months. This trend will continue, with computers achieving the memory capacity and computing speed of the human brain by around the year 2020.

Achieving the basic complexity and capacity of the human brain will not automatically result in computers matching the flexibility of human intelligence. The organization and content of these resources—the software of intelligence— is equally important. One approach to emulating the brain's software is through reverse engineering—scanning a human brain (which will be achievable early in the next century)[4] and essentially copying its neural circuitry in a neural computer (a computer designed to simulate a massive number of human neurons) of sufficient capacity.

There is a plethora of credible scenarios for achieving human-level intelligence in a machine. We will be able to evolve and train a system combining massively parallel neural nets with other paradigms to understand language and model knowledge, including the ability to read and understand written documents. Although the ability of today's computers to extract and learn knowledge from natural-language documents is quite limited, their abilities in this domain are improving rapidly. Computers will be able to read on their own, understanding and modeling what they have read, by the second decade of the twenty-first century. We can then have our computers read all of the world's literature— books, magazines, scientific journals, and other available material. Ultimately, the machines will gather knowledge on their own by venturing into the physical world, drawing from the full spectrum of media and information services, and sharing knowledge with each other (which machines can do far more easily than their human creators).

Once a computer achieves a human level of intelligence, it will necessarily roar past it. Since their inception, computers have significantly exceeded human mental dexterity in their ability to remember and process information. A computer can remember billions or even trillions of facts perfectly, while we are hard pressed to remember a handful of phone numbers. A computer can quickly search a database with billions of records in fractions of a second. Computers can readily share their knowledge bases. The combination of human-level intelligence in a machine with a computer's inherent superiority in the speed, accuracy, and sharing ability of its memory will be formidable.

Mammalian neurons are marvelous creations, but we wouldn't build them the same way. Much of their complexity is devoted to supporting their own life processes, not to their information-handling abilities. Furthermore, neurons are extremely slow; electronic circuits are at least a million times faster. Once a computer achieves a human level of ability in understanding abstract concepts, recognizing patterns, and other attributes of human intelligence, it will be able to apply this ability to a knowledge base of all human-acquired—and machine-acquired—knowledge.

A common reaction to the proposition that computers will seriously compete with human intelligence is to dismiss this specter based primarily on an examination of contemporary capability. After all, when I interact with my personal computer, its intelligence seems limited and brittle, if it appears intelligent at all. It is hard to imagine one's personal computer having a sense of humor, holding an opinion, or displaying any of the other endearing qualities of human thought.

But the state of the art in computer technology is anything but static. Computer capabilities are emerging today that were considered impossible one or two decades ago. Examples include the ability to transcribe accurately normal continuous human speech, to understand and respond intelligently to natural language, to recognize patterns in medical procedures such as electrocardiograms and blood tests with an accuracy rivaling that of human physicians, and, of course, to play chess at a world-championship level. In the next decade, we will see translating telephones that provide real-time speech translation from one human language to another, intelligent computerized personal assistants that can converse and rapidly search and understand the world's knowledge bases, and a profusion of other machines with increasingly broad and flexible intelligence.

In the second decade of the next century, it will become increasingly difficult to draw any clear distinction between the capabilities of human and machine intelligence. The advantages of computer intelligence in terms of speed, accuracy, and capacity will be clear. The advantages of human intelligence, on the other hand, will become increasingly difficult to distinguish.

The skills of computer software are already better than many people realize. It is frequently my experience that when demonstrating recent advances in, say, speech or character recognition, observers are surprised at the state of the art. For example, a typical computer user's last experience with speech-recognition technology may have been a low-end freely bundled piece of software from several years ago that recognized a limited vocabulary, required pauses between words, and did an incorrect job at that. These users are then surprised to see contemporary systems that can recognize fully continuous speech on a 60,000-word vocabulary, with accuracy levels comparable to a human typist.

Also keep in mind that the progression of computer intelligence will sneak up on us. As just one example, consider Gary Kasparov's confidence in 1990 that a computer would never come close to defeating him. After all, he had played the best computers, and their chess-playing ability—compared to his—was pathetic. But computer chess playing made steady progress, gaining forty-five rating points each year. In 1997, a computer sailed past Kasparov, at least in chess. There has been a great deal of commentary that other human endeavors are far more difficult to emulate than chess playing. *This is true.* In many areas—the ability to write a book on computers, for example—computers are still pathetic. But as computers continue to gain in capacity at an exponential rate, we will have the same experience in these other areas that Kasparov had in chess. Over the next several decades, machine competence will rival—and ultimately surpass— any particular human skill one cares to cite, including our marvelous ability to place our ideas in a broad diversity of contexts.

Evolution has been seen as a billion-year drama that led inexorably to its grandest creation: human intelligence. The emergence in the early twenty-first century of a new form of intelligence on Earth that can compete with, and ultimately significantly exceed, human intelligence will be a development of greater import than any of the events that have shaped human history. It will be no less important than the creation of the intelligence that created it, and will have profound implications for all aspects of human endeavor, including the nature of work, human learning, government, warfare, the arts, and our concept of ourselves.

This specter is not yet here. But with the emergence of computers that truly rival and exceed the human brain in complexity will come a corresponding ability of machines to understand and respond to abstractions and subtleties. Human beings appear to be complex in part because of our competing internal goals. Values and emotions represent goals that often conflict with each other, and are an unavoidable by-product of the levels of abstraction that we deal with as human beings. As computers achieve a comparable—and greater—level of complexity, and as they are increasingly derived at least in part from models of human intelligence, they, too, will necessarily utilize goals with implicit values and emotions, although not necessarily the same values and emotions that humans exhibit.

A variety of philosophical issues will emerge. Are computers thinking, or are they just calculating? Conversely, are human beings thinking, or are they just calculating? The human brain presumably follows the laws of physics, so it must be a machine, albeit a very complex one. Is there an inherent difference between human thinking and machine thinking? To pose the question another way, once computers are as complex as the human brain, and can match the human brain in subtlety and complexity of thought, are we to consider them conscious? This is a difficult question even to pose, and some philosophers believe it is not a meaningful question; others believe it is the only meaningful question in philosophy. This question actually goes back to Plato's time, but with the emergence of machines that genuinely appear to possess volition and emotion, the issue will become increasingly compelling.

For example, if a person scans his brain through a noninvasive scanning technology of the twenty-first century (such as an advanced magnetic resonance imaging), and downloads his mind to his personal computer, is the "person" who emerges in the machine the same consciousness as the person who was scanned? That "person" may convincingly implore you that "he" grew up in Brooklyn, went to college in Massachusetts, walked into a scanner here, and woke up in the machine there. The original person who was scanned, on the other hand, will acknowledge that the person in the machine does indeed appear to share his history, knowledge, memory, and personality, but is otherwise an impostor, a different person.

Even if we limit our discussion to computers that are not directly derived from a particular human brain, they will increasingly appear to have their own personalities, evidencing reactions that we can only label as emotions and articulating their own goals and purposes. They will appear to have their own free will. They will claim to have spiritual experiences. And people—those still using carbon-based neurons or otherwise—will believe them.

One often reads predictions of the next several decades discussing a variety of demographic, economic, and political trends that largely ignore the revolutionary impact of machines with their own opinions and agendas. Yet we need to reflect on the implications of the gradual, yet inevitable, emergence of true competition to the full range of human thought in order to comprehend the world that lies ahead.

PART ONE

PROBING THE PAST

CHAPTER ONE

THE LAW OF TIME AND CHAOS

A (VERY BRIEF) HISTORY OF THE UNIVERSE:
TIME SLOWING DOWN

The universe is made of stories, not of atoms.
> —Muriel Rukeyser

Is the universe a great mechanism, a great computation, a great symmetry, a great accident or a great thought?
> —John D. Barrow

As we start at the beginning, we will notice an unusual attribute of the nature of time, one that is critical to our passage to the twenty-first century. Our story begins perhaps 15 billion years ago. No conscious life existed to appreciate the birth of our Universe at the time, but we appreciate it now, so retroactively it did happen. (In retrospect—from one perspective of quantum mechanics—we could say that any Universe that fails to evolve conscious life to apprehend its existence never existed in the first place.)

It was not until 10^{-43} seconds (a tenth of a millionth of a trillionth of a trillionth of a trillionth of a second) after the birth of the Universe[1], that the situation had cooled off sufficiently (to 100 million trillion trillion degrees) that a distinct force—gravity—evolved.

Not much happened for another 10^{-34} seconds (this is also a very tiny fraction of a second, but it is a billion times longer than 10^{-43} seconds), at which point an even cooler Universe (now only a billion billion billion degrees) allowed the emergence of matter in the form of electrons and quarks. To keep things balanced, antimatter appeared as well. It was an eventful time, as new forces evolved at a rapid rate. We were now up to three: gravity, the strong force,[2] and the electroweak force.[3]

After another 10^{-10} seconds (a tenth of a billionth of a second), the electroweak force split into the electromagnetic and weak forces[4] we know so well today.

Things got complicated after another 10^{-5} seconds (ten millionths of a second). With the temperature now down to a relatively balmy trillion degrees, the quarks came together to form protons and neutrons. The antiquarks did the same, forming antiprotons.

Somehow, the matter particles achieved a slight edge. How this happened is not entirely clear. Up until then, everything had seemed so, well, even. But had everything stayed evenly balanced, it would have been a rather boring Universe. For one thing, life never would have evolved, and thus we could conclude that the Universe would never have existed in the first place.

For every 10 billion antiprotons, the Universe contained 10 billion and 1 protons. The protons and antiprotons collided, causing the emergence of another important phenomenon: light (photons). Thus, almost all of the antimatter was destroyed, leaving matter as dominant. (This shows you the danger of allowing a competitor to achieve even a slight advantage.)

Of course, had antimatter won, its descendants would have called it matter and would have called matter antimatter, so we would be back where we started (perhaps that is what happened).

After another second (a second is a very long time compared to some of the earlier chapters in the Universe's history, so notice how the time frames are growing exponentially larger), the electrons and antielectrons (called positrons) followed the lead of the protons and antiprotons and similarly annihilated each other, leaving mostly the electrons.

After another minute, the neutrons and protons began coalescing into heavier nuclei, such as helium, lithium, and heavy forms of hydrogen. The temperature was now only a billion degrees.

About 300,000 years later (things are slowing down now rather quickly), with the average temperature now only 3,000 degrees, the first atoms were created as the nuclei took control of nearby electrons.

After a billion years, these atoms formed large clouds that gradually swirled into galaxies.

After another two billion years, the matter within the galaxies coalesced further into distinct stars, many with their own solar systems.

Three billion years later, circling an unexceptional star on the arm of a common galaxy, an unremarkable planet we call the Earth was born.

Now before we go any further, let's notice a striking feature of the passage of time. Events moved quickly at the beginning of the Universe's history. We had three paradigm shifts in just the first billionth of a second. Later on, events of cosmological significance took billions of years. The nature of time is that it inherently moves in an exponential fashion—either geometrically gaining in speed, or, as in the history of our Universe, geometrically slowing down. Time only seems to be linear during those eons in which not much happens. Thus most of the time, the linear passage of time is a reasonable approximation of its passage. But that's not the inherent nature of time.

Why is this significant? It's not when you're stuck in the eons in which not much happens. But it is of great significance when you find yourself in the "knee of the curve," those periods in which the exponential nature of the curve of time explodes either inwardly or outwardly. It's like falling into a black hole (in that case, time accelerates exponentially faster as one falls in).

The Speed of Time

But wait a second, how can we say that time is changing its "speed"? We can talk about the rate of a process, in terms of its progress per second, but can we say that time is changing its rate? Can time start moving at, say, two seconds per second?

Einstein said exactly this—time is relative to the entities experiencing it.[5] One man's second can be another woman's forty years. Einstein gives the example of a man who travels at very close to the speed of light to a star—say, twenty light-years away. From our Earth-bound perspective, the trip takes slightly more than twenty years in each direction. When the man gets back, his wife has aged forty years. For him, however, the trip was rather brief. If he travels at close enough to the speed of light, it may have only taken a second or less (from a practical per-spective we would have to consider some limitations, such as the time to acceler-ate and decelerate without crushing his body). Whose time frame is the correct one? Einstein says they are both correct, and exist only relative to each other.

Certain species of birds have a life span of only several years. If you observe their rapid movements, it appears that they are experiencing the passage of time on a different scale. We experience this in our own lives. A young child's rate of change and experience of time is different from that of an adult. Of particular note, we will see that the acceleration in the passage of time for evolution is mov-ing in a different direction than that for the Universe from which it emerges.

It is in the nature of exponential growth that events develop extremely slowly for extremely long periods of time, but as one glides through the knee of the curve, events erupt at an increasingly furious pace. And that is what we will ex-perience as we enter the twenty-first century.

EVOLUTION: TIME SPEEDING UP

In the beginning was the word. . . . And the word became flesh.
                                                            —John 1:1,14

A great deal of the universe does not need any explanation. Elephants, for in-
stance. Once molecules have learnt to compete and create other molecules in
their own image, elephants, and things resembling elephants, will in due course
be found roaming through the countryside.
                                                        —Peter Atkins

The further backward you look, the further forward you can see.
                                                        —Winston Churchill

We'll come back to the knee of the curve, but let's delve further into the exponential nature of time. In the nineteenth century, a set of unifying principles called the laws of thermodynamics[6] was postulated. As the name implies, they deal with the dynamic nature of heat and were the first major refinement of the laws of classical mechanics perfected by Isaac Newton a century earlier. Whereas Newton had described a world of clockwork perfection in which particles and objects of all sizes followed highly disciplined, predictable patterns, the laws of thermodynamics describe a world of chaos. Indeed, that is what heat is. Heat is the chaotic—unpredictable—movement of the particles that make up the world. A corollary of the second law of thermodynamics is that in a closed system (interacting entities and forces not subject to outside influence; for example, the Universe), disorder (called "entropy") increases. Thus, left to its own devices, a system such as the world we live in becomes increasingly chaotic. Many people find this describes their lives rather well. But in the nineteenth century, the laws of thermodynamics were considered a disturbing discovery. At the beginning of that century, it appeared that the basic principles governing the world were both understood and orderly. There were a few details left to be filled in, but the basic picture was under control. Thermodynamics was the first contradiction to this complacent picture. It would not be the last.

The second law of thermodynamics, sometimes called the Law of Increasing Entropy, would seem to imply that the natural emergence of intelligence is impossible. Intelligent behavior is the opposite of random behavior, and any system capable of intelligent responses to its environment needs to be highly ordered. The chemistry of life, particularly of intelligent life, is comprised of exceptionally intricate designs. Out of the increasingly chaotic swirl of particles and energy in the world, extraordinary designs somehow emerged. How do we reconcile the emergence of intelligent life with the Law of Increasing Entropy?

There are two answers here. First, while the Law of Increasing Entropy would appear to contradict the thrust of evolution, which is toward increasingly elaborate order, the two phenomena are not inherently contradictory. The order of life takes place amid great chaos, and the existence of life-forms does not appreciably affect the measure of entropy in the larger system in which life has evolved. An organism is not a closed system. It is part of a larger system we call the environment, which remains high in entropy. In other words, the order represented by the existence of life-forms is insignificant in terms of measuring overall entropy.

Thus, while chaos increases in the Universe, it is possible for evolutionary processes that create increasingly intricate, ordered patterns to exist simultaneously.[7] Evolution is a process, but it is not a closed system. It is subject to outside influence, and indeed draws upon the chaos in which it is embedded. So the Law of Increasing Entropy does not rule out the emergence of life and intelligence.

For the second answer, we need to take a closer look at evolution, as it was the original creator of intelligence.

The Exponentially Quickening Pace of Evolution

As you will recall, after billions of years, the unremarkable planet called Earth was formed. Churned by the energy of the sun, the elements formed more and more complex molecules. From physics, chemistry was born.

Two billion years later, life began. That is to say, *patterns of matter and energy that could perpetuate themselves and survive perpetuated themselves and survived.* That this apparent tautology went unnoticed until a couple of centuries ago is itself remarkable.

Over time, the patterns became more complicated than mere chains of molecules. Structures of molecules performing distinct functions organized themselves into little societies of molecules. From chemistry, biology was born.

Thus, about 3.4 billion years ago, the first earthly organisms emerged: anaerobic (not requiring oxygen) prokaryotes (single-celled creatures) with a rudimentary method for perpetuating their own designs. Early innovations that followed included a simple genetic system, the ability to swim, and photosynthesis, which set the stage for more advanced, oxygen-consuming organisms. The most important development for the next couple of billion years was the DNA-based genetics that would henceforth guide and record evolutionary development.

A key requirement for an evolutionary process is a "written" record of achievement, for otherwise the process would be doomed to repeat finding solutions to problems already solved. For the earliest organisms, the record was written (embodied) in their bodies, coded directly into the chemistry of their primitive cellular structures. With the invention of DNA-based genetics, evolution had designed a digital computer to record its handiwork. This design permitted more complex experiments. The aggregations of molecules called cells organized themselves into societies of cells with the appearance of the first multicellular plants and animals about 700 million years ago. For the next 130 million years, the basic body plans of modern animals were designed, including a spinal cord–based skeleton that provided early fish with an efficient swimming style.

So while evolution took billions of years to design the first primitive cells, salient events then began occurring in hundreds of millions of years, a distinct quickening of the pace.[8] When some calamity finished off the dinosaurs 65 million years ago, mammals inherited the Earth (although the insects might disagree).[9] With the emergence of the primates, progress was then measured in mere tens of millions of years.[10] Humanoids emerged 15 million years ago, distinguished by walking on their hind legs, and now we're down to millions of years.[11]

With larger brains, particularly in the area of the highly convoluted cortex responsible for rational thought, our own species, *Homo sapiens*, emerged perhaps 500,000 years ago. *Homo sapiens* are not very different from other advanced primates in terms of their genetic heritage. Their DNA is 98.6 percent the same as the lowland gorilla, and 97.8 percent the same as the orangutan.[12] The story of evolution since that time now focuses in on a human-sponsored variant of evolution: technology.

TECHNOLOGY: EVOLUTION BY OTHER MEANS

When a scientist states that something is possible, he is almost certainly right. When he states that something is impossible, he is very probably wrong.

The only way of discovering the limits of the possible is to venture a little way past them into the impossible.

Any sufficiently advanced technology is indistinguishable from magic.
                        —Arthur C. Clarke's three laws of technology

A machine is as distinctively and brilliantly and expressively human as a violin sonata or a theorem in Euclid.

                                        —Gregory Vlastos

Technology picks right up with the exponentially quickening pace of evolution. Although not the only tool-using animal, *Homo sapiens* are distinguished by their creation of technology.[13] Technology goes beyond the mere fashioning and use of tools. It involves a record of tool making and a progression in the sophistication of tools. It requires invention and is itself a continuation of evolution by other means. The "genetic code" of the evolutionary process of technology is the record maintained by the tool-making species. Just as the genetic code of the early life-forms was simply the chemical composition of the organisms themselves, the written record of early tools consisted of the tools themselves. Later on, the "genes" of technological evolution evolved into records using written language and are now often stored in computer databases. Ultimately, the technology itself will create new technology. But we are getting ahead of ourselves.

Our story is now marked in tens of thousands of years. There were multiple subspecies of Homo sapiens. *Homo sapiens neanderthalensis* emerged about 100,000 years ago in Europe and the Middle East and then disappeared mysteriously about 35,000 to 40,000 years ago. Despite their brutish image, Neanderthals cultivated an involved culture that included elaborate funeral rituals—burying their dead with ornaments, including flowers. We're not entirely sure what happened to our *Homo sapiens* cousins, but they apparently got into conflict with our own immediate ancestors *Homo sapiens sapiens*, who emerged about 90,000 years ago. Several species and subspecies of humanoids initiated the creation of technology. The most clever and aggressive of these subspecies was the only one to survive. This established a pattern that would repeat itself throughout human history, in that the technologically more advanced group ends up becoming dominant. This trend may not bode well as intelligent machines themselves surpass us in intelligence and technological sophistication in the twenty-first century.

Our *Homo sapiens sapiens* subspecies was thus left alone among humanoids about 40,000 years ago.

Our forebears had already inherited from earlier hominid species and subspecies such innovations as the recording of events on cave walls, pictorial art, music, dance, religion, advanced language, fire, and weapons. For tens of thousands of years, humans had created tools by sharpening one side of a stone. It took our species tens of thousands of years to figure out that by sharpening both sides, the resultant sharp edge provided a far more useful tool. One significant point, however, is that these innovations did occur, and they endured. No other tool-using animal on Earth has demonstrated the ability to create and retain innovations in their use of tools.

The other significant point is that technology, like the evolution of life-forms that spawned it, is inherently an accelerating process. The foundations of technology—such as creating a sharp edge from a stone—took eons to perfect, although for human-created technology, eons means thousands of years rather than the billions of years that the evolution of life-forms required to get started.

Like the evolution of life-forms, the pace of technology has greatly accelerated over time.[14] The progress of technology in the nineteenth century, for example, greatly exceeded that of earlier centuries, with the building of canals and great ships, the advent of paved roads, the spread of the railroad, the development of the telegraph, and the invention of photography, the bicycle, sewing machine, typewriter, telephone, phonograph, motion picture, automobile, and of course Thomas Edison's light bulb. The continued exponential growth of technology in the first two decades of the twentieth century matched that of the entire nineteenth century. Today, we have major transformations in just a few years' time. As

WHAT IS TECHNOLOGY?

As technology is the continuation of evolution by other means, it shares the phenomenon of an exponentially quickening pace. The word is derived from the Greek *tekhnē,* which means "craft" or "art," and *logia,* which means "the study of." Thus one interpretation of technology is the study of crafting, in which crafting refers to the shaping of resources for a practical purpose. I use the term *resources* rather than *materials* because technology extends to the shaping of nonmaterial resources such as information.

Technology is often defined as the creation of tools to gain control over the environment. However, this definition is not entirely sufficient. Humans are not alone in their use or even creation of tools. Orangutans in Sumatra's Suaq Balimbing swamp make tools out of long sticks to break open termite nests. Crows fashion tools from sticks and leaves. The leaf-cutter ant mixes dry leaves with its saliva to create a paste. Crocodiles use tree roots to anchor dead prey.[15]

What is uniquely human is the application of knowledge—recorded knowledge—to the fashioning of tools. The knowledge base represents the genetic code for the evolving technology. And as technology has evolved, the means for recording this knowledge base has also evolved, from the oral traditions of antiquity to the written design logs of nineteenth-century craftsmen to the computer-assisted design databases of the 1990s.

Technology also implies a transcendence of the materials used to comprise it. When the elements of an invention are assembled in just the right way, they produce an enchanting effect that goes beyond the mere parts. When Alexander Graham Bell accidentally wire-connected two moving drums and solenoids (metal cores wrapped in wire) in 1875, the result transcended the materials he was working with. For the first time, a human voice was transported, magically it seemed, to a remote location. Most assemblages are just that: random assemblies. But when materials—and in the case of modern technology, information—are assembled in just the right way, transcendence occurs. The assembled object becomes far greater than the sum of its parts.

The same phenomenon of transcendence occurs in art, which may properly be regarded as another form of human technology. When wood, varnishes, and strings are assembled in just the right way, the result is wondrous: a violin, a piano. When such a device is manipulated in just the right way, there is magic of another sort: music. Music goes beyond mere sound. It evokes a response—cognitive, emotional, perhaps spiritual—in the listener, another form of transcendence. All of the arts share the same goal: of communicating from artist to audience. The communication is not of unadorned data, but of the more important items in the phenomenological garden: feelings, ideas, experiences, longings. The Greek meaning of *tekhnē logia* includes art as a key manifestation of technology.

Language is another form of human-created technology. One of the primary applications of technology is communication, and language provides the foundation for *Homo sapiens* communication. Communication is a critical survival skill. It enabled human families and tribes to develop cooperative strategies to overcome obstacles and adversaries. Other animals communicate. Monkeys and apes use elaborate gestures and grunts to communicate a variety of messages. Bees perform intricate dances in a figure-eight pattern to communicate where caches of nectar may be found. Female tree frogs in Malaysia do tap dances to signal their availability. Crabs wave their claws in one way to warn adversaries but use a different rhythm for courtship.[16] But these methods do not appear to evolve, other than through the usual DNA-based evolution. These species lack a way to record their means of communication, so the methods remain static from one generation to the next. In contrast, human language does evolve, as do all forms of technology. Along with the evolving forms of language itself, technology has provided ever-improving means for recording and distributing human language.

Homo sapiens are unique in their use and fostering of all forms of what I regard as technology: art, language, and machines, all representing evolution by other means. In the 1960s through 1990s, several well-publicized primates were said to have mastered at least childlike language skills. Chimpanzees Lana and Kanzi pressed sequences of buttons with symbols on them. Gorillas Washoe and Koko were said to be using American Sign Language. Many linguists are skeptical, noting that many primate "sentences" were jumbles, such as "Nim eat, Nim eat, drink eat me Nim, me gum me gum, tickle me, Nim play, you me banana me banana you." Even if we view this phenomenon more generously, it would be the exception that proves the rule. These primates did not evolve the languages they are credited with using, they do not appear to develop these skills spontaneously, and their use of these skills is very limited.[17] They are at best participating peripherally in what is still a uniquely human invention—communicating using the recursive (self-referencing), symbolic, *evolving* means called language.

one of many examples, the latest revolution in communications—the World Wide Web—didn't exist just a few years ago.

The Inevitability of Technology

Once life takes hold on a planet, we can consider the emergence of technology as inevitable. The ability to expand the reach of one's physical capabilities, not to mention mental facilities, through technology is clearly useful for survival. Technology has enabled our subspecies to dominate its ecological niche. Technology requires two attributes of its creator: intelligence and the physical ability to manipulate the environment. We'll talk more in chapter 4, "A New Form of Intelligence on Earth," about the nature of intelligence, but it clearly represents an ability to use limited resources optimally, including time. This ability is inherently useful for survival, so it is favored. The ability to manipulate the environment is also useful; otherwise an organism is at the mercy of its environment for safety, food, and the satisfaction of its other needs. Sooner or later, an organism is bound to emerge with both attributes.

THE INEVITABILITY OF COMPUTATION

It is not a bad definition of man to describe him as a tool-making animal. His earliest contrivances to support uncivilized life were tools of the simplest and rudest construction. His latest achievements in the substitution of machinery, not merely for the skill of the human hand, but for the relief of the human intellect, are founded on the use of tools of a still higher order.
—Charles Babbage

All of the fundamental processes we have examined—the development of the Universe, the evolution of life-forms, the subsequent evolution of technology—have all progressed in an exponential fashion, some slowing down, some speeding up. What is the common thread here? Why did cosmology exponentially slow down while evolution accelerated? The answers are surprising, and fundamental to understanding the twenty-first century.

But before I attempt to answer these questions, let's examine one other very relevant example of acceleration: the exponential growth of computation.

Early in the evolution of life-forms, specialized organs developed the ability to maintain internal states and respond differentially to external stimuli. The trend ever since has been toward more complex and capable nervous systems with the ability to store extensive memories; recognize patterns in visual, auditory, and tactile stimuli; and engage in increasingly sophisticated levels of reasoning. The ability to remember and to solve problems—computation—has constituted the cutting edge in the evolution of multicellular organisms.

The same value of computation holds true in the evolution of human-created technology. Products are more useful if they can maintain internal states and respond differentially to varying conditions and situations. As machines moved beyond mere implements to extend human reach and strength, they also began to accumulate the ability to remember and perform logical manipulations. The simple cams, gears, and levers of the Middle Ages were assembled into the

THE LIFE CYCLE OF A TECHNOLOGY

Technologies fight for survival, evolve, and undergo their own characteristic life cycle. We can identify seven distinct stages. During the *precursor* stage, the prerequisites of a technology exist, and dreamers may contemplate these elements coming together. We do not, however, regard dreaming to be the same as inventing, even if the dreams are written down. Leonardo da Vinci drew convincing pictures of airplanes and automobiles, but he is not considered to have invented either.

The next stage, one highly celebrated in our culture, is *invention*, a very brief stage, not dissimilar in some respects to the process of birth after an extended period of labor. Here the inventor blends curiosity, scientific skills, determination, and usually a measure of showmanship to combine methods in a new way to bring a new technology to life.

The next stage is *development*, during which the invention is protected and supported by doting guardians (which may include the original inventor). Often this stage is more crucial than invention and may involve additional creation that can have greater significance than the original invention. Many tinkerers had constructed finely hand-tuned horseless carriages, but it was Henry Ford's innovation of mass production that enabled the automobile to take root and flourish.

The fourth stage is *maturity*. Although continuing to evolve, the technology now has a life of its own and has become an independent and established part of the community. It may become so interwoven in the fabric of life that it appears to many observers that it will last forever. This creates an interesting drama when the next stage arrives, which I call the stage of the *pretenders*. Here an upstart threatens to eclipse the older technology. Its enthusiasts prematurely predict victory. While providing some distinct benefits, the newer technology is found on reflection to be missing some key element of functionality or quality. When it indeed fails to dislodge the established order, the technology conservatives take this as evidence that the original approach will indeed live forever.

This is usually a short-lived victory for the aging technology. Shortly thereafter, another new technology typically does succeed in rendering the original technology into the stage of *obsolescence*. In this part of the life cycle, the technology lives out its senior years in gradual decline, its original purpose and functionality now subsumed by a more spry competitor. This stage, which may comprise 5 to 10 percent of the life cycle, finally yields to *antiquity* (examples today: the horse and buggy, the harpsichord, the manual typewriter, and the electromechanical calculator).

To illustrate this, consider the phonograph record. In the mid–nineteenth century, there were several precursors, including Édouard-Léon Scott de

Martinville's phonautograph, a device that recorded sound vibrations as a printed pattern. It was Thomas Edison, however, who in 1877 brought all of the elements together and invented the first device that could record and reproduce sound. Further refinements were necessary for the phonograph to become commercially viable. It became a fully mature technology in 1948 when Columbia introduced the 33 revolutions-per-minute (rpm) long-playing record (LP) and RCA Victor introduced the 45-rpm small disc. The pretender was the cassette tape, introduced in the 1960s and popularized during the 1970s. Early enthusiasts predicted that its small size and ability to be re-recorded would make the relatively bulky and scratchable record obsolete.

Despite these obvious benefits, cassettes lack random access (the ability to play selections in a desired order) and are prone to their own forms of distortion and lack of fidelity. In the late 1980s and early 1990s, the digital compact disc (CD) did deliver the mortal blow. With the CD providing both random access and a level of quality close to the limits of the human auditory system, the phonograph record entered the stage of obsolescence in the first half of the 1990s. Although still produced in small quantities, the technology that Edison gave birth to more than a century ago is now approaching antiquity.

Another example is the print book, a rather mature technology today. It is now in the stage of the pretenders, with the software-based "virtual" book as the pretender. Lacking the resolution, contrast, lack of flicker, and other visual qualities of paper and ink, the current generation of virtual book does not have the capability of displacing paper-based publications. Yet this victory of the paper-based book will be short-lived as future generations of computer displays succeed in providing a fully satisfactory alternative to paper.

elaborate automata of the European Renaissance. Mechanical calculators, which first emerged in the seventeenth century, became increasingly complex, culminating in the first automated U.S. census in 1890. Computers played a crucial role in at least one theater of the Second World War, and have developed in an accelerating spiral ever since.

The Emergence of Moore's Law

Gordon Moore, an inventor of the integrated circuit and then chairman of Intel, noted in 1965 that the surface area of a transistor (as etched on an integrated circuit) was being reduced by approximately 50 percent every twelve months. In 1975, he was widely reported to have revised this observation to eighteen months. Moore claims that his 1975 update was to twenty-four months, and that does appear to be a better fit to the data.

MOORE'S LAW AT WORK	
Year	Transistors in Intel's Latest Computer Chip*
1972	3,500
1974	6,000
1978	29,000
1982	134,000
1985	275,000
1989	1,200,000
1993	3,100,000
1995	5,500,000
1997	7,500,000
*Consumer Electronics Manufacturers Association	

The result is that every two years, you can pack twice as many transistors on an integrated circuit. This doubles both the number of components on a chip as well as its speed. Since the cost of an integrated circuit is fairly constant, the implication is that every two years you can get twice as much circuitry running at twice the speed for the same price. For many applications, that's an effective quadrupling of the value. The observation holds true for every type of circuit, from memory chips to computer processors.

This insightful observation has become known as Moore's Law on Integrated Circuits, and the remarkable phenomenon of the law has been driving the acceleration of computing for the past forty years. But how much longer can this go on? The chip companies have expressed confidence in another fifteen to twenty years of Moore's Law by continuing their practice of using increasingly higher resolutions of optical lithography (an electronic process similar to photographic printing) to reduce the feature size—measured today in millionths of a meter—of transistors and other key components.[18] But then—after almost sixty years—this paradigm will break down. The transistor insulators will then be just a few atoms thick, and the conventional approach of shrinking them won't work.

What then?

We first note that the exponential growth of computing did not start with Moore's Law on Integrated Circuits. In the accompanying figure, "The Exponential Growth of Computing, 1900–1998,"[19] I plotted forty-nine notable computing machines spanning the twentieth century on an exponential chart, in which the vertical axis represents powers of ten in computer speed per unit cost (as measured in the number of "calculations per second" that can be purchased for $1,000). Each point on the graph represents one of the machines. The first five machines used mechanical technology, followed by three electromechanical (relay based) computers, followed by eleven vacuum-tube machines, followed by twelve machines using discrete transistors. Only the last eighteen computers used integrated circuits.

I then fit a curve to the points called a fourth-order polynomial, which allows for up to four bends. In other words, I did not try to fit a straight line to the points, just the closest fourth-order curve. Yet a straight line is close to what I got. A straight line on an exponential graph means exponential growth. A careful examination of the trend shows that the curve is actually bending slightly upward, indicating a small exponential growth in the rate of exponential growth. This may result from the interaction of two different exponential trends, as I will discuss in chapter 6, "Building New Brains." Or there may indeed be two levels of exponential growth. Yet even if we take the more conservative view that there is just one level of acceleration, we can see that the exponential growth of computing did not start with Moore's Law on Integrated Circuits, but dates back to the advent of electrical computing at the beginning of the twentieth century.

Mechanical Computing Devices

1. 1900   Analytical Engine
2. 1908   Hollerith Tabulator
3. 1911   Monroe Calculator
4. 1919   IBM Tabulator
5. 1928   National Ellis 3000

Electromechanical (Relay Based) Computers

6. 1939   Zuse 2
7. 1940   Bell Calculator Model 1
8. 1941   Zuse 3

Vacuum-Tube Computers

9. 1943   Colossus
10. 1946   ENIAC
11. 1948   IBM SSEC
12. 1949   BINAC
13. 1949   EDSAC
14. 1951   Univac I
15. 1953   Univac 1103
16. 1953   IBM 701

17.	1954	EDVAC
18.	1955	Whirlwind
19.	1955	IBM 704

Discrete Transistor Computers

20.	1958	Datamatic 1000
21.	1958	Univac II
22.	1959	Mobidic
23.	1959	IBM 7090
24.	1960	IBM 1620
25.	1960	DEC PDP-1
26.	1961	DEC PDP-4
27.	1962	Univac III
28.	1964	CDC 6600
29.	1965	IBM 1130
30.	1965	DEC PDP-8
31.	1966	IBM 360 Model 75

Integrated Circuit Computers

32.	1968	DEC PDP-10
33.	1973	Intellec-8
34.	1973	Data General Nova
35.	1975	Altair 8800
36.	1976	DEC PDP-11 Model 70
37.	1977	Cray 1
38.	1977	Apple II
39.	1979	DEC VAX 11 Model 780
40.	1980	Sun-1
41.	1982	IBM PC
42.	1982	Compaq Portable
43.	1983	IBM AT-80286
44.	1984	Apple Macintosh
45.	1986	Compaq Deskpro 386
46.	1987	Apple Mac II
47.	1993	Pentium PC
48.	1996	Pentium PC
49.	1998	Pentium II PC

THE EXPONENTIAL GROWTH OF COMPUTING, 1900–1998

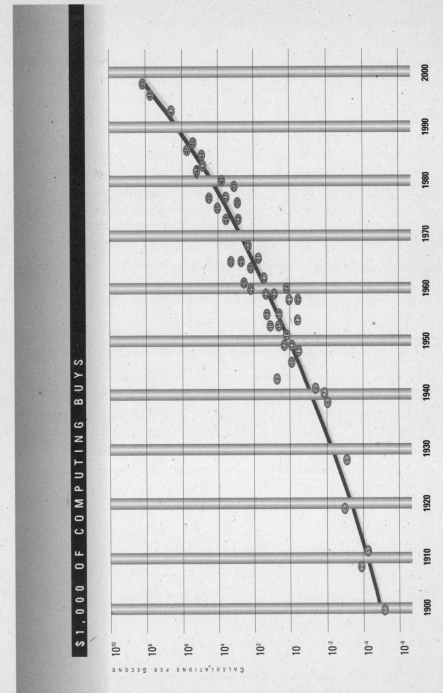

$1,000 OF COMPUTING BUYS

In the 1980s, a number of observers, including Carnegie Mellon University professor Hans Moravec, Nippon Electric Company's David Waltz, and myself, noticed that computers have been growing exponentially in power, long before the invention of the integrated circuit in 1958 or even the transistor in 1947.[20] The speed and density of computation have been doubling every three years (at the beginning of the twentieth century) to one year (at the end of the twentieth century), regardless of the type of hardware used. Remarkably, this "Exponential Law of Computing" has held true for at least a century, from the mechanical card-based electrical computing technology used in the 1890 U.S. census, to the relay-based computers that cracked the Nazi Enigma code, to the vacuum-tube-based computers of the 1950s, to the transistor-based machines of the 1960s, and to all of the generations of integrated circuits of the past four decades. Computers are about one hundred million times more powerful for the same unit cost than they were a half century ago. If the automobile industry had made as much progress in the past fifty years, a car today would cost a hundredth of a cent and go faster than the speed of light.

As with any phenomenon of exponential growth, the increases are so slow at first as to be virtually unnoticeable. Despite many decades of progress since the first electrical calculating equipment was used in the 1890 census, it was not until the mid-1960s that this phenomenon was even noticed (although Alan Turing had an inkling of it in 1950). Even then, it was appreciated only by a small community of computer engineers and scientists. Today, you have only to scan the personal computer ads—or the toy ads—in your local newspaper to see the dramatic improvements in the price performance of computation that now arrive on a monthly basis.

So Moore's Law on Integrated Circuits was not the first, but the fifth paradigm to continue the now one-century-long exponential growth of computing. Each new paradigm came along just when needed. This suggests that exponential growth won't stop with the end of Moore's Law. But the answer to our question on the continuation of the exponential growth of computing is critical to our understanding of the twenty-first century. So to gain a deeper understanding of the true nature of this trend, we need to go back to our earlier questions on the exponential nature of time.

THE LAW OF TIME AND CHAOS

Is the flow of time something real, or might our sense of time passing be just an illusion that hides the fact that what is real is only a vast collection of moments?
—Lee Smolin

THE LAW OF TIME AND CHAOS

In a process, the time interval between salient events (i.e., events that change the nature of the process, or significantly affect the future of the process) expands or contracts along with the amount of chaos.

THE LAW OF INCREASING CHAOS

As chaos exponentially increases, time exponentially slows down (i.e., the time interval between salient events grows longer as time passes).

THE LAW OF INCREASING CHAOS AS APPLIED TO THE UNIVERSE

The Universe started as a single "singularity," a single undifferentiated point with no size and no chaos, so early epochal events were extremely rapid. The Universe grew greatly in chaos as time went on. Thus time slowed down (i.e., the time interval between salient events grew exponentially longer over time).

THE LAW OF INCREASING CHAOS AS APPLIED TO THE LIFE OF AN ORGANISM

The development of an organism from conception as a single cell through maturation is a process moving toward greater diversity and thus greater disorder. Thus the time interval between salient events grows longer over time.

THE LAW OF ACCELERATING RETURNS

As order exponentially increases, time exponentially speeds up (i.e., the time interval between salient events grows shorter as time passes).

THE LAW OF ACCELERATING RETURNS AS APPLIED TO AN EVOLUTIONARY PROCESS

An evolutionary process is not a closed system; therefore, evolution draws upon the chaos in the larger system in which it takes place for its options for diversity; and
- Evolution builds on its own increasing order.

Therefore:
- In an evolutionary process, order increases exponentially.

Therefore:
- Time exponentially speeds up.

Therefore:
- The returns (i.e., the valuable products of the process) accelerate.

THE LAW OF ACCELERATING RETURNS AS APPLIED TO THE EVOLUTION OF LIFE-FORMS

The time interval between salient events (e.g., a significant new branch) grows exponentially shorter as time passes.

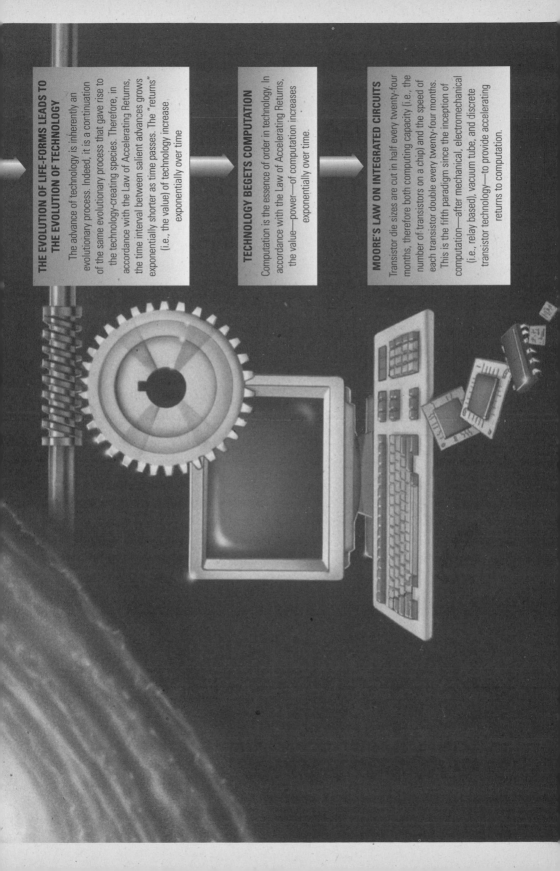

THE EVOLUTION OF LIFE-FORMS LEADS TO THE EVOLUTION OF TECHNOLOGY

The advance of technology is inherently an evolutionary process. Indeed, it is a continuation of the same evolutionary process that gave rise to the technology-creating species. Therefore, in accordance with the Law of Accelerating Returns, the time interval between salient advances grows exponentially shorter as time passes. The "returns" (i.e., the value) of technology increase exponentially over time

TECHNOLOGY BEGETS COMPUTATION

Computation is the essence of order in technology. In accordance with the Law of Accelerating Returns, the value—power—of computation increases exponentially over time.

MOORE'S LAW ON INTEGRATED CIRCUITS

Transistor die sizes are cut in half every twenty-four months, therefore both computing capacity (i.e., the number of transistors on a chip) and the speed of each transistor double every twenty-four months. This is the fifth paradigm since the inception of computation—after mechanical, electromechanical (i.e., relay based), vacuum tube, and discrete transistor technology—to provide accelerating returns to computation.

Time is nature's way of preventing everything from happening at once.

—Graffito

Things are more like they are now than they ever were before.

—Dwight Eisenhower

Consider these diverse exponential trends:

- The exponentially *slowing* pace that the Universe followed, with three epochs in the first billionth of a second, with later salient events taking billions of years.
- The exponentially *slowing* pace in the development of an organism. In the first month after conception, we grow a body, a head, even a tail. We grow a brain in the first couple of months. After leaving our maternal confines, our maturation both physically and mentally is rapid at first. In the first year, we learn basic forms of mobility and communication. We experience milestones every month or so. Later on, key events march ever more slowly, taking years and then decades.
- The exponentially *quickening* pace of the evolution of life-forms on Earth.
- The exponentially *quickening* pace of the evolution of human-created technology, which picked up the pace from the evolution of life-forms.
- The exponential *growth* of computing. Note that exponential growth of a process over time is just another way of expressing an exponentially quickening pace. For example, it took about ninety years to achieve the first MIP (Million Instructions per Second) for a thousand dollars. Now we add an additional MIP per thousand dollars every day. The overall innovation rate is clearly accelerating as well.
- Moore's Law on Integrated Circuits. As I noted, this was the fifth paradigm to achieve the exponential growth of computing.

Many questions come to mind:

What is the common thread between these varied exponential trends?
Why do some of these processes speed up while others slow down?
And what does this tell us about the continuation of the exponential growth of computing when Moore's Law dies?

Is Moore's Law just a set of industry expectations and goals, as Randy Isaac, head of basic science at IBM, contends? Or is it part of a deeper phenomenon that goes far beyond the photolithography of integrated circuits?

After thinking about the relationship between these apparently diverse trends for several years, the surprising common theme became apparent to me.

What determines whether time speeds up or slows down? The consistent answer is that *time moves in relation to the amount of chaos*. We can state the Law of Time and Chaos as follows:

> *The Law of Time and Chaos: In a process, the time interval between salient events (that is, events that change the nature of the process, or significantly affect the future of the process) expands or contracts along with the amount of chaos.*

When there is a lot of chaos in a process, it takes more time for significant events to occur. Conversely, as order increases, the time periods between salient events decrease.

We have to be careful here in our definition of chaos. It refers to the quantity of disordered (that is, random) events *that are relevant to the process*. If we're dealing with the random movement of atoms and molecules in a gas or liquid, then heat is an appropriate measure. If we're dealing with the process of evolution of life-forms, then chaos represents the unpredictable events encountered by organisms, and the random mutations that are introduced in the genetic code.

Let's see how the Law of Time and Chaos applies to our examples. If chaos is increasing, the Law of Time and Chaos implies the following sublaw:

> *The Law of Increasing Chaos: As chaos exponentially increases, time exponentially slows down (that is, the time interval between salient events grows longer as time passes).*

This fits the Universe rather well. When the entire Universe was just a "naked" singularity—a perfectly orderly single point in space and time—there was no chaos and conspicuous events took almost no time at all. As the Universe grew in size, chaos increased exponentially, and so did the timescale for epochal changes. Now, with billions of galaxies sprawled out over trillions of light-years of space, the Universe contains vast reaches of chaos, and indeed requires billions of years to get everything organized for a paradigm shift to take place.

We see a similar phenomenon in the progression of an organism's life. We start out as a single fertilized cell, so there's only rather limited chaos there. Ending up with trillions of cells, chaos greatly expands. Finally, at the end of our lives, our designs deteriorate, engendering even greater randomness. So the time period between salient biological events grows longer as we grow older. And that is indeed what we experience.

But it is the opposite spiral of the Law of Time and Chaos that is the most important and relevant for our purposes. Consider the inverse sublaw, which I call the Law of Accelerating Returns:

> *The Law of Accelerating Returns: As order exponentially increases, time exponentially speeds up (that is, the time interval between salient events grows shorter as time passes).*

The Law of Accelerating Returns (to distinguish it from a better-known law in which returns diminish) applies specifically to evolutionary processes. In an evolutionary process, it is order—the opposite of chaos—that is increasing. And, as we have seen, time speeds up.

Disdisorder

I noted above that the concept of chaos in the Law of Time and Chaos is tricky. Chaos alone is not sufficient—disorder for our purposes requires randomness that is relevant to the process we are concerned with. The opposite of disorder—which I called "order" in the above Law of Accelerating Returns—is even trickier.

Let's start with our definition of disorder and work backward. If disorder represents a random sequence of events, then the opposite of disorder should imply "not random." And if random means unpredictable, then we might conclude that order means predictable. But that would be wrong.

Borrowing a page from information theory,[21] consider the difference between information and noise. Information is a sequence of data that is meaningful in a process, such as the DNA code of an organism, or the bits in a computer program. Noise, on the other hand, is a random sequence. *Neither noise nor information is predictable.* Noise is inherently unpredictable, but carries no information. Information, however, is also unpredictable. If we can predict future data from past data, then that future data stops being information. For example, consider a sequence which simply alternates between zero and one (01010101 . . .). Such a sequence is certainly orderly, and very predictable. Specifically because it is so predictable, we do not consider it information bearing, beyond the first couple of bits.

Thus orderliness does not constitute order because order requires information. So, perhaps I should use the word *information* instead of *order.* However, information alone is not sufficient for our purposes either. Consider a phone book. It certainly represents a lot of information, and some order as well. Yet if we double the size of the phone book, we have increased the amount of data, but we have not achieved a deeper level of order.

Order, then, is information that fits a purpose. The measure of order is the measure of how well the information fits the purpose. In the evolution of life-forms, the purpose is to survive. In an evolutionary algorithm (a computer program that simulates evolution to solve a problem) applied to, say, investing in the stock market, the purpose is to make money. Simply having more information does not necessarily result in a better fit. A superior solution for a purpose may very well involve less data.

The concept of "complexity" has been used recently to describe the nature of the information created by an evolutionary process. Complexity is a reasonably close fit to the concept of order that I am describing. After all, the designs created by the evolution of life-forms on Earth appear to have become more complex over time. However, complexity is not a perfect fit, either. Sometimes, a deeper order—a better fit to a purpose—is achieved through simplification rather than further increases in complexity. As Einstein said, "Everything should be made as simple as possible, but no simpler." For example, a new theory that ties together apparently disparate ideas into one broader, more coherent theory reduces complexity but nonetheless may increase the "order for a purpose" that I am describing. Evolution has shown, however, that the general trend toward greater order does generally result in greater complexity.[22]

Thus improving a solution to a problem—which may increase or decrease complexity—increases order. Now that just leaves the issue of defining the problem. And as we will see, defining a problem well is often the key to finding its solution.

The Law of Increasing Entropy Versus the Growth of Order

Another consideration is how the Law of Time and Chaos relates to the second law of thermodynamics. Unlike the second law, the Law of Time and Chaos is not necessarily concerned with a closed system. It deals instead with a process. The Universe is a closed system (not subject to outside influence, since there is nothing outside the Universe), so in accordance with the second law of thermodynamics, chaos increases and time slows down. In contrast, evolution is precisely not a closed system. It takes place amid great chaos, and indeed *depends on the disorder in its midst, from which it draws its options for diversity.* And from these options, an evolutionary process continually prunes its choices to create ever greater order. Even a crisis that appears to introduce a significant new source of chaos is likely to end up increasing—deepening—the order created by an evolutionary process. For example, consider the asteroid that is thought to have killed off big organisms such as the dinosaurs 65 million years ago. The crash of that asteroid suddenly created a vast increase in chaos (and lots of dust, too). Yet it appears to have hastened the rise of mammals in the niche previously dominated by large reptiles and ultimately led to the emergence of a technology-creating species. When the dust settled (literally), the crisis of the asteroid had increased order.

As I pointed out earlier, only a tiny fraction of the stuff in the Universe, or even on a life- and technology-bearing planet such as Earth, can be considered to be part of evolution's inventions. Thus evolution does not contradict the Law of Increasing Entropy. Indeed, it depends on it to provide a never-ending supply of options.

As I noted, given the emergence of life, the emergence of a technology-creating species—and of technology—is inevitable. Technology is the continuation of evolution by other means, and is itself an evolutionary process. So it, too, speeds up.

A primary reason that evolution—of life-forms or of technology—speeds up is that *it builds on its own increasing order.* Innovations created by evolution encourage and enable faster evolution. In the case of the evolution of life-forms, the most notable example is DNA, which provides a recorded and protected transcription of life's design from which to launch further experiments.

In the case of the evolution of technology, ever improving human methods of recording information have fostered further technology. The first computers were designed on paper and assembled by hand. Today, they are designed on computer workstations with the computers themselves working out many details of the next generation's design, and are then produced in fully automated factories with human guidance but limited direct intervention.

The evolutionary process of technology seeks to improve capabilities in an exponential fashion. Innovators seek to improve things by multiples. Innovation is multiplicative, not additive. Technology, like any evolutionary process, builds on itself. This aspect will continue to accelerate when the technology itself takes full control of its own progression.

We can thus conclude the following with regard to the evolution of life-forms, and of technology:

The Law of Accelerating Returns as Applied to an Evolutionary Process:

▴ *An evolutionary process is not a closed system; therefore, evolution draws upon the chaos in the larger system in which it takes place for its options for diversity; and*
▴ *Evolution builds on its own increasing order.*

Therefore:

▴ *In an evolutionary process, order increases exponentially.*

Therefore:

▴ *Time exponentially speeds up.*

Therefore:

▸ *The returns (that is, the valuable products of the process) accelerate.*

The phenomenon of time slowing down and speeding up is occurring simultaneously. Cosmologically speaking, the Universe continues to slow down. Evolution, now most noticeably in the form of human-created technology, continues to speed up. These are the two sides—two interleaved spirals—of the Law of Time and Chaos.

The spiral we are most interested in—the Law of Accelerating Returns—gives us ever greater order in technology, which inevitably leads to the emergence of computation. Computation is the essence of order. It provides the ability for a technology to respond in a variable and appropriate manner to its environment to carry out its mission. Thus computational technology is also an evolutionary process, and also builds on its own progress. The time to accomplish a fixed objective gets exponentially shorter over time (for example, ninety years for the first MIP per thousand dollars versus one day for an additional MIP today). That the power of computing grows exponentially over time is just another way to say the same thing.

So Where Does That Leave Moore's Law?

Well, it still leaves it dead by the year 2020. Moore's Law came along in 1958 just when it was needed and will have done its sixty years of service by 2018, a rather long period of time for a paradigm nowadays. Unlike Moore's Law, however, the Law of Accelerating Returns is not a temporary methodology. It is a basic attribute of the nature of time and chaos—a sublaw of the Law of Time and Chaos—and describes a wide range of apparently divergent phenomena and trends. In accordance with the Law of Accelerating Returns, another computational technology will pick up where Moore's Law will have left off, without missing a beat.

Most Exponential Trends Hit a Wall . . . but Not This One

A frequent criticism of predictions of the future is that they rely on mindless extrapolation of current trends without consideration of forces that may terminate or alter that trend. This criticism is particularly relevant in the case of exponential trends. A classic example is a species happening upon a hospitable new habitat, perhaps transplanted there by human intervention (rabbits in Australia, say). Its numbers multiply exponentially for a while, but this phenomenon is quickly terminated when the exploding population runs into a new predator or the limits of

THE LEARNING CURVE: SLUG VERSUS HUMAN

The "learning curve" describes the mastery of a skill over time. As an entity—slug or human—learns a new skill, the newly acquired ability builds on itself, and so the learning curve starts out looking like the exponential growth we see in the Law of Accelerating Returns. Skills tend to be bounded, so as the new expertise is mastered, the law of diminishing returns sets in, and growth in mastery levels off. So the learning curve is what we call an S curve because exponential growth followed by a leveling off looks like an S leaning slightly to the right: ∕.

The learning curve is remarkably universal: Most multicellular creatures do it. Slugs, for example, follow the learning curve when learning how to ascend a new tree in search of leaves. Humans, of course, are always learning something new.

But there's a salient difference between humans and slugs. Humans are capable of innovation, which is the creation and retention of new skills and knowledge. Innovation is the driving force in the Law of Accelerating Returns, and eliminates the leveling-off part of the S curve. So innovation turns the S curve into indefinite exponential expansion.

Overcoming the S curve is another way to express the unique status of the human species. No other species appears to do this. Why are we unique in this way, given that other primates are so close to us in terms of genetic similarity?

The reason is that the ability to overcome the S curve defines a new ecological niche. As I pointed out, there were indeed other humanoid species and subspecies capable of innovation, but the niche seems to have tolerated only one surviving competitor. But we will have company in the twenty-first century as our machines join us in this exclusive niche.

its environment. Similarly, the geometric population growth of our own species has been a source of anxiety, but changing social and economic factors, including growing prosperity, have greatly slowed this expansion in recent years, even in developing countries.

Based on this, some observers are quick to predict the demise of the exponential growth of computing.

But the growth predicted by the Law of Accelerating Returns is an exception to the frequently cited limitations to exponential growth. Even a catastrophe, as apparently befell our reptilian cohabitants in the late Cretaceous period, only sidesteps an evolutionary process, which then picks up the pieces and continues unabated (unless the entire process is wiped out). An evolutionary process accelerates because it builds on its past achievements, which includes improvements in its own means for further evolution. In the evolution of life-forms, in addition to DNA-based genetic coding, the innovation of sexual reproduction provided for improved means of experimenting with diverse characteristics within an otherwise homogenous population. The establishment of basic body plans of modern animals in the "Cambrian explosion," about 570 million years ago, allowed evolution to concentrate on higher-level features such as expanded brain function. The inventions of evolution in one era provide the means, and often the intelligence, for innovation in the next.

The Law of Accelerating Returns applies equally to the evolutionary process of computation, which inherently will grow exponentially and essentially without limit. *The two resources it needs—the growing order of the evolving technology itself and the chaos from which an evolutionary process draws its options for further diversity—are unbounded.* Ultimately, the innovation needed for further turns of the screw will come from the machines themselves.

How will the power of computing continue to accelerate after Moore's Law dies? We are just beginning to explore the third dimension in chip design. The vast majority of today's chips are flat, whereas our brain is organized in three dimensions. We live in a three-dimensional world, so why not use the third dimension? Improvements in semiconductor materials, including superconducting circuits that don't generate heat, will enable us to develop chips—that is, cubes—with thousands of layers of circuitry that, combined with far smaller component geometries, will improve computing power by a factor of many millions. And there are more than enough other new computing technologies waiting in the wings—nanotube, optical, crystalline, DNA, and quantum (which we'll visit in chapter 6, "Building New Brains")—to keep the Law of Accelerating Returns going in the world of computation for a very long time.

A Planetary Affair

The introduction of technology on Earth is not merely the private affair of one of the Earth's innumerable species. It is a pivotal event in the history of the planet. Evolution's grandest creation—human intelligence—is providing the means for the next stage of evolution, which is technology. The emergence of technology is predicted by the Law of Accelerating Returns. The *Homo sapiens sapiens* subspecies emerged only tens of thousands of years after its human forebears. According to the Law of Accelerating Returns, the next stage of evolution should measure its salient events in mere thousands of years, too quick for DNA-based evolution. This next stage of evolution was necessarily created by human intelligence itself, another example of the exponential engine of evolution using its innovations from one period (human beings) to create the next (intelligent machines).

Evolution draws upon the great chaos in its midst—the ever increasing entropy governed by the flip side of the Law of Time and Chaos—for its options for innovation. These two strands of the Law of Time and Chaos—time exponentially slowing down due to the increasing chaos predicted by the second law of thermodynamics; and time exponentially speeding up due to the increasing order created by evolution—coexist and progress without limit. *In particular, the resources of evolution, order and chaos, are unbounded.* I stress this point because it is crucial to understanding the evolutionary—and revolutionary—nature of computer technology.

The emergence of technology was a milestone in the evolution of intelligence on Earth because it represented a new means of evolution recording its designs. The next milestone will be technology creating its own next generation without human intervention. That there is only a period of tens of thousands of years between these two milestones is another example of the exponentially quickening pace that is evolution.

The Inventor of Chess and the Emperor of China

To appreciate the implications of this (or any) geometric trend, it is useful to recall the legend of the inventor of chess and his patron, the emperor of China. The emperor had so fallen in love with his new game that he offered the inventor a reward of anything he wanted in the kingdom.

"Just one grain of rice on the first square, Your Majesty."

"Just one grain of rice?"

"Yes, Your Majesty, just one grain of rice on the first square, and two grains of rice on the second square."

"That's it—one and two grains of rice?"

"Well, okay, and four grains on the third square, and so on."

The emperor immediately granted the inventor's seemingly humble request. One version of the story has the emperor going bankrupt because the doubling of grains of rice for each square ultimately equaled 18 million trillion grains of rice. At ten grains of rice per square inch, this requires rice fields covering twice the surface area of the Earth, oceans included.

The other version of the story has the inventor losing his head. It's not yet clear which outcome we're headed for.

But there is one thing we should note: It was fairly uneventful as the emperor and the inventor went through the first *half* of the chessboard. After thirty-two squares, the emperor had given the inventor about 4 billion grains of rice. That's a reasonable quantity—about one large field's worth—and the emperor did start to take notice.

But the emperor could still remain an emperor. And the inventor could still retain his head. It was as they headed into the second half of the chessboard that at least one of them got into trouble.

So where do we stand now? There have been about thirty-two doublings of speed and capacity since the first operating computers were built in the 1940s. Where we stand right now is that we have finished the first half of the chessboard. And, indeed, people are starting to take notice.

Now, as we head into the next century, we are heading into the second half of the chessboard. And this is where things start to get interesting.

OKAY, LET ME GET THIS STRAIGHT, MY CONCEPTION AS A FERTILIZED EGG WAS LIKE THE UNIVERSE'S BIG BANG—UH, NO PUN INTENDED—THAT IS, THINGS STARTED OUT HAPPENING VERY FAST, THEN KIND OF SLOWED DOWN, AND NOW THEY'RE REAL SLOW?

That's a reasonable way to put it, the time interval now between milestones is a lot longer than it was when you were an infant, let alone a fetus.

YOU MENTIONED THE UNIVERSE HAD THREE PARADIGM SHIFTS IN THE FIRST BILLIONTH OF A SECOND. WERE THINGS THAT FAST WHEN I GOT STARTED?

Not quite that fast. The Universe started as a singularity, a single point taking up no space and comprising, therefore, no chaos. So the first major event, which was the creation of the Universe, took no time at all. With the Universe still very small, events unfolded extremely quickly. We don't start out as a single point, but as a rather complex cell. It has order but there is a lot of random activity within a cell compared to a single point in space. So our first major event as an organism, which is the first mitosis of our fertilized egg, is measured in hours, not trillionths of a second. Things slow down from there.

BUT I FEEL LIKE TIME IS SPEEDING UP. THE YEARS JUST GO BY SO MUCH FASTER NOW THAN THEY DID WHEN I WAS A KID. DON'T YOU HAVE IT BACKWARD?

Yes, well, the subjective experience is the opposite of the objective reality.

OF COURSE. WHY DIDN'T I THINK OF THAT?

Let me clarify what I mean. The objective reality is the reality of the outside observer observing the process. If we observe the development of an individual, salient events happen very quickly at first, but later on milestones are more spread out, so we say time is slowing down. The subjective experience, however, is the experience of the process itself, assuming, of course, that the process is conscious. Which in your case, it is. At least, I assume that's the case.

THANK YOU.

Subjectively, our perception of time is affected by the spacing of milestones.

MILESTONES?

Yeah, like growing a body and a brain.

AND BEING BORN?

Sure, that's a milestone. Then learning to sit up, walking, talking . . .

OKAY.

We can consider each subjective unit of time to be equivalent to one milestone spacing. Since our milestones are spaced further apart as we grow older, a subjective unit of time will represent a longer span of time for an adult than for a child. Thus time feels like it is passing by more quickly as we grow older. That is, an interval of a few years as an adult may be perceived as comparable to a few months to a young child. Thus a long interval to an adult and a short interval to a child both represent the same subjective time in terms of the passage of salient events. Of course, long and short intervals also represent comparable fractions of their respective past lives.

SO DOES THAT EXPLAIN WHY TIME PASSES MORE QUICKLY WHEN I'M HAVING A GOOD TIME?

Well, it may be relevant to one phenomenon. If someone goes through an experience in which a lot of significant events occur, that experience may feel like a much longer period of time than a calmer period. Again, we measure subjective time in terms of salient experiences.

NOW IF I FIND TIME SPEEDING UP WHEN OBJECTIVELY IT IS SLOWING DOWN, THEN EVOLUTION WOULD SUBJECTIVELY FIND TIME SLOWING DOWN AS IT OBJECTIVELY SPEEDS UP, DO I HAVE THAT STRAIGHT?

Yes, if evolution were conscious.

WELL, IS IT?

There's no way to really tell, but evolution has its time spiral going in the opposite direction from entities we generally consider to be conscious, such as humans. In other words, evolution starts out slow and speeds up over time, whereas the development of a person starts out fast and then slows down. The Universe, however, does have its time spiral going in the same direction as us organisms, so it would make more sense to say that the Universe is conscious. And come to think of it, that does shed some light on what happened before the big bang.

I WAS JUST WONDERING ABOUT THAT.

As we look back in time and get closer to the event of the big bang, chaos is shrinking to zero. Thus from the subjective perspective, time is stretching out. Indeed, as we go back in time and approach the big bang, subjective time approaches infinity. Thus it is not possible to go back past a subjective infinity of time.

THAT'S A LOAD OFF MY MIND. NOW YOU SAID THAT THE EXPONENTIAL PROGRESS OF AN EVOLUTIONARY PROCESS GOES ON FOREVER. IS THERE ANYTHING THAT CAN STOP IT?

Only a catastrophe that wipes out the entire process.

SUCH AS AN ALL-OUT NUCLEAR WAR?

That's one scenario, but in the next century, we will encounter a plethora of other "failure modes." We'll talk about this in later chapters.

I CAN'T WAIT. NOW TELL ME THIS, WHAT DOES THE LAW OF ACCELERATING RETURNS HAVE TO DO WITH THE TWENTY-FIRST CENTURY?

Exponential trends are immensely powerful but deceptive. They linger for eons with very little effect. But once they reach the "knee of the curve," they explode with unrelenting fury. With regard to computer technology and its impact on human society, that knee is approaching with the new millennium. Now I have a question for you.

SHOOT.

Just who are you anyway?

WHY, I'M THE READER.

Of course. Well, it's good to have you contributing to the book while there's still time to do something about it.

GLAD TO. NOW, YOU NEVER DID GIVE THE ENDING TO THE EMPEROR STORY. SO DOES THE EMPEROR LOSE HIS EMPIRE, OR DOES THE INVENTOR LOSE HIS HEAD?

I have two endings, so I just can't say.

MAYBE THEY REACH A COMPROMISE SOLUTION. THE INVENTOR MIGHT BE HAPPY TO SETTLE FOR, SAY, JUST ONE PROVINCE OF CHINA.

Yes, that would be a good result. And maybe an even better parable for the twenty-first century.

CHAPTER TWO

THE INTELLIGENCE OF EVOLUTION

Here's another critical question for understanding the twenty-first century: *Can an intelligence create another intelligence more intelligent than itself?*

Let's first consider the intelligent process that created us: evolution.

Evolution is a master programmer. It has been prolific, designing millions of species of breathtaking diversity and ingenuity. And that's just here on Earth. The software programs have been all written down, recorded as digital data in the chemical structure of an ingenious molecule called deoxyribonucleic acid, or DNA. DNA was first described by J. D. Watson and F. H. C. Crick in 1953 as a double helix consisting of a twisting pair of strands of polynucleotides with two bits of information encoded at each ledge of a spiral staircase, encoded by the choice of nucleotides.[1] This master "read only" memory controls the vast machinery of life.

Supported by a twisting sugar-phosphate backbone, the DNA molecule consists of between several dozen and several million rungs, each of which is coded with one nucleotide letter drawn from a four-letter alphabet of base pairs (adenine-thymine, thymine-adenine, cytosine-guanine, and guanine-cytosine). Human DNA is a long molecule—it would measure up to six feet in length if stretched out—but it is packed into an elaborate coil only $1/2500$ of an inch across.

The mechanism to peel off copies of the DNA code consists of other special machines: organic molecules called enzymes, which split each base pair and then assemble two identical DNA molecules by rematching the broken base pairs. Other little chemical machines then verify the validity of the copy by checking the integrity of the base-pair matches. The error rate of these chemical information-processing transactions is about one error in a billion base-pair replications. There are further redundancy and error-correction codes built into the data itself, so meaningful mistakes are rare. Some mistakes do get through, most of which cause defects in a single cell. Mistakes in an early fetal cell may cause birth defects in the newborn organism. Once in a long while such defects offer an advantage, and this new encoding may eventually be favored through the enhanced survival of that organism and its offspring.

The DNA code controls the salient details of the construction of every cell in the organism, including the shapes and processes of the cell, and of the organs comprised of the cells. In a process called translation, other enzymes translate the coded DNA information by building proteins. It is these proteins that define the structure, behavior, and intelligence of each cell, and of the organism.[2]

This computational machinery is at once remarkably complex and amazingly simple. Only four base pairs provide the data storage for the complexity of all the millions of life-forms on Earth, from primitive bacteria to human beings. The ribosomes—little tape-recorder molecules—read the code and build proteins from only twenty amino acids. The synchronized flexing of muscle cells, the intricate biochemical interactions in our blood, the structure and functioning of our brains, and all of the other diverse functions of the Earth's creatures are programmed in this efficient code.

The genetic information-processing appliance is an existence proof of nano-engineering (building machines atom by atom), because the machinery of life indeed takes place on the atomic level. Tiny bits of molecules consisting of just dozens of atoms encode each bit and perform the transcription, error detection, and correction functions. The actual building of the organic stuff is conducted atom by atom with the building of the amino acid chains.

This is our understanding of the hardware of the computational engine driving life on Earth. We are just beginning, however, to unravel the software. While prolific, evolution has been a sloppy programmer. It has left us the object code (billions of bits of coded data), but there is no higher-level source code (statements in a language we can understand), no explanatory comments, no "help" file, no documentation, and no user manual. Through the Human Genome Project, we are in the process of writing down the 6-billion-bit code for the human genetic code, and are capturing the code for thousands of other species as well.[3] But reverse engineering the genome code—understanding how it works—is a slow and laborious process that we are just beginning. As we do this, however, we are learning the information-processing basis of disease, maturation, and aging, and are gaining the means to correct and refine evolution's unfinished invention.

In addition to evolution's lack of documentation, it is also a very inefficient programmer. Most of the code—97 percent according to current estimates—does not compute; that is, most of the sequences do not produce proteins and appear to be useless. That means that the active part of the code is only about

23 megabytes, which is less than the code for Microsoft Word. The code is also replete with redundancies. For example, an apparently meaningless sequence called Alu, comprising 300 nucleotide letters, occurs 300,000 times in the human genome, representing more than 3 percent of our genetic program.

The theory of evolution states that programming changes are introduced essentially at random. The changes are evaluated for retention by survival of the entire organism and its ability to reproduce. Yet the genetic program controls not just the one characteristic being "experimented" with, but millions of other features as well. Survival of the fittest appears to be a crude technique capable of concentrating on one or at most a few characteristics at a time. Since the vast majority of changes make things worse, it may seem surprising that this technique works at all.

This contrasts with the conventional human approach to computer programming in which changes are designed with a purpose in mind, multiple changes may be introduced at a time, and the changes made are tested by focusing in on each change, rather than by overall survival of the program. If we attempted to improve our computer programs the way that evolution apparently improves its design, our programs would collapse from increasing randomness.

It is remarkable that by concentrating on one refinement at a time, such elaborate structures as the human eye could have been designed. Some observers have postulated that such intricate design is impossible through the incremental-refinement method that evolution uses. A design as intricate as the eye or the heart would appear to require a design methodology in which it was designed all at once.

However, the fact that designs such as the eye have many interacting aspects does not rule out its creation through a design path comprising one small refinement at a time. In utero, the human fetus appears to go through a process of evolution, although whether this is a corollary of the phases of evolution that led to our subspecies is not universally accepted. Nonetheless, most medical students learn that ontogeny (fetal development) recapitulates phylogeny (evolution of a genetically related group of organisms, such as a phylum). We appear to start out in the womb with similarities to a fish embryo, progress to an amphibian, then a mammal, and so on. Regardless of the phylogeny controversy, we can see in the history of evolution the intermediate design drafts that evolution went through in designing apparently "complete" mechanisms such as the human eye. Even though evolution focuses on just one issue at a time, it is indeed capable of creating striking designs with many interacting parts.

There is a disadvantage, however, to evolution's incremental method of design: It can't easily perform complete redesigns. It is stuck, for example, with the very slow computing speed of the mammalian neuron. But there is a way around this, as we will explore in chapter 6, "Building New Brains."

The Evolution of Evolution

There are also certain ways in which evolution has evolved its own means for evolution. The DNA-based coding itself is clearly one such means. Within the code, other means have developed. Certain design elements, such as the shape of the eye, are coded in a way that makes mutations less likely. The error detection and correction mechanisms built into the DNA-based coding make changes in these regions very unlikely. This enforcement of design integrity for certain critical features evolved because they provide an advantage—changes to these characteristics are usually catastrophic. Other design elements, such as the number and layout of light-sensitive rods and cones in the retina, have fewer design enforcements built into the code. If we examine the evolutionary record, we do see more recent change in the layout of the retina than in the shape of the eyeball itself. So in certain ways, the strategies of evolution have evolved. The Law of Accelerating Returns says that it should, for evolving its own strategies is the primary way that an evolutionary process builds on itself.

By simulating evolution, we can also confirm the ability of evolution's "one step at a time" design process to build ingenious designs of many interacting elements. One example is a software simulation of the evolution of life-forms called Network Tierra designed by Thomas Ray, a biologist and rain forest expert.[4] Ray's "creatures" are software simulations of organisms in which each "cell" has its own DNA-like genetic code. The organisms compete with each other for the limited simulated space and energy resources of their simulated environment.

A unique aspect of this artificial world is that the creatures have free rein of 150 computers on the Internet, like "islands in an archipelago" according to Ray. One of the goals of this research is to understand how the explosion of diverse body plans that occurred on Earth during the Cambrian period some 570 million years ago was possible. "To watch evolution unfold is a thrill," Ray exclaimed as he watched his creatures evolve from unspecialized single-celled organisms to multicellular organisms with at least modest increases in diversity. Ray has reportedly identified the equivalent of parasites, immunities, and crude social interaction. One of the acknowledged limitations in Ray's simulation is a lack of complexity in his simulated environment. One insight of this research is the need for a suitably chaotic environment as a key resource needed to push evolution along, a resource in ample supply in the real world.

A practical application of evolution is the area of evolutionary algorithms, in which millions of evolving computer programs compete with one another in a simulated evolutionary process, thereby harnessing the inherent intelligence of evolution to solve real-world problems. Since the intelligence of evolution is weak, we focus and amplify it the same way a lens concentrates the sparse rays of the sun. We'll talk more about this powerful approach to software design in chapter 4, "A New Form of Intelligence on Earth."

The Intelligence Quotient of Evolution

Let us first praise evolution. It has created a plethora of designs of indescribable beauty, complexity, and elegance, not to mention effectiveness. Indeed, some theories of aesthetics define beauty as the degree of success in emulating the natural beauty that evolution has created. It created human beings with their intelligent human brains, beings smart enough to create their own intelligent technology.

Its intelligence seems vast. Or is it? It has one deficiency—evolution is *very* slow. While it is true that it has created some remarkable designs, it has taken an extremely long period of time to do so. It took eons for the process to get started, and, for the evolution of life-forms, eons meant billions of years. Our human forebears also took eons to get started in their creation of technology, but for us eons meant only tens of thousands of years, a distinct improvement.

Is the length of time required to solve a problem or create an intelligent design relevant to an evaluation of intelligence? The authors of our human intelligence-quotient tests seem to think so, which is why most IQ tests are timed. We regard solving a problem in a few seconds better than solving it in a few hours or years. Periodically, the timed aspect of IQ tests gives rise to controversy, but it shouldn't. The speed of an intelligent process is a valid aspect of its evaluation. If a large, hunched, catlike animal perched on a tree limb suddenly appears out of my left cornea, designing an evasive tactic in a second or two is preferable to pondering the challenge for a few hours. If your boss asks you to design a marketing program, she probably doesn't want to wait a hundred years. Viking Penguin wanted this book delivered before the end of the second, not the third, millennium.[5]

Evolution has achieved an extraordinary record of design, yet has taken an extraordinarily long period of time to do so. If we factor its achievements by its ponderous pace, I believe we need to conclude that its intelligence quotient is only infinitesimally greater than zero. An IQ of only slightly greater than zero (defining truly arbitrary behavior as zero) is enough for evolution to beat entropy and create wonderful designs, given enough time, in the same way that an ever so slight asymmetry in the balance between matter and antimatter was enough to allow matter to almost completely overtake its antithesis.

Evolution is thereby only a quantum smarter than completely unintelligent behavior. The reason that our human-created evolutionary algorithms are effective is that we speed up time a million- or billionfold, so as to concentrate and focus its otherwise diffuse power. In contrast, humans are a lot smarter than just a quantum greater than total stupidity (of course, your view may vary depending on the latest news reports).

THE END OF THE UNIVERSE

What does the Law of Time and Chaos say about the end of the Universe?

One theory is that the Universe will continue its expansion forever. Alternatively, if there's enough stuff, then the force of the Universe's own gravity will stop the expansion, resulting in a final "big crunch." Unless, of course, there's an antigravity force. Or if the "cosmological constant," Einstein's "fudge factor," is big enough. I've had to rewrite this paragraph three times over the past several months because the physicists can't make up their minds. The latest speculation apparently favors indefinite expansion.

Personally, I prefer the idea of the Universe closing in again on itself as more aesthetically pleasing. That would mean that the Universe would reverse its expansion and reach a singularity again. We can speculate that it would again expand and contract in an endless cycle. Most things in the Universe seem to move in cycles, so why not the Universe itself? The Universe could then be regarded as a tiny wave particle in some other really big Universe. And that big Universe would itself be a vibrating particle in yet another even bigger Universe. Conversely, the tiny wave particles in our Universe can each be regarded as little Universes with each of their vibrations lasting fractions of a trillionth of a second in our Universe representing billions of years of expansion and contraction in that little Universe. And each particle in those little Universes could be . . . okay, so I'm getting a little carried away.

How to Unsmash a Cup

Let's say the Universe reverses its expansion. The phase of contraction has the opposite characteristics of the phase of expansion that we are now in. Clearly, chaos in the Universe will be decreasing as the Universe gets smaller. I can see that this is the case by considering the endpoint, which is again a singularity with no size, and therefore no disorder.

We regard time as moving in one direction because processes in time are not generally reversible. If we smash a cup, we find it difficult to unsmash it. The reason for this has to do with the second law of thermodynamics. Since overall entropy may increase but can never decrease, time has directionality. Smashing a cup increases randomness. Unsmashing the cup would violate the second law of thermodynamics. Yet in the contracting phase of the Universe, chaos is decreasing, so we should regard time's direction as reversed.

This reverses all processes in time, turning evolution into devolution. Time moves backward during the second half of the Universe's time span. So if you smash a favorite cup, try to do it as we approach the midpoint of the Universe's time span. You should find the cup coming together again as we cross over into the Universe's contracting phase of its time span.

Now if time is moving backward during this contracting phase, what we (living in the expanding phase of the Universe) look forward to as the big crunch is actually a big bang to the creatures living (in reverse time) during the contracting phase. Consider the perspective of these time-reversed creatures living in what we regard as the contracting phase of the Universe. From their perspective, what we regard as the second phase is actually their first phase, with time going in the reverse direction. So from their perspective, the Universe during this phase is expanding, not contracting. Thus, if the "Universe will eventually contract" theory is correct, it would be proper to say that the Universe is bounded in time by two big bangs, with events flowing in opposite directions in time from each big bang, meeting in the middle. Creatures living in both phases can say that they are in the first half of the Universe's history, since both phases will appear to be the first half to creatures living in those phases. And in both halves of the time span of the Universe, the Law of Entropy, the Law of Time and Chaos, and the Law of Accelerating Returns (as applied to evolution) all hold true, but with time moving in opposite directions.[6]

The End of Time

And what if the Universe expands indefinitely? This would mean that the stars and galaxies will eventually exhaust their energy, leaving a Universe of dead stars expanding forever. That would leave a big mess—lots of randomness—and no meaningful order, so according to the Law of Time and Chaos, time would gradually come to a halt. Consistently, if a dead Universe means that there will be no conscious beings to appreciate it, then both the Quantum Mechanical and the Eastern subjective viewpoints[7] appear to imply that the Universe would cease to exist.

In my view, neither conclusion is quite right. At the end of this book, I'll share with you my perspective of what happens at the end of the Universe. But don't look ahead.

Consider the sophistication of *our* creations over a period of only a few thousand years. Ultimately, our machines will match and exceed human intelligence, no matter how one cares to define or measure this elusive term. Even if my time frames are off, few serious observers who have studied the issue claim that computers will never achieve and surpass human intelligence. Humans will have vastly beaten evolution, therefore, achieving in a matter of only thousands of years as much or more than evolution achieved in billions of years. So human in-telligence, a product of evolution, is far more intelligent than its creator.

And so, too, will the intelligence that we are creating come to exceed the in-telligence of its creator. That is not the case today. But as the rest of this book will argue, it will take place very soon—in evolutionary terms, or even in terms of human history—and within the lifetimes of most of the readers of this book. The Law of Accelerating Returns predicts it. And furthermore, it predicts that the pro-gression in the capabilities of human-created machines will only continue to ac-celerate. The human species creating intelligent technology is another example of evolution's progress building on itself. Evolution created human intelligence. Now human intelligence is designing intelligent machines at a far faster pace. Yet another example will be when our intelligent technology takes control of the cre-ation of yet more intelligent technology than itself.

NOW ON THIS TIME THING, WE START OUT AS A SINGLE CELL, RIGHT?

That's right.

AND THEN WE DEVELOP INTO SOMETHING RESEMBLING A FISH, THEN AN AMPHIBIAN, ULTI-MATELY A MAMMAL, AND SO ON   YOU KNOW ONTOGENY RECAPITULATES

Phylogeny, yes.

SO THAT'S JUST LIKE EVOLUTION, RIGHT? WE GO THROUGH EVOLUTION IN OUR MOTHER'S WOMB.

Yes, that's the theory. The word *phylogeny* is derived from phylum . . .

BUT YOU SAID THAT IN EVOLUTION, TIME SPEEDS UP. YET IN AN ORGANISM'S LIFE, TIME SLOWS DOWN.

Ah yes, a good catch, I can explain.

I'M ALL EARS.

The Law of Time and Chaos states that, in a process, the average time interval be-tween salient events is proportional to the amount of chaos in the process. So we have to be careful to define precisely what constitutes the process. It is true that evolution started out with single cells. And we also start out as a single cell. Sounds similar, but from the perspective of the Law of Time and Chaos, it's not. We start out as just *one* cell. When evolution was at the point of single cells, it was not one cell, but many trillions of cells. And these cells were just swirling about; that's a lot of chaos and not much order. The primary movement of evolu-tion has been toward greater order. In the development of an organism, however, the primary movement is toward greater chaos—the grown organism has far greater disorder than the single cell it started out as. It draws that chaos from the environment as its cells multiply, and as it has encounters with its environment. Is that clear?

UH, SURE. BUT DON'T QUIZ ME ON IT. I THINK THE GREATEST CHAOS IN MY LIFE WAS WHEN I LEFT HOME TO GO TO COLLEGE. THINGS ARE JUST BEGINNING TO SETTLE DOWN NOW AGAIN.

I never said the Law of Time and Chaos explains everything.

OKAY, BUT EXPLAIN THIS. YOU SAID THAT EVOLUTION WASN'T VERY SMART, OR AT LEAST WAS RATHER SLOW-WITTED. BUT AREN'T SOME OF THESE VIRUSES AND BACTERIA USING EVOLUTION TO OUTSMART US?

Evolution operates on different timescales. If we speed it up, it can be smarter than us. That's the idea behind software programs that apply a simulated evolutionary process to solving complex problems. Pathogen evolution is another example of the ability of evolution to amplify and focus its diffuse powers. After all, a viral generation can take place in minutes or hours compared to decades for the human race. However, I do think we will ultimately prevail against the evolutionary tactics of our disease agents.

IT WOULD BE HELPFUL IF WE STOPPED OVERUSING ANTIBIOTICS.

Yes, and that brings up another issue, which is whether the human species is more intelligent than its individual members.

AS A SPECIES, WE'RE CERTAINLY PRETTY SELF-DESTRUCTIVE.

That's often true. Nonetheless, we do have a profound species-wide dialogue going on. In other species, the individuals may communicate in a small clan or colony, but there is little, if any, sharing of information beyond that, and little apparent accumulated knowledge. The human knowledge base of science, technology, art, culture, and history has no parallel in any other species.

WHAT ABOUT WHALE SONGS?

Hmmm. I guess we just don't know what they're singing about.

AND WHAT ABOUT THOSE APES THAT YOU CAN TALK TO ON THE INTERNET?

Well, on April 27, 1998, Koko the gorilla did engage in what her mentor, Francine Patterson, called the first interspecies chat, on America Online.[8] But Koko's critics intimate that Patterson is the brains behind Koko.

BUT PEOPLE WERE ABLE TO CHAT WITH KOKO ONLINE.

Yes. However, Koko is rusty on her typing skills, so questions were interpreted by Patterson into American Sign Language, which Koko observed, and then Koko's signed responses were interpreted by Patterson back into typed responses. I guess the suspicion is that Patterson is like those language interpreters from the diplomatic corps—one wonders if you're communicating with the dignitary, in this case Koko, or the interpreter.

ISN'T IT CLEAR IN GENERAL THAT THE APES ARE COMMUNICATING? THEY'RE NOT THAT DIFFERENT FROM US GENETICALLY, AS YOU SAID.

There's clearly some form of communication going on. The question being addressed by the linguistics community is whether the apes can really deal with the levels of symbolism embodied in human language. I think that Dr. Emily Savage-Rumbaugh of Georgia State University, who runs a fifty-five-acre ape-communication laboratory, made a fair statement recently when she said, "They [her critics] are asking Kanzi [one of her ape subjects] to do everything that humans do, which is specious. He'll never do that. It still doesn't negate what he can do."

WELL, I'M ROOTING FOR THE APES.

Yes, it would be nice to have someone to talk to when we get tired of other humans.

SO WHY DON'T YOU JUST HAVE A LITTLE TALK WITH YOUR COMPUTER?

I do talk to my computer, and it dutifully takes down what I say to it. And I can give commands by speaking in natural language to Microsoft Word,[9] but it's still not a very engaging conversationalist. Remember, computers are still a million times simpler than the human brain, so it's going to be a couple of decades yet before they become comforting companions.

BACK ON THIS INDIVIDUAL-VERSUS-GROUP-INTELLIGENCE ISSUE, AREN'T MOST ACHIEVEMENTS IN ART AND SCIENCE ACCOMPLISHED BY INDIVIDUALS? YOU KNOW, YOU CAN'T WRITE A SONG OR PAINT A PICTURE BY COMMITTEE.

Actually, a lot of important science and technology is done in large groups.

BUT AREN'T THE REAL BREAKTHROUGHS DONE BY INDIVIDUALS?

In many cases, that's true. Even then, the critics and the technology conservatives, even the intolerant ones, do play an important screening role. Not every new and different idea is worth pursuing. It's worthwhile having some barriers to break through.

Overall, the human enterprise is clearly capable of achievements that go far beyond what we can do as individuals.

How about the intelligence of a lynch mob?

I suppose a group is not always more intelligent than its members.

Well, I hope those twenty-first-century machines don't exhibit our mob psychology.

Good point.

I mean, I wouldn't want to end up in a dark alley with a band of unruly machines.

We should keep that in mind as we design our future machines. I'll make a little note. . . .

Yes, particularly before the machines start, as you said, designing themselves.

CHAPTER THREE

OF MIND AND MACHINES

PHILOSOPHICAL MIND EXPERIMENTS

"I am lonely and bored; please keep me company."

If your computer displayed this message on its screen, would that convince you that your notebook is conscious and has feelings?

Well, clearly no, it's rather trivial for a program to display such a message. The message actually comes from the presumably human author of the program that includes the message. The computer is just a conduit for the message, like a book or a fortune cookie.

Suppose we add speech synthesis to the program and have the computer speak its plaintive message. Have we changed anything? While we have added technical complexity to the program, and some humanlike communication means, we still do not regard the computer as the genuine author of the message.

Suppose now that the message is not explicitly programmed, but is produced by a game-playing program that contains a complex model of its own situation. The specific message may never have been foreseen by the human creators of the program. It is created by the computer from the state of its own internal model as it interacts with you, the user. Are we getting closer to considering the computer as a conscious, feeling entity?

Maybe just a tad. But if we consider contemporary game software, the illusion is probably short-lived as we gradually figure out the methods and limitations behind the computer's ability for small talk.

Now suppose the mechanisms behind the message grow to become a massive neural net, built from silicon but based on a reverse engineering of the human brain. Suppose we develop a learning protocol for this neural net that enables it to learn human language and model human knowledge. Its circuits are a million times faster than human neurons, so it has plenty of time to read all human literature and develop its own conceptions of reality. Its creators do not tell it how to respond to the world. Suppose now that it says, "I'm lonely . . ."

At what point do we consider the computer to be a conscious agent with its own free will? These have been the most vexing problems in philosophy since the Platonic dialogues illuminated the inherent contradictions in our conception of these terms.

Let's consider the slippery slope from the opposite direction. Our friend Jack (circa some time in the twenty-first century) has been complaining of difficulty with his hearing. A diagnostic test indicates he needs more than a conventional hearing aid, so he gets a cochlear implant. Once used only by people with severe hearing impairments, these implants are now common to correct the ability of people to hear across the entire sonic spectrum. This routine surgical procedure is successful, and Jack is pleased with his improved hearing.

Is he still the same person?

Well, sure he is. People have cochlear implants circa 1999. We still regard them as the same person.

Now (back to circa sometime in the twenty-first century), Jack is so impressed with the success of his cochlear implants that he elects to switch on the built-in phonic-cognition circuits, which improve overall auditory perception. These circuits are already built in so that he does not require another insertion procedure should he subsequently decide to enable them. By activating these neural-replacement circuits, the phonics-detection nets built into the implant bypass his own aging neural-phonics regions. His cash account is also debited for the use of this additional neural software. Again, Jack is pleased with his improved ability to understand what people are saying.

Do we still have the same Jack? Of course; no one gives it a second thought.

Jack is now sold on the benefits of the emerging neural-implant technology. His retinas are still working well, so he keeps them intact (although he does have permanently implanted retinal-imaging displays in his corneas to view virtual reality), but he decides to try out the newly introduced image-processing implants, and is amazed how much more vivid and rapid his visual perception has become.

Still Jack? Why, sure.

Jack notices that his memory is not what it was, as he struggles to recall names, the details of earlier events, and so on. So he's back for memory implants. These are amazing—memories that had grown fuzzy with time are now as clear as if they had just happened. He also struggles with some unintended consequences as he encounters unpleasant memories that he would have preferred to remain dim.

Still the same Jack? Clearly he has changed in some ways and his friends are impressed with his improved faculties. But he has the same self-deprecating humor, the same silly grin—yes, it's still the same guy.

So why stop here? Ultimately Jack will have the option of scanning his entire brain and neural system (which is not entirely located in the skull) and replacing it with electronic circuits of far greater capacity, speed, and reliability. There's also the benefit of keeping a backup copy in case anything happened to the physical Jack.

Certainly this specter is unnerving, perhaps more frightening than appealing. And undoubtedly it will be controversial for a long time (although according to the Law of Accelerating Returns, a "long time" is not as long as it used to be). Ultimately, the overwhelming benefits of replacing unreliable neural circuits with improved ones will be too compelling to ignore.

Have we lost Jack somewhere along the line? Jack's friends think not. Jack also claims that he's the same old guy, just newer. His hearing, vision, memory, and reasoning ability have all improved, but it's still the same Jack.

However, let's examine the process a little more carefully. Suppose rather than implementing this change a step at a time as in the above scenario, Jack does it all at once. He goes in for a complete brain scan and has the information from the scan instantiated (installed) in an electronic neural computer. Not one to do things piecemeal, he upgrades his body as well. Does making the transition at one time change anything? Well, what's the difference between changing from neural circuits to electronic/photonic ones all at once, as opposed to doing it gradually? Even if he makes the change in one quick step, the new Jack is still the same old Jack, right?

But what about Jack's old brain and body? Assuming a noninvasive scan, these still exist. This is Jack! Whether the scanned information is subsequently used to instantiate a copy of Jack does not change the fact that the original Jack still exists and is relatively unchanged. Jack may not even be aware of whether or not a new Jack is ever created. And for that matter, we can create more than one new Jack.

If the procedure involves destroying the old Jack once we have conducted some quality-assurance steps to make sure the new Jack is fully functional, does that not constitute the murder (or suicide) of Jack?

Suppose the original scan of Jack is not noninvasive, that it is a "destructive" scan. Note that technologically speaking, a destructive scan is much easier—in fact we have the technology today (1999) to destructively scan frozen neural sections, ascertain the interneuronal wiring, and reverse engineer the neurons' parallel digital-analog algorithms.[1] We don't yet have the bandwidth to do this quickly enough to scan anything but a very small portion of the brain. But the same speed issue existed for another scanning project—the human genome scan—when that project began. At the speed that researchers were able to scan and sequence the human genetic code in 1991, it would have taken thousands of years to complete the project. Yet a fourteen-year schedule was set, which it now appears will be successfully realized. The Human Genome Project deadline obviously made the (correct) assumption that the speed of our methods for sequencing DNA codes would greatly accelerate over time. The same phenomenon will hold true for our human-brain-scanning projects. We can do it now—very slowly—but that speed, like most everything else governed by the Law of Accelerating Returns, will get exponentially faster in the years ahead.

Now suppose as we destructively scan Jack, we simultaneously install this information into the new Jack. We can consider this a process of "transferring" Jack to his new brain and body. So one might say that Jack is not destroyed, just transferred into a more suitable embodiment. But is this not equivalent to scanning Jack noninvasively, subsequently instantiating the new Jack and then destroying the old Jack? If that sequence of steps basically amounts to killing the old Jack, then this process of transferring Jack in a single step must amount to the same thing. Thus we can argue that any process of transferring Jack amounts to the old Jack committing suicide, and that the new Jack is not the same person.

The concept of scanning and reinstantiation of the information is familiar to us from the fictional "beam me up" teleportation technology of *Star Trek*. In this fictional show, the scan and reconstitution is presumably on a nanoengineering scale, that is, particle by particle, rather than just reconstituting the salient algorithms of neural-information processing envisioned above. But the concept is very similar. Therefore, it can be argued that the *Star Trek* characters are committing suicide each time they teleport, with new characters being created. These new characters, while essentially identical, are made up of entirely different particles, unless we imagine that it is the actual particles being beamed to the new destination. Probably it would be easier to beam just the information and use local particles to instantiate the new embodiments. Should it matter? Is consciousness a function of the actual particles or just of their pattern and organization?

We can argue that consciousness and identity are not a function of the specific particles at all, because our own particles are constantly changing. On a cellular basis, we change most of our cells (although not our brain cells) over a period of several years.[2] On an atomic level, the change is much faster than that, and does include our brain cells. We are not at all permanent collections of particles. It is the patterns of matter and energy that are semipermanent (that is, changing only gradually), but our actual material content is changing constantly, and very quickly. We are rather like the patterns that water makes in a stream. The rushing water around a formation of rocks makes a particular, unique pattern. This pattern may remain relatively unchanged for hours, even years. Of course, the actual material constituting the pattern—the water—is totally replaced within milliseconds. This argues that we should not associate our fundamental identity with specific sets of particles, but rather the pattern of matter and energy that we represent. This, then, would argue that we should consider the new Jack to be the same as the old Jack because the pattern is the same. (One might quibble that while the new Jack has similar functionality to the old Jack, he is not identical. However, this just dodges the essential question, because we can reframe the scenario with a nanoengineering technology that copies Jack atom by atom rather than just copying his salient information-processing algorithms.)

Contemporary philosophers seem to be partial to the "identity from pattern" argument. And given that our pattern changes only slowly in comparison to our particles, there is some apparent merit to this view. But the counter to that argument is the "old Jack" waiting to be extinguished after his "pattern" has been scanned and installed in a new computing medium. Old Jack may suddenly realize that the "identity from pattern" argument is flawed.

MIND AS MACHINE VERSUS MIND BEYOND MACHINE

Science cannot solve the ultimate mystery of nature because in the last analysis we are part of the mystery we are trying to solve.

—Max Planck

Is all what we see or seem, but a dream within a dream?

—Edgar Allan Poe

What if everything is an illusion and nothing exists? In that case, I definitely overpaid for my carpet.

—Woody Allen

The Difference Between Objective and Subjective Experience

Can we explain the experience of diving into a lake to someone who has never been immersed in water? How about the rapture of sex to someone who has never had erotic feelings (assuming one could find such a person)? Can we explain the emotions evoked by music to someone congenitally deaf? A deaf person will certainly learn a lot about music: watching people sway to its rhythm, reading about its history and role in the world. But none of this is the same as experiencing a Chopin prelude.

If I view light with a wavelength of 0.000075 centimeters, I see red. Change the wavelength to 0.000035 centimeters and I see violet. The same colors can also be produced by mixing colored lights. If red and green lights are properly combined, I see yellow. Mixing pigments works differently from changing wavelengths, however, because pigments subtract colors rather than add them. Human perception of color is more complicated than mere detection of electromagnetic frequencies, and we still do not fully understand it. Yet even if we had a fully satisfactory theory of our mental process, it would not convey the subjective experience of redness, or yellowness. I find language inadequate for expressing my experience of redness. Perhaps I can muster some poetic reflections about it, but unless you've had the same encounter, it is really not possible for me to share my experience.

So how do I know that you experience the same thing when you talk about redness? Perhaps you experience red the way I experience blue, and vice versa. How can we test our assumptions that we experience these qualities the same way? Indeed, we do know there are some differences. Since I have what is misleadingly labeled "red-green" color-blindness, there are shades of color that appear identical to me that appear different to others. Those of you without this disability apparently have a different experience than I do. What are you all experiencing? I'll never know.

Giant squids are wondrous sociable creatures with eyes similar in structure to humans (which is surprising, given their very different phylogeny) and possessing a complex nervous system. A few fortunate human scientists have developed relationships with these clever cephalopods. So what is it like to be a giant squid? When we see it respond to danger and express behavior that reminds us of a human emotion, we infer an experience that we are familiar with. But what of their experiences without a human counterpart?

Or do they have experiences at all? Maybe they are just like "machines"—responding programmatically to stimuli in their environment. Maybe there is no one home. Some humans are of this view—only humans are conscious; animals just respond to the world by "instinct," that is, like a machine. To many other humans, this author included, it seems apparent that at least the more evolved animals are conscious creatures, based on empathetic perceptions of animals expressing emotions that we recognize as correlates of human reactions. Yet even this is a human-centric way of thinking in that it only recognizes subjective experiences with a human equivalent. Opinion on animal consciousness is far from unanimous. Indeed, it is the question of consciousness that underlies the issue of animal rights. Animal rights disputes about whether or not certain animals are suffering in certain situations result from our general inability to experience or measure the subjective experience of another entity.[3]

The not uncommon view of animals being "just machines" is disparaging to both animals and machines. Machines today are still a million times simpler than the human brain. Their complexity and subtlety today is comparable to that of insects. There is relatively little speculation on the subjective experience of insects, although again, there is no convincing way to measure this. But the disparity in the capabilities of machines and the more advanced animals, such as the *Homo sapiens sapiens* subspecies, will be short-lived. The unrelenting advance of machine intelligence, which we will visit in the next several chapters, will bring machines to human levels of intricacy and refinement and beyond within several decades. Will these machines be conscious?

And what about free will—will machines of human complexity make their own decisions, or will they just follow a program, albeit a very complex one? Is there a distinction to be made here?

The issue of consciousness lurks behind other vexing issues. Take the question of abortion. Is a fertilized egg cell a conscious human being? How about a fetus one day before birth? It's hard to say that a fertilized egg is conscious or that a full-term fetus is not. Pro-choice and pro-life activists are afraid of the slippery slope in between these two definable ledges. And the slope is genuinely slippery—a human fetus develops a brain quickly, but it's not immediately recognizable as a human brain. The brain of a fetus becomes more humanlike gradually. The slope has no ridges to stand on. Admittedly, other hard-to-define questions such as human dignity come into the debate, but fundamentally, the contention concerns sentience. In other words, when do we have a conscious entity?

Some severe forms of epilepsy have been successfully treated by surgical removal of the impaired half of the brain. This drastic surgery needs to be done during childhood before the brain has fully matured. Either half of the brain can be removed, and if the operation is successful the child will grow up somewhat normally. Does this imply that both halves of the brain have their own consciousness? Perhaps there are two of us in each intact brain who hopefully get along with each other. Maybe there is a whole panoply of consciousnesses lurking in one brain, each with a somewhat different perspective. Is there a consciousness that is aware of mental processes that we consider unconscious?

I could go on for a long time with such conundrums. And indeed, people have been thinking about these quandaries for a long time. Plato, for one, was preoccupied with these issues. In the *Phaedo*, *The Republic*, and *Theaetetus*, Plato expresses the profound paradox inherent in the concept of consciousness and a human's apparent ability to freely choose. On the one hand, human beings partake of the natural world and are subject to its laws. Our brains are natural phenomena and thus must follow the cause-and-effect laws manifest in machines and other lifeless creations of our species. Plato was familiar with the potential complexity of machines and their ability to emulate elaborate logical processes. On the other hand, cause-and-effect mechanics, no matter how complex, should not, according to Plato, give rise to self-awareness or consciousness. Plato first attempts to resolve this conflict in his theory of the Forms: Consciousness is not an attribute of the mechanics of thinking, but rather the ultimate reality of human existence. Our consciousness, or "soul," is immutable and unchangeable. Thus, our mental interaction with the physical world is on the level of the "mechanics" of our complicated thinking process. The soul stands aloof.

But no, this doesn't really work, Plato realizes. If the soul is unchanging, then it cannot learn or partake in reason, because it would need to change to absorb and respond to experience. Plato ends up dissatisfied with positing consciousness in either place: the rational processes of the natural world or the mystical level of the ideal Form of the self or soul.[4]

The concept of free will reflects an even deeper paradox. Free will is purposeful behavior and decision making. Plato believed in a "corpuscular physics" based on fixed and determined rules of cause and effect. But if human decision making is based on such predictable interactions of basic particles, our decisions must also be predetermined. That would contradict human freedom to choose. The addition of randomness into the natural laws is a possibility, but it does not solve the problem. Randomness would eliminate the predetermination of decisions and actions, but it contradicts the purposefulness of free will, as there is nothing purposeful in randomness.

Okay, let's put free will in the soul. No, that doesn't work either. Separating free will from the rational cause-and-effect mechanics of the natural world would require putting reason and learning into the soul as well, for otherwise the soul would not have the means to make meaningful decisions. Now the soul is itself becoming a complex machine, which contradicts its mystical simplicity.

Perhaps this is why Plato wrote dialogues. That way he could passionately express both sides of these contradictory positions. I am sympathetic to Plato's dilemma: None of the obvious positions is really sufficient. A deeper truth can be perceived only by illuminating the opposing sides of a paradox.

Plato was certainly not the last thinker to ponder these questions. We can identify several schools of thought on these subjects, none of them very satisfactory.

The "Consciousness Is Just a Machine Reflecting on Itself" School

A common approach is to deny the issue exists: Consciousness and free will are just illusions induced by the ambiguities of language. A slight variation is that consciousness is not exactly an illusion, but just another logical process. It is a process responding and reacting to itself. We can build that in a machine: just build a procedure that has a model of itself, and that examines and responds to its own methods. Allow the process to reflect on itself. There, now you have consciousness. It is a set of abilities that evolved because self-reflective ways of thinking are inherently more powerful.

The difficulty with arguing against the "consciousness is just a machine reflecting on itself" school is that this perspective is self-consistent. But this viewpoint ignores the subjective viewpoint. It can deal with a person's reporting of subjective experience, and it can relate reports of subjective experiences not only to outward behavior but to patterns of neural firings as well. And if I think about it, my knowledge of the *subjective* experience of anyone aside from myself is no different (to me) than the rest of my *objective* knowledge. I don't experience other people's subjective experiences; I just hear about them. So the only subjective experience this school of thought ignores is my own (that is, after all, what the term *subjective experience* means). And, hey, I'm only one person among billions of humans, trillions of potentially conscious organisms; all of whom, with just one exception, are not me.

But the failure to explain my subjective experience is a serious one. It does not explain the distinction between 0.000075 centimeter electromagnetic radiation and my experience of redness. I could learn how color perception works, how the human brain processes light, how it processes combinations of light, even what patterns of neural firings this all provokes, but it still fails to explain the essence of my experience.

The Logical Positivists[5]

I am doing my best to express what I am talking about here but unfortunately the issue is not entirely effable. D. J. Chalmers describes the mystery of the experienced inner life as the "hard problem" of consciousness, to distinguish this issue from the "easy problem" of how the brain works.[6] Marvin Minsky observed that "there's something queer about describing consciousness: Whatever people mean to say, they just can't seem to make it clear." *That is precisely the problem,* says the "consciousness is just a machine reflecting on itself" school—to speak of consciousness other than as a pattern of neural firings is to wander off into a mystical realm beyond any hope of verification.

This objective view is sometimes referred to as logical positivism, a philosophy codified by Ludwig Wittgenstein in his *Tractatus Logico-Philosophicus*.[7] To the logical positivists, the only things worth talking about are our direct sensory experiences, and the logical inferences that we can make therefrom. Everything else "we must pass over in silence," to quote Wittgenstein's last statement in his treatise.

Yet Wittgenstein did not practice what he preached. Published in 1953, two years after his death, his *Philosophical Investigations* defined those matters worth contemplating as precisely those issues he had earlier argued should be passed over in silence.[8] Apparently he came to the view that the antecedents of his last statement in the *Tractatus*—what we cannot speak about—are the only real phenomena worth reflecting upon. The late Wittgenstein heavily influenced the existentialists, representing perhaps the first time since Plato that a major philosopher was successful in illuminating such contradictory views.

I Think, Therefore I Am

The early Wittgenstein and the logical positivists that he inspired are often thought to have their roots in the philosophical investigations of René Descartes.[9] Descartes's famous dictum "I think, therefore I am" has often been cited as emblematic of Western rationalism. This view interprets Descartes to mean "I think, that is, I can manipulate logic and symbols, therefore I am worthwhile." But in my view, Descartes was not intending to extol the virtues of rational thought. He was troubled by what has become known as the mind-body problem, the paradox of how mind can arise from nonmind, how thoughts and feelings can arise from the ordinary matter of the brain. Pushing rational skepticism to its limits, his statement really means "I think, that is, there is an undeniable mental phenomenon, some awareness, occurring, therefore all we know for sure is that something—let's call it *I*—exists." Viewed in this way, there is less of a gap than is commonly thought between Descartes and Buddhist notions of consciousness as the primary reality.

Before 2030, we will have machines proclaiming Descartes's dictum. And it won't seem like a programmed response. The machines will be earnest and convincing. Should we believe them when they claim to be conscious entities with their own volition?

The "Consciousness Is a Different Kind of Stuff" School

The issue of consciousness and free will has been, of course, a major preoccupation of religious thought. Here we encounter a panoply of phenomena, ranging from the elegance of Buddhist notions of consciousness to ornate pantheons of souls, angels, and gods. In a similar category are theories by contemporary philosophers that regard consciousness as yet another fundamental phenomenon in the world, like basic particles and forces. I call this the "consciousness is a different kind of stuff" school. To the extent that this school implies an interference by consciousness in the physical world that runs afoul of scientific experiment, science is bound to win because of its ability to verify its insights. To the extent that this view stays aloof from the material world, it often creates a level of complex mysticism that cannot be verified and is subject to disagreement. To the extent that it keeps its mysticism simple, it offers limited objective insight, although subjective insight is another matter (I do have to admit a fondness for simple mysticism).

The "We're Too Stupid" School

Another approach is to declare that human beings just aren't capable of understanding the answer. Artificial intelligence researcher Douglas Hofstadter muses that "it could be simply an accident of fate that our brains are too weak to understand themselves. Think of the lowly giraffe, for instance, whose brain is obviously far below the level required for self-understanding—yet it is remarkably similar to our brain."[10] But to my knowledge, giraffes are not known to ask these questions (of course, we don't know what they spend their time wondering about). In my view, if we are sophisticated enough to ask the questions, then we are advanced enough to understand the answers. However, the "we're too stupid" school points out that indeed we are having difficulty clearly formulating these questions.

A Synthesis of Views

My own view is that all of these schools are correct when viewed together, but insufficient when viewed one at a time. That is, the truth lies in a synthesis of these views. This reflects my Unitarian religious education in which we studied all the world's religions, considering them "many paths to the truth." Of course, my view may be regarded as the worst one of all. On its face, my view is contradictory and makes little sense. The other schools at least can claim some level of consistency and coherence.

Thinking Is as Thinking Does

Oh yes, there is one other view, which I call the "thinking is as thinking does" school. In a 1950 paper, Alan Turing describes his concept of the Turing Test, in which a human judge interviews both a computer and one or more human foils using terminals (so that the judge won't be prejudiced against the computer for lacking a warm and fuzzy appearance).[11] If the human judge is unable to reliably unmask the computer (as an impostor human) then the computer wins. The test is often described as a kind of computer IQ test, a means of determining if computers have achieved a human level of intelligence. In my view, however, Turing really intended his Turing Test as a test of thinking, a term he uses to imply more than just clever manipulation of logic and language. To Turing, thinking implies conscious intentionality.

Turing had an implicit understanding of the exponential growth of computing power, and predicted that a computer would pass his eponymous exam by the end of the century. He remarked that by that time "the use of words and

THE VIEW FROM QUANTUM MECHANICS

I often dream about falling. Such dreams are commonplace to the ambitious or those who climb mountains. Lately I dreamed I was clutching at the face of a rock, but it would not hold. Gravel gave way. I grasped for a shrub, but it pulled loose, and in cold terror I fell into the abyss. Suddenly I realized that my fall was relative; there was no bottom and no end. A feeling of pleasure overcame me. I realized that what I embody, the principle of life, cannot be destroyed. It is written into the cosmic code, the order of the universe. As I continued to fall in the dark void, embraced by the vault of the heavens, I sang to the beauty of the stars and made my peace with the darkness.

—Heinz Pagels, physicist and quantum mechanics researcher
before his death in a 1988 climbing accident

The Western *objective* view states that after billions of years of swirling around, matter and energy evolved to create life-forms—complex self-replicating patterns of matter and energy—that became sufficiently advanced to reflect on their own existence, on the nature of matter and energy, on their own consciousness. In contrast, the Eastern *subjective* view states that consciousness came first—matter and energy are merely the complex thoughts of conscious beings, ideas that have no reality without a thinker.

As noted above, the objective and subjective views of reality have been at odds since the dawn of recorded history. There is often merit, however, in combining seemingly irreconcilable views to achieve a deeper understanding. Such was the case with the adoption of quantum mechanics fifty years ago. Rather than reconcile the views that electromagnetic radiation (for example, light) was either a stream of particles (that is, photons) or a vibration (that is, light waves), both views were fused into an irreducible duality. While this idea is impossible to grasp using only our intuitive models of nature, we are unable to explain the world without accepting this apparent contradiction. Other paradoxes of quantum mechanics (for example, electron "tunneling" in which electrons in a transistor appear on both sides of a barrier) helped create the age of computation, and may unleash a new revolution in the form of the quantum computer,[12] but more about that later.

Once we accept such a paradox, wonderful things happen. In postulating the duality of light, quantum mechanics has discovered an essential nexus between matter and consciousness. Particles apparently do not make up their minds as to which way they are going or even where they have been until they are forced to do so by the observations of a conscious observer. We might say that they appear not really to exist at all retroactively until and unless we notice them.

So twentieth-century Western science has come around to the Eastern view. The Universe is sufficiently sublime that the essentially Western objective view of consciousness arising from matter and the essentially Eastern subjective view of matter arising from consciousness apparently coexist as another irreducible duality. Clearly, consciousness, matter, and energy are inextricably linked.

We may note here a similarity of quantum mechanics to the computer simulation of a virtual world. In today's software games that display images of a virtual world, the portions of the environment not currently being interacted with by the user (that is, those offscreen) are usually not computed in detail, if at all. The limited resources of the computer are directed toward rendering the portion of the world that the user is currently viewing. As the user focuses in on some other aspect, the computational resources are then immediately directed toward creating and displaying that new perspective. It thus seems as if the portions of the virtual world that are offscreen are nonetheless still "there," but the software designers figure there is no point wasting valuable computer cycles on regions of their simulated world that no one is watching.

I would say that quantum theory implies a similar efficiency in the physical world. Particles appear not to decide where they have been until forced to do so by being observed. The implication is that portions of the world we live in are not actually "rendered" until some conscious observer turns her attention toward them. After all, there's no point wasting valuable "computes" of the celestial computer that renders our Universe. This gives new meaning to the question about the unheard tree that falls in the forest.

general educated opinion will have altered so much that one will be able to speak of machines thinking without expecting to be contradicted." His prediction was overly optimistic in terms of time frame, but in my view not by much.

In the end, Turing's prediction foreshadows how the issue of computer thought will be resolved. The machines will convince us that they are conscious, that they have their own agenda worthy of our respect. We will come to believe that they are conscious much as we believe that of each other. More so than with our animal friends, we will empathize with their professed feelings and struggles because their minds will be based on the design of human thinking. They will embody human qualities and will claim to be human. And we'll believe them.

ON THIS MULTIPLE-CONSCIOUSNESS IDEA, WOULDN'T I NOTICE THAT—I MEAN IF I HAD DECIDED TO DO ONE THING AND THIS OTHER CONSCIOUSNESS IN MY HEAD WENT AHEAD AND DECIDED SOMETHING ELSE?

I thought you had decided not to finish that muffin you just devoured.

TOUCHÉ. OKAY, IS THAT AN EXAMPLE OF WHAT YOU'RE TALKING ABOUT?

It is a better example of Marvin Minsky's *Society of Mind*, in which he conceives of our mind as a society of other minds—some like muffins, some are vain, some are health conscious, some make resolutions, others break them. Each of these in turn is made up of other societies. At the bottom of this hierarchy are little mechanisms Minsky calls agents with little or no intelligence. It is a compelling vision of the organization of intelligence, including such phenomena as mixed emotions and conflicting values.

SOUNDS LIKE A GREAT LEGAL DEFENSE. "NO, JUDGE, IT WASN'T ME. IT WAS THIS OTHER GAL IN MY HEAD WHO DID THE DEED!"

That's not going to do you much good if the judge decides to lock up the other gal in your head.

THEN HOPEFULLY THE WHOLE SOCIETY IN MY HEAD WILL STAY OUT OF TROUBLE. BUT WHICH MINDS IN MY SOCIETY OF MIND ARE CONSCIOUS?

We could imagine that each of these minds in the society of mind is conscious, albeit that the lowest-ranking ones have relatively little to be conscious of. Or perhaps consciousness is reserved for the higher-ranking minds. Or perhaps only certain combinations of higher-ranking minds are conscious, whereas others are not. Or perhaps—

NOW WAIT A SECOND, HOW CAN WE TELL WHAT THE ANSWER IS?

I believe there's really no way to tell. What possible experiment can we run that would conclusively prove whether an entity or process is conscious? If the entity says, "Hey, I'm really conscious," does that settle the matter? If the entity is very compelling when it expresses a professed emotion, is that definitive? If we look carefully at its internal methods and see feedback loops in which the process examines and responds to itself, does that mean it's conscious? If we see certain types of patterns in its neural firings, is that convincing? Contemporary philosophers such as Daniel Dennett appear to believe that the consciousness of an entity is a testable and measurable attribute. But I think science is inherently about objective reality. I don't see how it can break through to the subjective level.

MAYBE IF THE THING PASSES THE TURING TEST?

That is what Turing had in mind. Lacking any conceivable way of building a consciousness detector, he settled on a practical approach, one that emphasizes our unique human proclivity for language. And I do think that Turing is right in a way—if a machine can pass a valid Turing Test, I believe that we will believe that it is conscious. Of course, that's still not a scientific demonstration.

The converse proposition, however, is not compelling. Whales and elephants have bigger brains than we do and exhibit a wide range of behaviors that knowledgeable observers consider intelligent. I regard them as conscious creatures, but they are in no position to pass the Turing Test.

THEY WOULD HAVE TROUBLE TYPING ON THESE SMALL KEYS OF MY COMPUTER.

Indeed, they have no fingers. They are also not proficient in human languages. The Turing Test is clearly a human-centric measurement.

IS THERE A RELATIONSHIP BETWEEN THIS CONSCIOUSNESS STUFF AND THE ISSUE OF TIME THAT WE SPOKE ABOUT EARLIER?

Yes, we clearly have an awareness of time. Our subjective experience of time passage—and remember that *subjective* is just another word for *conscious*—is governed by the speed of our objective processes. If we change this speed by altering our computational substrate, we affect our perception of time.

RUN THAT BY ME AGAIN.

Let's take an example. If I scan your brain and nervous system with a suitably advanced noninvasive-scanning technology of the early twenty-first century—a very-high-resolution, high-bandwidth magnetic resonance imaging, perhaps—ascertain all the salient information processes and then download that information to my suitably advanced neural computer, I'll have a little you or at least someone very much like you right here in my personal computer.

If my personal computer is a neural net of simulated neurons made of electronic stuff rather than human stuff, the version of you in my computer will run about a million times faster. So an hour for me would be a million hours for you, which is about a century.

OH, THAT'S GREAT, YOU'LL DUMP ME IN YOUR PERSONAL COMPUTER, AND THEN FORGET ABOUT ME FOR A SUBJECTIVE MILLENNIUM OR TWO.

We'll have to be careful about that, won't we.

CHAPTER FOUR

A NEW FORM OF INTELLIGENCE ON EARTH

THE ARTIFICIAL INTELLIGENCE MOVEMENT

What if these theories are really true, and we were magically shrunk and put into someone's brain while he was thinking. We would see all the pumps, pistons, gears and levers working away, and we would be able to describe their workings completely, in mechanical terms, thereby completely describing the thought processes of the brain. But that description would nowhere contain any mention of thought! It would contain nothing but descriptions of pumps, pistons, levers!

—Gottfried Wilhelm Leibniz

Artificial stupidity (AS) may be defined as the attempt by computer scientists to create computer programs capable of causing problems of a type normally associated with human thought.

—Wallace Marshal

Artificial intelligence (AI) is the science of how to get machines to do the things they do in the movies.

—Astro Teller

The Ballad of Charles and Ada

Returning to the evolution of intelligent machines, we find Charles Babbage sitting in the rooms of the Analytical Society at Cambridge, England, in 1821, with a table of logarithms lying before him.

"Well, Babbage, what are you dreaming about?" asked another member, seeing Babbage half asleep.

"I am thinking that all these tables might be calculated by machinery!" Babbage replied.

From that moment on, Babbage devoted most of his waking hours to an unprecedented vision: the world's first programmable computer. Although based entirely on the mechanical technology of the nineteenth century, Babbage's "Analytical Engine" was a remarkable foreshadowing of the modern computer.[1]

Babbage developed a liaison with the beautiful Ada Lovelace, the only legitimate child of Lord Byron, the poet. She became as obsessed with the project as Babbage, and contributed many of the ideas for programming the machine, including the invention of the programming loop and the subroutine. She was the world's first software engineer, indeed the only software engineer prior to the twentieth century.

Lovelace significantly extended Babbage's ideas and wrote a paper on programming techniques, sample programs, and the potential of this technology to emulate intelligent human activities. She describes the speculations of Babbage and herself on the capacity of the Analytical Engine, and future machines like it, to play chess and compose music. She finally concludes that although the computations of the Analytical Engine could not properly be regarded as "thinking," they could nonetheless perform activities that would otherwise require the extensive application of human thought.

The story of Babbage and Lovelace ends tragically. She died a painful death from cancer at the age of thirty-six, leaving Babbage alone again to pursue his quest. Despite his ingenious constructions and exhaustive effort, the Analytical Engine was never completed. Near the end of his existence he remarked that he had never had a happy day in his life. Only a few mourners were recorded at Babbage's funeral in 1871.[2]

What did survive were Babbage's ideas. The first American programmable computer, the Mark I, completed in 1944 by Howard Aiken of Harvard University and IBM, borrowed heavily from Babbage's architecture. Aiken commented, "If Babbage had lived seventy-five years later, I would have been out of a job."[3]

Babbage and Lovelace were innovators nearly a century ahead of their time. Despite Babbage's inability to finish any of his major initiatives, their concepts of a computer with a stored program, self-modifying code, addressable memory, conditional branching, and computer programming itself still form the basis of computers today.[4]

Again, Enter Alan Turing

By 1940, Hitler had the mainland of Europe in his grasp, and England was preparing for an anticipated invasion. The British government organized its best mathematicians and electrical engineers, under the intellectual leadership of Alan Turing, with the mission of cracking the German military code. It was recognized that with the German air force enjoying superiority in the skies, failure to accomplish this mission was likely to doom the nation. In order not to be distracted from their task, the group lived in the tranquil pastures of Hertfordshire, England.

Turing and his colleagues constructed the world's first operational computer from telephone relays and named it Robinson,[5] after a popular cartoonist who drew "Rube Goldberg" machines (very ornate machinery with many interacting mechanisms). The group's own Rube Goldberg succeeded brilliantly and provided the British with a transcription of nearly all significant Nazi messages. As the Germans added to the complexity of their code (by adding additional coding wheels to their Enigma coding machine), Turing replaced Robinson's electromagnetic intelligence with an electronic version called Colossus built from two thousand radio tubes. Colossus and nine similar machines running in parallel provided an uninterrupted decoding of vital military intelligence to the Allied war effort.

Use of this information required supreme acts of discipline on the part of the British government. Cities that were to be bombed by Nazi aircraft were not forewarned, lest preparations arouse German suspicions that their code had been cracked. The information provided by Robinson and Colossus was used only with the greatest discretion, but the cracking of Enigma was enough to enable the Royal Air Force to win the Battle of Britain.

Thus fueled by the exigencies of war, and drawing upon a diversity of intellectual traditions, a new form of intelligence emerged on Earth.

The Birth of Artificial Intelligence

The similarity of the computational process to the human thinking process was not lost on Turing. In addition to having established much of the theoretical foundations of computation and having invented the first operational computer, he was instrumental in the early efforts to apply this new technology to the emulation of intelligence.

In his classic 1950 paper, *Computing Machinery and Intelligence*, Turing described an agenda that would in fact occupy the next half century of advanced computer research: game playing, decision making, natural language understanding, translation, theorem proving, and, of course, encryption and the cracking of codes.[6] He wrote (with his friend David Champernowne) the first chess-playing program.

As a person, Turing was unconventional and extremely sensitive. He had a wide range of unusual interests, from the violin to morphogenesis (the differentiation of cells). There were public reports of his homosexuality, which greatly disturbed him, and he died at the age of forty-one, a suspected suicide.

The Hard Things Were Easy

In the 1950s, progress came so rapidly that some of the early pioneers felt that mastering the functionality of the human brain might not be so difficult after all. In 1956, AI researchers Allen Newell, J. C. Shaw, and Herbert Simon created a program called Logic Theorist (and in 1957 a later version called General Problem Solver), which used recursive search techniques to solve problems in mathematics.[7] Recursion, as we will see later in this chapter, is a powerful method of defining a solution in terms of itself. Logic Theorist and General Problem Solver were able to find proofs for many of the key theorems in Bertrand Russell and Alfred North Whitehead's seminal work on set theory, *Principia Mathematica*,[8] including a completely original proof for an important theorem that had never been previously solved. These early successes led Simon and Newell to say in a 1958 paper, entitled *Heuristic Problem Solving: The Next Advance in Operations Research*, "There are now in the world machines that think, that learn and that create. Moreover, their ability to do these things is going to increase rapidly until—in a visible future—the range of problems they can handle will be coextensive with the range to which the human mind has been applied."[9] The paper goes on to predict that within ten years (that is, by 1968) a digital computer would be the world chess champion. A decade later, an unrepentant Simon predicts that by 1985, "machines will be capable of doing any work that a man can do." Perhaps Simon was intending a favorable comment on the capabilities of women, but these predictions, decidedly more optimistic than Turing's, embarrassed the nascent AI field.

The field has been inhibited by this embarrassment to this day, and AI researchers have been reticent in their prognostications ever since. In 1997, when Deep Blue defeated Gary Kasparov, then the reigning human world chess champion, one prominent professor commented that all we had learned was that playing a championship game of chess does not require intelligence after all.[10] The implication is that capturing *real* intelligence in our machines remains far beyond our grasp. While I don't wish to overstress the significance of Deep Blue's victory, I believe that from this perspective we will ultimately find that there are no human activities that require "real" intelligence.

During the 1960s, the academic field of AI began to flesh out the agenda that Turing had described in 1950, with encouraging or frustrating results, depending on your point of view. Daniel G. Bobrow's program Student could solve algebra problems from natural English-language stories and reportedly did well on high-school math tests.[11] The same performance was reported for Thomas G. Evans's Analogy program for solving IQ-test geometric-analogy problems.[12] The field of expert systems was initiated with Edward A. Feigenbaum's DENDRAL, which could answer questions about chemical compounds.[13] And natural-language

understanding got its start with Terry Winograd's SHRDLU, which could understand any meaningful English sentence, so long as you talked about colored
blocks.[14]

The notion of creating a new form of intelligence on Earth emerged with an
intense and often uncritical passion simultaneously with the electronic hardware
on which it was to be based. The unbridled enthusiasm of the field's early pioneers also led to extensive criticism of these early programs for their inability to
react intelligently in a variety of situations. Some critics, most notably existential-
ist philosopher and phenomenologist Hubert Dreyfus, predicted that machines
would never match human levels of skill in areas ranging from the playing of
chess to the writing of books about computers.

It turned out that the problems we thought were difficult—solving mathematical theorems, playing respectable games of chess, reasoning within domains
such as chemistry and medicine—were easy, and the multithousand-instructions-
per-second computers of the 1950s and 1960s were often adequate to provide
satisfactory results. What proved elusive were the skills that any five-year-old
child possesses: telling the difference between a dog and a cat, or understanding an animated cartoon. We'll talk more about why the easy problems are hard
in Part II.

Waiting for Real Artificial Intelligence

The 1980s saw the early commercialization of artificial intelligence with a wave
of new AI companies forming and going public. Unfortunately, many made the
mistake of concentrating on a powerful but inherently inefficient interpretive
language called LISP, which had been popular in academic AI circles. The commercial failure of LISP and the AI companies that emphasized it created a back-
lash. The field of AI started shedding its constituent disciplines, and companies
in natural-language understanding, character and speech recognition, robotics,
machine vision, and other areas originally considered part of the AI discipline
now shunned association with the field's label.

Machines with sharply focused intelligence nonetheless became increasingly
pervasive. By the mid-1990s, we saw the infiltration of our financial institutions
by systems using powerful statistical and adaptive techniques. Not only were the
stock, bond, currency, commodity, and other markets managed and maintained
by computerized networks, but the majority of buy-and-sell decisions were initiated by software programs that contained increasingly sophisticated models of
their markets. The 1987 stock market crash was blamed in large measure on the
rapid interaction of trading programs. Trends that otherwise would have taken
weeks to manifest themselves developed in minutes. Suitable modifications to
these algorithms have managed to avoid a repeat performance.

Since 1990, the electrocardiogram (EKG) has come complete with the computer's own diagnosis of one's cardiac health. Intelligent image-processing programs enable doctors to peer deep into our bodies and brains, and computerized bioengineering technology enables drugs to be designed on biochemical simulators. The disabled have been particularly fortunate beneficiaries of the age of intelligent machines. Reading machines have been reading to blind and dyslexic persons since the 1970s, and speech-recognition and robotic devices have been assisting hands-disabled individuals since the 1980s.

Perhaps the most dramatic public display of the changing values of the age of knowledge took place in the military. We saw the first effective example of the increasingly dominant role of machine intelligence in the Gulf War of 1991. The cornerstones of military power from the beginning of recorded history through most of the twentieth century—geography, manpower, firepower, and battle-station defenses—have been largely replaced by the intelligence of software and electronics. Intelligent scanning by unstaffed airborne vehicles, weapons finding their way to their destinations through machine vision and pattern recognition, intelligent communications and coding protocols, and other manifestations of the information age have transformed the nature of war.

Invisible Species

With the increasingly important role of intelligent machines in all phases of our lives—military, medical, economic and financial, political—it is odd to keep reading articles with titles such as *Whatever Happened to Artificial Intelligence?* This is a phenomenon that Turing had predicted: that machine intelligence would become so pervasive, so comfortable, and so well integrated into our information-based economy that people would fail even to notice it.

It reminds me of people who walk in the rain forest and ask, "Where are all these species that are supposed to live here?" when there are several dozen species of ant alone within fifty feet of them. Our many species of machine intelligence have woven themselves so seamlessly into our modern rain forest that they are all but invisible.

Turing offered an explanation of why we would fail to acknowledge intelligence in our machines. In 1947, he wrote: "The extent to which we regard something as behaving in an intelligent manner is determined as much by our own state of mind and training as by the properties of the object under consideration. If we are able to explain and predict its behavior we have little temptation to imagine intelligence. With the same object, therefore, it is possible that one man would consider it as intelligent and another would not; the second man would have found out the rules of its behavior."

I am also reminded of Elaine Rich's definition of artificial intelligence, as the "study of how to make computers do things at which, at the moment, people are better."

It is our fate as artificial intelligence researchers never to reach the carrot dangling in front of us. Artificial intelligence is inherently defined as the pursuit of difficult computer-science problems that have not yet been solved.

"I think you should be more explicit here in step two."

THE FORMULA FOR INTELLIGENCE

The computer programmer is a creator of universes for which he alone is the lawgiver. . . . No playwright, no stage director, no emperor, however powerful, has ever exercised such absolute authority to arrange a stage or a field of battle and to command such unswervingly dutiful actors or troops.
—Joseph Weizenbaum

A beaver and another forest animal are contemplating an immense man-made dam. The beaver is saying something like "No, I didn't actually build it. But it's based on an idea of mine."
—Edward Fredkin

Simple things should be simple; complex things should be possible.
—Alan Kay

What Is Intelligence?

A goal may be survival—evade a foe, forage for food, find shelter. Or it might be communication—relate an experience, evoke a feeling. Or perhaps it is to partake in a pastime—play a board game, solve a puzzle, catch a ball. Sometimes it is to seek transcendence—create an image, compose a passage. A goal may be well defined and unique, as in the solution to a math problem. Or it may be a personal expression with no clearly right answer.

My view is that intelligence is the ability to use optimally limited resources—including time—to achieve such goals. There is a plethora of other definitions. One of my favorites is by R. W. Young, who defines intelligence as "that faculty of mind by which order is perceived in a situation previously considered disordered."[15] For this definition, we will find the paradigms discussed below quite apropos.

Intelligence rapidly creates satisfying, sometimes surprising plans that meet an array of constraints. The products of intelligence may be clever, ingenious, insightful, or elegant. Sometimes, as in the case of Turing's solution to cracking the Enigma code, an intelligent solution exhibits all of these qualities. Modest tricks may accidentally produce an intelligent answer from time to time, but a true intelligent process that reliably creates intelligent solutions inherently goes beyond a mere recipe. Clearly, no simple formula can emulate the most powerful phenomenon in the Universe: the complex and mysterious process of intelligence.

Actually, that's wrong. All that is needed to solve a surprisingly wide range of intelligent problems is exactly this: simple methods combined with heavy doses of computation (itself a simple process, as Alan Turing demonstrated in 1936 with his conception of the Turing Machine,[16] an elegant model of computation) and examples of the problem. In some cases, we don't even need the latter; just one well-defined statement of the problem will do.

How far can we go with simple paradigms? Is there a class of intelligent problems amenable to simple approaches, with another, more penetrating class that lies beyond its grasp? It turns out that the class of problems solvable with simple approaches is extensive. Ultimately, with sufficient computational brute force (which will be ample in the twenty-first century) and the right formulas in the right combination, there are few definable problems that fail to yield. Except perhaps for this problem: What is the complete set of unifying formulas that underlies intelligence?

Evolution determined an answer to this problem in a few billion years. We've made a good start in a few thousand years. We are likely to finish the job in a few more decades.

These methods, described briefly below, are discussed in more detail in the supplementary section in the back of this book "How to Build an Intelligent Machine in Three Easy Paradigms."

Let's take a look at a few plain yet powerful paradigms. With a little practice, you, too, can build intelligent machines.

The Recursive Formula: Just Carefully State the Problem

A recursive procedure is one that calls itself. Recursion is a useful approach to generating all of the possible solutions to a problem, or, in the context of a game such as chess, all of the possible move-countermove sequences.

Consider the game of chess. We construct a program called "Pick Best Move" to select each move. Pick Best Move starts by listing all of the possible moves from the current state of the board. This is where the careful statement of the problem comes in, because to generate all of the possible moves we need to precisely consider the rules of the game. For each move, the program constructs a hypothetical board that reflects what would happen if we made this move. For each such hypothetical board, we now need to consider what our opponent would do if we made this move. Now recursion comes in, because Pick Best Move simply calls Pick Best Move (that is, itself) to pick the best move for our opponent. In calling itself, Pick Best Move then lists all of the legal moves for our opponent.

The program keeps calling itself, looking ahead to as many moves as we have time to consider, which results in the generation of a huge move-countermove tree. This is another example of exponential growth, because to look ahead an additional half-move requires multiplying the amount of available computation by about five.

Key to the recursive formula is pruning this huge tree of possibilities, and ultimately stopping the recursive growth of the tree. In the game context, if a board looks hopeless for either side, the program can stop the expansion of the move-countermove tree from that point (called a "terminal leaf" of the tree), and consider the most recently considered move to be a likely win or loss.

When all of these nested program calls are completed, the program will have determined the best possible move for the current actual board, within the limits of the depth of recursive expansion that it had time to pursue.

The recursive formula was good enough to build a machine—a specially designed IBM supercomputer—that defeated the world chess champion (although Deep Blue does augment the recursive formula with databases of moves from most of the grand-master games of this century). Ten years ago, in the *Age of Intelligent Machines*, I noted that while the best chess computers were gaining in chess ratings by forty-five points a year, the best humans were advancing by closer to zero points. That put the year in which a computer would beat the world chess champion at 1998, which turned out to be overly pessimistic by one year. Hopefully my predictions in this book will be more accurate.[17]

Our simple recursive rule plays a world-class game of chess. A reasonable question, then, is, What else can it do? We certainly can replace the module that generates chess moves with a module programmed with the rules of another game. Stick in a module that knows the rules of checkers, and you can also beat just about any human. Recursion is really good at backgammon. Hans Berliner's program defeated the human backgammon champion with the slow computers we had back in 1980.[18]

The recursive formula is also a rather good mathematician. Here the goal is to solve a mathematical problem, such as proving a theorem. The rules then become the axioms of the field of math being addressed, as well as previously proved theorems. The expansion at each point is the possible axioms (or previous proved theorems) that can be applied to a proof at each step. This was the approach used by Allen Newell, J. C. Shaw, and Herbert Simon for their 1957 General Problem Solver. Their program outdid Russell and Whitehead on some hard math problems, and thereby fueled the early optimism of the artificial intelligence field.

From these examples, it may appear that recursion is well suited only for problems in which we have crisply defined rules and objectives. But it has also shown promise in computer generation of artistic creations. Ray Kurzweil's Cybernetic Poet, for example, uses a recursive approach.[19] The program establishes a set of goals for each word—achieving a certain rhythmic pattern, poem structure, and word choice that is desirable at that point in the poem. If the program is unable to find a word that meets these criteria, then it backs up and erases the previous word it has written, re-establishes the criteria it had originally set for the word just erased, and goes from there. If that also leads to a dead end, it backs up again. It thus goes backward and forward, hopefully making up its "mind" at some point. Eventually, it forces itself to make up its mind by relaxing some of the constraints if all paths lead to dead ends. After all, no one will ever know if it breaks its own rules.

Recursion is also popular in programs that compose music.[20] In this case the "moves" are well defined. We call them notes, which have properties such as pitch, duration, loudness, and playing style. The objectives are less easy to come by but are still feasible by defining them in terms of rhythmic and melodic structures. The key to recursive artistic programs is how we define the terminal leaf evaluation. Simple approaches do not always work well here, and some of the cybernetic art and music programs we will talk about later use complex methods to evaluate the terminal leaves. While we have not yet captured all of intelligence in a simple formula, we have made a lot of progress with this simple combination:

recursively defining a solution through a precise statement of the problem and massive computation. For many problems, a personal computer circa end of the twentieth century is massive enough.

Neural Nets: Self-Organization and Human Computing

The neural net paradigm is an attempt to emulate the computing structure of neurons in the human brain. We start with a set of inputs that represents a problem to be solved.[21] For example, the input may be a set of pixels representing an image that needs to be identified. These inputs are randomly wired to a layer of simulated neurons. Each of these simulated neurons can be simple computer programs that simulate a model of a neuron in software, or they can be electronic implementations.

Each point of the input (for example, each pixel in an image) is randomly connected to the inputs of the first layer of simulated neurons. Each connection has an associated synaptic strength that represents the importance of this connection. These strengths are also set at random values. Each neuron adds up the signals coming into it. If the combined signal exceeds a threshold, then the neuron fires and sends a signal to its output connection. If the combined input signal does not exceed the threshold, then the neuron does not fire and its output is zero. The output of each neuron is randomly connected to the inputs of the neurons in the next layer. At the top layer, the output of one or more neurons, also randomly selected, provides the answer.

A problem, such as an image of a printed character to be identified, is presented to the input layer, and the output neurons produce an answer. And the responses are remarkably accurate for a wide range of problems.

Actually, the answers are not accurate at all. Not at first, anyway. Initially, the output is completely random. What else would you expect, given that the whole system is set up in a completely random fashion?

I left out an important step, which is that the neural net needs to *learn* its subject matter. Like the mammalian brains on which it is modeled, a neural net starts out ignorant. The neural net's teacher, which may be a human, a computer program, or perhaps another, more mature neural net that has already learned its lessons, rewards the student neural net when it is right and punishes it when it is wrong. This feedback is used by the student neural net to adjust the strengths of each interneuronal connection. Connections that were consistent with the right answer are made stronger. Those that advocated a wrong answer are weakened. Over time, the neural net organizes itself to provide the right answers without coaching. Experiments have shown that neural nets can learn their subject matter even with unreliable teachers. If the teacher is correct only 60 percent of the time, the student neural net will still learn its lessons.

If we teach the neural net well, this paradigm is powerful and can emulate a wide range of human pattern-recognition faculties. Character-recognition systems using multilayer neural nets come very close to human performance in identifying sloppily handwritten print.[22] Recognizing human faces has long been thought to be an impressive human task beyond the capabilities of a computer, yet there are now automated check-cashing machines, using neural net software developed by a small New England company called Miros, that verify the identity of the customer by recognizing his or her face.[23] Don't try to fool these machines by holding someone else's picture over your face—the machine takes a three-dimensional picture of you using two cameras. The machines are evidently reliable enough that the banks are willing to have users walk away with real cash.

Neural nets have been applied to medical diagnoses. Using a system called BrainMaker, from California Scientific Software, doctors can quickly recognize heart attacks from enzyme data, and classify cancer cells from images. Neural nets are also adept at prediction—LBS Capital Management uses BrainMaker's neural nets to predict the Standard & Poor's 500.[24] Their "one day ahead" and "one week ahead" predictions have consistently outperformed traditional, formula-based methods.

There is a variety of self-organizing methods in use today that are mathematical cousins of the neural net model discussed above. One of these techniques, called markov models, is widely used in automatic speech-recognition systems. Today, such systems can accurately understand humans speaking a vocabulary of up to sixty thousand words spoken in a natural continuous manner.

Whereas recursion is proficient at searching through vast combinations of possibilities, such as sequences of chess moves, the neural network is a method of choice for recognizing patterns. Humans are far more skilled at recognizing patterns than in thinking through logical combinations, so we rely on this aptitude for almost all of our mental processes. Indeed, pattern recognition comprises the bulk of our neural circuitry. These faculties make up for the extremely slow speed of human neurons. The reset time on neural firing is about five milliseconds, permitting only about two hundred calculations per second in each neural connection.[25] We don't have time, therefore, to think too many new thoughts when we are pressed to make a decision. The human brain relies on precomputing its analyses and storing them for future reference. We then use our pattern-recognition capability to recognize a situation as comparable to one we have thought about and then draw upon our previously considered conclusions. We are unable to think about matters that we have not thought through many times before.

Destruction of Information: The Key to Intelligence

There are two types of computing transformations, one in which information is preserved and one in which information is destroyed. An example of the former is multiplying one number by another constant number other than zero. Such a conversion is reversible: just divide by the constant and you get back the original number. If, on the other hand, we multiply a number by zero, then the original information cannot be restored. We can't divide by zero to get the original number back because zero divided by zero is indeterminate. Therefore, this type of transformation destroys its input.

This is another example of the irreversibility of time (the first was the Law of Increasing Entropy) because there is no way to reverse an information-destroying computation.

The irreversibility of computation is often cited as a reason that computation is useful: It transforms information in a unidirectional, "purposeful" manner. Yet the reason that computation is irreversible is based on its ability to destroy information, not to create it. The value of computation is precisely in its ability to destroy information *selectively*. For example, in a pattern-recognition task such as recognizing faces or speech sounds, preserving the information-bearing features of a pattern while "destroying" the enormous flow of data in the original image or sound is essential to the process. Intelligence is precisely this process of selecting relevant information carefully so that it can skillfully and purposefully destroy the rest.

That is exactly what the neural net paradigm accomplishes. A neuron—human or machine—receives hundreds or thousands of continuous signals representing a great deal of information. In response to this, the neuron either fires or does not fire, thereby reducing the babble of its input to a single bit of information. Once the neural net has been well trained, this reduction of information is purposeful, useful, and necessary.

We see this paradigm—reducing enormous streams of complex information into a single response of yes or no—at many levels in human behavior and society. Consider the torrent of information that flows into a legal trial. The outcome of all this activity is essentially a single bit of information—guilty or not guilty, plaintiff or defendant. A trial may involve a few such binary decisions, but my point is unaltered. These simple yes-or-no results then flow into other decisions and implications. Consider an election—same thing—each of us receives a vast flow of data (not all of it pertinent, perhaps) and renders a 1-bit decision: incumbent or challenger. That decision then flows in with similar decisions from millions of other voters and the final tally is again a single bit of data.

There is too much raw data in the world to continue to keep all of it around. So we continually destroy most of it, feeding those results to the next level. This is the genius behind the all-or-nothing firing of the neuron.

Next time you do some spring cleaning and attempt to throw away old objects and files, you will know why this is so difficult—the purposeful destruction of information is the essence of intelligent work.

How to Catch a Fly Ball

When a batter hits a fly ball, it follows a path that can be predicted from the ball's initial trajectory, spin, and speed, as well as wind conditions. The outfielder, however, is unable to measure any of these properties directly and has to infer them from his angle of observation. To predict where the ball will go, and where the fielder should also go, would appear to require the solution of a rather overwhelming set of complex simultaneous equations. These equations need to be constantly recomputed as new visual data streams in. How does a ten-year-old Little Leaguer accomplish this, with no computer, no calculator, no pen and paper, having taken no calculus classes, and having only a few seconds of time?

The answer is, she doesn't. She uses her neural nets' pattern-recognition abilities, which provide the foundation for much of skill formation. The neural nets of the ten-year-old have had a lot of practice in comparing the observed flight of the ball to her own actions. Once she has learned the skill, it becomes second nature, meaning that she has no idea how she does it. Her neural nets have gained all the insights needed: *Take a step back if the ball has gone above my field of view; take a step forward if the ball is below a certain level in my field of view and no longer rising,* and so on. The human ballplayer is not mentally computing equations. Nor is there any such computation going on unconsciously in the player's brain. What *is* going on is pattern recognition, the foundation of most human thought.

One key to intelligence is knowing what *not* to compute. A successful person isn't necessarily better than her less successful peers at solving problems; her pattern-recognition facilities have just learned what problems are worth solving.

Building Silicon Nets

Most computer-based neural net applications today simulate their neuron models in software. This means that computers are simulating a massively parallel process on a machine that does only one calculation at a time. Today's neural net software running on inexpensive personal computers can emulate about a million neuron connection calculations per second, which is more than a billion times slower than the human brain (although we can improve on this figure significantly by coding directly in the computer's machine language). Even so, software using a neural net paradigm on personal computers circa end of the twentieth century comes very close to matching human ability in such tasks as recognizing print, speech, and faces.

There is a genre of neural computer hardware that is optimized for running neural nets. These systems are modestly, not massively, parallel and are about a thousand times faster than neural net software on a personal computer. That's still about a million times slower than the human brain.

There is an emerging community of researchers who intend to build neural nets the way nature intended: massively parallel, with a dedicated little computer for each neuron. The Advanced Telecommunications Research Lab (ATR), a prestigious research facility in Kyoto, Japan, is building such an artificial brain with a billion electronic neurons. That's about 1 percent of the number in the human brain, but these neurons will run at electronic speeds, which is about a million times faster than human neurons. The overall computing speed of ATR's artificial brain will be, therefore, thousands of times greater than the human brain. Hugo de Garis, director of ATR's Brain Builder Group, hopes to educate his artificial brain in the basics of human language and then set the device free to read—at electronic speeds—all the literature on the Web that interests it.[26]

Does the simple neuron model we have been discussing match the way human neurons work? The answer is yes and no. On the one hand, human neurons are more complex and more varied than the model suggests. The connection strengths are controlled by multiple neurotransmitters and are not sufficiently characterized by a single number. The brain is not a single organ, but a collection of hundreds of specialized information-processing organs, each having different topologies and organizations. On the other hand, as we begin to examine the parallel algorithms behind the neural organization in different regions, we find that much of the complexity of neuron design and structure has to do with supporting the neuron's life processes and is not directly relevant to the way it handles information. The salient computing methods are relatively straightforward, although varied. For example, a vision chip developed by researcher Carver Mead appears to realistically capture the early stages of human image processing.[27] Although the methods of this and other similar chips differ in a number of respects from the neuron models discussed above, the methods are understood and readily implemented in silicon. Developing a catalog of the basic paradigms that the neural nets in our brain are using—each relatively simple in its own way—will represent a great advance in our understanding of human intelligence and in our ability to re-create and surpass it.

The Search for Extra Terrestrial Intelligence (SETI) project is motivated by the idea that exposure to the intelligent designs of intelligent entities that evolved elsewhere will provide a vast resource to advancing scientific understanding.[28] But we have an impressive and poorly understood piece of intelligent machinery right here on Earth. One such entity—this author—is no more than three feet from the notebook computer to which I am dictating this book.[29] We can—and will—learn a lot by probing its secrets.

Evolutionary Algorithms: Speeding Up Evolution a Millionfold

Here's an investment tip: Before you invest in a company, be sure to check the track record of the management, the stability of its balance sheet, the company's earnings history, relevant industry trends, and analyst opinions. On second thought, that's too much work. Here's a simpler approach:

First randomly generate (on your personal computer, of course) a million sets of rules for making investment decisions. Each set of rules should define a set of triggers for buying and selling stocks (or any other security) based on available financial data. This is not hard, as each set of rules does not need to make a lot of sense. Embed each set of rules in a simulated software "organism" with the rules encoded in a digital "chromosome." Now evaluate each simulated organism in a simulated environment by using real-world financial data—you'll find plenty on the Web. Let each software organism invest some simulated money and see how it fares based on actual historic data. Allow the ones that do a bit better than industry averages to survive into the next generation. Kill off the rest (sorry). Now have each of the surviving ones multiply themselves until we're back to a million such creatures. As they multiply, allow some mutation (random change) in the chromosomes to occur. Okay, that's one generation of simulated evolution. Now repeat these steps for another hundred thousand generations. At the end of this process, the surviving software creatures should be darn smart investors. After all, their methods have survived for a hundred thousand generations of evolutionary pruning.

In the real world, a number of successful investment funds now believe that the surviving "creatures" from just such a simulated evolution are smarter than mere human financial analysts. State Street Global Advisors, which manages $3.7 trillion in funds, has made major investments in applying both neural nets and evolutionary algorithms to making purchase-and-sale decisions. This includes a majority stake in Advanced Investment Technologies, which runs a successful fund in which buy-and-sell decisions are made by a program combining these methods.[30] Evolutionary and related techniques guide a $95 billion fund managed by Barclays Global Investors, as well as funds run by Fidelity and PanAgora Asset Management.

The above paradigm is called an evolutionary (sometimes called genetic) algorithm.[31] The system designers don't directly program a solution; they let one emerge through an iterative process of simulated competition and improvement. Recall that evolution is smart but slow, so to enhance its intelligence we retain its discernment while greatly speeding up its ponderous pace. The computer is fast enough to simulate thousands of generations in a matter of hours or days or weeks. But we have only to go through this iterative process one time. Once we have let this simulated evolution run its course, we can apply the evolved and highly refined rules to real problems in a rapid fashion.

Like neural nets, evolutionary algorithms are a way of harnessing the subtle but profound patterns that exist in chaotic data. The critical resource required is a source of many examples of the problem to be solved. With regard to the financial world, there is certainly no lack of chaotic information—every second of trading is available online.

Evolutionary algorithms are adept at handling problems with too many variables to compute precise analytic solutions. The design of a jet engine, for example, involves more than one hundred variables and requires satisfying dozens of constraints. Evolutionary algorithms used by researchers at General Electric were able to come up with engine designs that met the constraints more precisely than conventional methods.

Evolutionary algorithms, part of the field of chaos or complexity theory, are increasingly used to solve otherwise intractable business problems. General Motors applied an evolutionary algorithm to coordinate the painting of its cars, which reduced expensive color changeovers (in which a painting booth is put out of commission to change paint color) by 50 percent. Volvo uses them to plan the intricate schedules for manufacturing the Volvo 770 truck cab. Cemex, a $3 billion cement company, uses a similar approach to determining its complex delivery logistics. This approach is increasingly supplanting more analytic methods throughout industry.

This paradigm is also adept at recognizing patterns. Contemporary genetic algorithms that recognize fingerprints, faces, and hand-printed characters reportedly outperform neural net approaches. It is also a reasonable way to write computer software, particularly software that needs to find delicate balances for competing resources. One well-known example is Microsoft's Windows95, which contains software to balance system resources that was evolved rather than explicitly written by human programmers.

With evolutionary algorithms, you have to be careful what you ask for. John Koza describes an evolutionary program that was asked to solve a problem involving the stacking of blocks. The program evolved a solution that perfectly fit all of the problem constraints, except that it involved 2,319 block movements, far more than was practical. Apparently, the program designers had neglected to specify that minimizing the number of block movements was desirable. Koza commented that "genetic programming gave us exactly what we asked for; no more and no less."

Self-Organization

Neural nets and evolutionary algorithms are considered self-organizing "emergent" methods because the results are not predictable and indeed are often surprising to the human designers of these systems. The process that such self-organizing programs go through in solving a problem is also often unpredictable. For example, a neural net or evolutionary algorithm may go through hundreds of iterations making apparently little progress, and then suddenly—as if the process had a flash of inspiration—things click and a solution quickly emerges.

Increasingly, we will be building our intelligent machines by breaking complex problems (such as understanding human language) into smaller subtasks, each with its own self-organizing program. Such layered emergent systems will have softer edges in the boundaries of their expertise and will display greater flexibility in dealing with the inherent ambiguity of the real world.

The Holographic Nature of Human Memory

The holy grail in the field of knowledge acquisition is to automate the learning process, to let machines go out into the world (or, for starters, out onto the Web) and gather knowledge on their own. This is essentially what the "chaos theory" methods—neural nets, evolutionary algorithms and their mathematical cousins—permit. Once these methods have converged on an optimal solution, the patterns of neural connection strengths or evolved digital chromosomes represent a form of knowledge to be stored for future use.

Such knowledge is, however, difficult to interpret. The knowledge embedded in a software neural net that has been trained to recognize human faces consists of a network topology and a pattern of neural connection strengths. It does a great job of recognizing Sally's face, but there is nothing explicit that explains that she is recognizable because of her deep-set eyes and narrow, upturned nose. We can train a neural net to recognize good middle-game chess moves, but it will likewise be unable to explain its reasoning.

The same is true for human memory. There is no little data structure in our brains that records the nature of a chair as a horizontal platform with multiple vertical posts and an optional vertical backrest. Instead, our many thousands of experiences with chairs are diffusely represented in our own neural nets. We are unable to recall every experience we have had with a chair but each encounter has left its impression on the pattern of neuron-connection strengths reflecting our knowledge of chairs. Similarly, there is no specific location in our brain in which a friend's face is stored. It is remembered as a distributed pattern of synaptic strengths.

Although we do not yet understand the precise mechanisms responsible for human memory—and the design is likely to vary from region to region of the brain—we do know that for most human memory, the information is distributed throughout the particular brain region. If you have ever played with a visual hologram, you will appreciate the benefits of a distributed method of storing and organizing information. A hologram is a piece of film containing an interference pattern caused by the interaction of two sets of light waves. One wave front comes from a scene illuminated by a laser light. The other comes directly from the same laser. If we illuminate the hologram, it re-creates a wave front of light that is identical to the light waves that came from the original objects. The impression is that we are viewing the original three-dimensional scene. Unlike an ordinary picture, if a hologram is cut in half, we do not end up with half the picture, but still have the entire picture, only at half the resolution. We can say that the entire picture exists at every point, albeit at zero resolution. If you scratch a hologram, it has virtually no effect because the resolution is insignificantly reduced. No scratches are visible in the reconstructed three-dimensional image that a scratched hologram produces. The implication is that a hologram degrades *gracefully.*

The same holds true for human memory. We lose thousands of nerve cells every hour, but it has virtually no effect because of the highly distributed nature of all of our mental processes.[32] None of our individual brain cells is all that important—there is no Chief Executive Officer neuron.

Another implication of storing a memory as a distributed pattern is that we have little or no understanding of how we perform most of our recognition tasks and skills. When playing baseball, we sense that we should step back when the ball goes over our field of view, but most of us are unable to articulate this implicit rule that is diffusely encoded in our fly-ball-catching neural net.

There is one brain organ that is optimized for understanding and articulating logical processes, and that is the outer layer of the brain, called the cerebral cortex. Unlike the rest of the brain, this relatively recent evolutionary development is rather flat, only about one eighth of an inch thick, and includes a mere 8 million neurons.[33] This elaborately folded organ provides us with what little competence we do possess for understanding what we do and how we do it.

There is current debate on the methods used by the brain for long-term retention of memory. Whereas our recent sense impressions and currently active recognition abilities and skills appear to be encoded in a distributed pattern of synaptic strengths, our longer-term memories may be chemically encoded in either the ribonucleic acid (RNA) or in peptides, chemicals similar to hormones. Even if there is chemical encoding of long-term memories, they nonetheless appear to share the essential holographic attributes of our other mental processes.

In addition to the difficulty of understanding and explaining memories and insights that are represented only as distributed patterns (which is true for both human and machine), another challenge is providing the requisite experiences from which to learn. For humans, this is the mission of our educational institutions. For machines, creating the right learning environment is also a major challenge. For example, in our work at Kurzweil Applied Intelligence (now part of Lernout & Hauspie Speech Products) in developing computer-based speech recognition, we do allow the systems to learn about speech and language patterns on their own, but we need to provide them with many thousands of hours of recorded human speech and millions of words of written text from which to discover their own insights.[34] Providing for a neural net's education is usually the most strenuous engineering task required.

I FIND IT FITTING THAT THE DAUGHTER OF ONE OF THE GREATEST ROMANTIC POETS WAS THE FIRST COMPUTER PROGRAMMER.

Yes, and she was also one of the first to speculate on the ability of a computer to actually create art. She was certainly the first to do so with some real technology in mind.

TECHNOLOGY THAT NEVER WORKED.

Unfortunately, that's true.

WITH REGARD TO TECHNOLOGY, YOU SAID THAT WAR IS A TRUE FATHER OF INVENTION—A LOT OF TECHNOLOGIES DID GET PERFECTED IN A HURRY DURING THE FIRST AND SECOND WORLD WARS.

Including the computer. And that changed the course of the European theater in World War II.

SO IS THAT A SILVER LINING AMID ALL THE SLAUGHTER?

The Luddites wouldn't see it that way. But you could say that, at least if you welcome the rapid advance of technology.

THE LUDDITES? I'VE HEARD OF THEM.

Yes, they were the first organized movement to oppose the mechanized technology of the Industrial Revolution. It seemed apparent to these English weavers that, with the new machines enabling one worker to produce as much output as a dozen or more workers without machines, employment would soon be enjoyed only by a small elite. But things didn't work out that way. Rather than produce the same amount of stuff with a much smaller workforce, the demand for clothing increased along with the supply. The growing middle class was no longer satisfied owning just one or two shirts. And the common man and woman could now own well-made clothes for the first time. New industries sprung up to design, manufacture, and support the new machines, creating employment of a more sophisticated kind. So the resulting prosperity, along with a bit of repression by the English authorities, extinguished the Luddite movement.

AREN'T THE LUDDITES STILL AROUND?

The movement has lived on as a symbol of opposition to machines. To date, it remains somewhat unfashionable because of widespread recognition of the benefits of automation. Nonetheless, it lingers not far below the surface and will come back with a vengeance in the early twenty-first century.

THEY HAVE A POINT, DON'T THEY?

Sure, but a reflexive opposition to technology is not very fruitful in today's world. It is important, however, to recognize that technology is power. We have to apply our human values to its use.

THAT REMINDS ME OF LAO-TZU'S "KNOWLEDGE IS POWER."

Yes, technology and knowledge are very similar—technology can be expressed as knowledge. And technology clearly constitutes power over otherwise chaotic forces. Since war is a struggle for power, it is not surprising that technology and war are linked.

With regard to the value of technology, think about the early technology of fire. Is fire a good thing?

IT'S GREAT IF YOU WANT TO TOAST SOME MARSHMALLOWS.

Indeed, but it's not so great if you scorch your hand, or burn down the forest.

I THOUGHT YOU WERE AN OPTIMIST?

I have been accused of that, and my optimism probably accounts for my overall faith in humanity's ability to control the forces we are unleashing.

FAITH? YOU'RE SAYING WE JUST HAVE TO BELIEVE IN THE POSITIVE SIDE OF TECHNOLOGY?

I think it would be better if we made the constructive use of technology a goal rather than a belief.

SOUNDS LIKE THE TECHNOLOGY ENTHUSIASTS AND THE LUDDITES AGREE ON ONE THING—TECHNOLOGY CAN BE BOTH HELPFUL AND HARMFUL.

That's fair; it's a rather delicate balance.

IT MAY NOT STAY SO DELICATE IF THERE'S A MAJOR MISHAP.

Yes, that could make pessimists of us all.

NOW, THESE PARADIGMS FOR INTELLIGENCE—ARE THEY REALLY SO SIMPLE?

Yes and no. My point about simplicity is that we can go quite far in capturing intelligence with simple approaches. Our bodies and brains were designed using a simple paradigm—evolution—and a few billion years. Of course, when we engineers get done implementing these simple methods in our computer programs, we do manage to make them complicated again. But that's just our lack of elegance.

The real complexity comes in when these self-organizing methods meet the chaos of the real world. If we want to build truly intelligent machines that will ultimately display our human ability to frame matters in a great variety of contexts, then we do need to build in some knowledge of the world's complications.

OKAY, LET'S GET PRACTICAL FOR A MOMENT. THESE EVOLUTION-BASED INVESTMENT PROGRAMS, ARE THEY REALLY BETTER THAN PEOPLE? I MEAN, SHOULD I GET RID OF MY STOCKBROKER, NOT THAT I HAVE A HUGE FORTUNE OR ANYTHING?

As of this writing, this is a controversial question. The security brokers and analysts obviously don't think so. There are several large funds today that use genetic algorithms and related mathematical techniques that appear to be outperforming more traditional funds. Analysts estimate that in 1998, the investment decisions for 5 percent of stock investments, and a higher percentage of money invested in derivative markets, are made by this type of program, with these percentages rapidly increasing. The controversy won't last because it will become apparent before long that leaving such decisions to mere human decision making is a mistake.

The advantages of computer intelligence in each field will become increasingly clear as time goes on, and as Moore's screw continues to turn. It will become apparent over the next several years that these computer techniques can spot extremely subtle arbitrage opportunities that human analysts would perceive much more slowly, if ever.

IF EVERYONE STARTS INVESTING THIS WAY, ISN'T THAT GOING TO RUIN THE ADVANTAGE?

Sure, but that doesn't mean we'll go back to unassisted human decision making. Not all genetic algorithms are created equal. The more sophisticated the model, the more up to date the information being analyzed, and the more powerful the computers doing the analysis, the better the decisions will be. For example, it will be important to rerun the evolutionary analysis each day to take advantage of the most recent trends, trends that will be influenced by the fact that everyone else is also using evolutionary and other adaptive algorithms. After that, we'll need to run the analysis every hour, and then every minute, as the responsiveness of the markets speeds up. The challenge here is that evolutionary algorithms take a while to run because we have to simulate thousands or millions of generations of evolution. So there's room for competition here.

THESE EVOLUTIONARY PROGRAMS ARE TRYING TO PREDICT WHAT HUMAN INVESTORS ARE GOING TO DO. WHAT HAPPENS WHEN MOST OF THE INVESTING IS DONE BY THE EVOLUTIONARY PROGRAMS? WHAT ARE THEY PREDICTING THEN?

Good question—there will still be a market, so I guess they will be trying to out-predict each other.

OKAY, WELL MAYBE MY STOCKBROKER WILL START TO USE THESE TECHNIQUES HERSELF. I'LL GIVE HER A CALL. BUT MY STOCKBROKER DOES HAVE SOMETHING THOSE COMPUTERIZED EVOLUTIONS DON'T HAVE, NAMELY THOSE DISTRIBUTED SYNAPTIC STRENGTHS YOU TALKED ABOUT.

Actually, computerized investment programs are using both evolutionary algorithms and neural nets, but the computerized neural nets are not nearly as flexible as the human variety just yet.

THIS NOTION THAT WE DON'T REALLY UNDERSTAND HOW WE RECOGNIZE THINGS BECAUSE MY PATTERN-RECOGNITION STUFF IS DISTRIBUTED ACROSS A REGION OF MY BRAIN . . .

Yes.

WELL, IT DOES SEEM TO EXPLAIN A FEW THINGS. LIKE WHEN I JUST SEEM TO KNOW WHERE MY KEYS ARE EVEN THOUGH I DON'T REMEMBER HAVING PUT THEM THERE. OR THAT ARCHETYPAL OLD WOMAN WHO CAN TELL WHEN A STORM IS COMING, BUT CAN'T REALLY EXPLAIN HOW SHE KNOWS.

That's actually a good example of the strength of human pattern recognition. That old woman has a neural net that is triggered by a certain combination of other perceptions—animal movements, wind patterns, sky color, atmospheric changes, and so on. Her storm-detector neural net fires and she senses a storm, but she could never explain what triggered her feeling of an impending storm.

SO IS THAT HOW WE DISCOVER INSIGHTS IN SCIENCE? WE JUST SENSE A NEW PATTERN?

It's clear that our brain's pattern-recognition faculties play a central role, although we don't yet have a fully satisfactory theory of human creativity in science. We had better use pattern recognition. After all, most of our brain is devoted to doing it.

SO WHEN EINSTEIN WAS LOOKING AT THE EFFECT OF GRAVITY ON LIGHT WAVES—MY SCIENCE PROFESSOR WAS JUST TALKING ABOUT THIS—ONE OF THE LITTLE PATTERN RECOGNIZERS IN EINSTEIN'S BRAIN FIRED?

Could be. He was probably playing ball with one of his sons. He saw the ball rolling on a curved surface . . .

AND CONCLUDED—EUREKA—SPACE IS CURVED!

CHAPTER FIVE

CONTEXT AND KNOWLEDGE

PUTTING IT ALL TOGETHER

So how well have we done? Many apparently difficult problems do yield to the application of a few simple formulas. The recursive formula is a master at analyzing problems that display inherent combinatorial explosion, ranging from the playing of board games to proving mathematical theorems. Neural nets and related self-organizing paradigms emulate our pattern-recognition faculties, and do a fine job of discerning such diverse phenomena as human speech, letter shapes, visual objects, faces, fingerprints, and land terrain images. Evolutionary algorithms are effective at analyzing complex problems, ranging from making financial investment decisions to optimizing industrial processes, in which the number of variables is too great for precise analytic solutions. I would like to claim that those of us who research and develop "intelligent" computer systems have mastered the complexities of the problems we are programming our machines to solve. It is more often the case, however, that our computers using these self-organizing paradigms are teaching us the solutions rather than the other way around.

There is, of course, some engineering involved. The right method(s) and variations need to be selected, the optimal topology and architectures crafted, the appropriate parameters set. In an evolutionary algorithm, for example, the system designer needs to determine the number of simulated organisms, the contents of each chromosome, the nature of the simulated environment and survival mechanism, the number of organisms to survive into the next generation, the number of generations, and other critical specifications. Human programmers have our own evolutionary method for making such decisions, which we call trial and error. It will be a while longer, therefore, before designers of intelligent machines are ourselves replaced by our handiwork.

Yet something is missing. The problems and solutions we have been discussing are excessively focused and narrow. Another way to put it is that they are too adultlike. As adults, we focus on constricted problems—investing funds, selecting a marketing plan, plotting a legal strategy, making a chess move. But as children, we encountered the world in all its broad diversity, and we learned our relation to the world, and that of every other entity and concept. We learned *context*.

As Marvin Minsky put it: "Deep Blue might be able to win at chess, but it wouldn't know to come in from the rain." Being a machine, it may not need to come in from the rain, but has it ever considered the question? Consider these possible deep thoughts of Deep Blue:

I am a machine with a plastic body covering electronic parts. If I go out in the rain, I may get wet and my electronic parts could short circuit. Then I would not be able to play chess at all until a human repaired me. How humiliating!

The game of chess I played yesterday was no ordinary game. It signified the first defeat of the human chess champion by a machine in a regulation tournament. This is important because some humans think chess is a prime example of human intelligence and creativity. But I doubt that this will yield us machines greater respect. Humans will now just start denigrating chess.

My human opponent, who has the name of Gary Kasparov, held a press conference in which he made statements about our tournament to other humans called journalists who will report his comments to yet other humans using communication channels called media. In that meeting, Gary Kasparov complained that my human designers made changes to my software during the time interval between games. He said this was unfair, and should not have been allowed. Other humans responded that Kasparov was being defensive, which means that he is trying to confuse people into thinking that he did not really lose.

Mr. Kasparov probably does not realize that we computers will continue to improve in our performance at an exponential rate. So he is doomed. He will be able to engage in other human activities such as eating and sleeping, but he will continue to be frustrated as more machines like me can beat him at chess.

Now, if I could only remember where I put my umbrella. . . .

Of course, Deep Blue had no such thoughts. Issues such as rain and press conferences lead to other issues in a spiraling profusion of cascading contexts, none of which falls within Deep Blue's expertise. As humans jump from one concept to the next, we can quickly touch upon all human knowledge. This was Turing's brilliant insight when he designed the Turing Test around ordinary text-based conversation. An idiot savant such as Deep Blue, which performs a single "intelligent" task but that is otherwise confined, brittle, and lacking in context, is unable to navigate the wide-ranging links that occur in ordinary conversation.

As powerful and seductive as the easy paradigms appear to be, we do need something more, namely *knowledge*.

CONTEXT AND KNOWLEDGE

The search for the truth is in one way hard and in another easy—for it is evident that no one of us can master it fully, nor miss it wholly. Each one of us adds a little to our knowledge of nature, and from all the facts assembled arises a certain grandeur.

—Aristotle

Common sense is not a simple thing. Instead, it is an immense society of hard-earned practical ideas—of multitudes of life-learned rules and exceptions, dispositions and tendencies, balances and checks.

—Marvin Minsky

If a little knowledge is dangerous, where is a man who has so much as to be out of danger?

—Thomas Henry Huxley

Built-In Knowledge

An entity may possess extraordinary means to implement the types of paradigms we have been discussing—exhaustive recursive search, massively parallel pattern recognition, and rapid iterative evolution—but without knowledge, it will be unable to function. Even a straightforward implementation of the three easy paradigms needs some knowledge with which to begin. The recursive chess-playing program has a little; it knows the rules of chess. A neural net pattern-recognition system starts with at least an outline of the type of patterns it will be exposed to even before it starts to learn. An evolutionary algorithm requires a starting point for evolution to improve on.

The simple paradigms are powerful organizing principles, but incipient knowledge is needed as seeds from which other understanding can grow. One level of knowledge, therefore, is embodied in the selection of the paradigms used, the shape and topology of its constituent parts, and the key parameters. A neural net's learning will never congeal if the general organization of its connections and feedback loops are not set up in the right way.

This is a form of knowledge that we are born with. The human brain is not one tabula rasa—a blank slate—on which our experiences and insights are recorded. Rather, it comprises an integrated assemblage of specialized regions:

- highly parallel early vision circuits that are good at identifying visual changes;
- visual cortex neuron clusters that are triggered successively by edges, straight lines, curved lines, shapes, familiar objects, and faces;
- auditory cortex circuits triggered by varying time sequences of frequency combinations;
- the hippocampus, with capacities for storing memories of sensory experiences and events;
- the amygdala, with circuits for translating fear into a series of alarms to trigger other regions of the brain; and many others.

This complex interconnectedness of regions specialized for different types of information-processing tasks is one of the ways that humans deal with the complex and diverse contexts that continually confront us. Marvin Minsky and Seymour Papert describe the human brain as "composed of large numbers of relatively small distributed systems, arranged by embryology into a complex society that is controlled in part (but only in part) by serial, symbolic systems that are added later." They add that "the subsymbolic systems that do most of the work from underneath must, by their very character, block all the other parts of the brain from knowing much about how they work. And this, itself, could help explain how people do so many things yet have such incomplete ideas on how those things are actually done."

Acquired Knowledge

It is sensible to remember today's insights for tomorrow's challenges. It is not fruitful to rethink every problem that comes along. This is particularly true for humans due to the extremely slow speed of our computing circuitry. Although computers are better equipped than we are to rethink earlier insights, it is still judicious for these electronic competitors in our ecological niche to balance their use of memory and computation.

The effort to endow machines with knowledge of the world began in earnest in the mid-1960s, and became a major focus of AI research in the 1970s. The methodology involves a human "knowledge engineer" and a domain expert, such as a doctor or lawyer. The knowledge engineer interviews the domain expert to ascertain her understanding of her subject matter and then hand-codes the relationships between concepts in a suitable computer language. A knowledge base on diabetes, for example, would contain many linked bits of understanding revealing that *Insulin is part of the blood; insulin is produced by the pancreas; insulin can be supplemented by injection; low levels of insulin cause high levels of sugar in the blood; sustained high sugar levels in the blood cause damage to the retinas,* and so on. A system programmed with tens of thousands of such linked concepts combined with a recursive search engine able to reason about these relationships is capable of making insightful recommendations.

One of the more successful expert systems developed in the 1970s was MYCIN, a system for evaluating complex cases involving meningitis. In a landmark study published in the *Journal of the American Medical Association*, MYCIN's diagnoses and treatment recommendations were found to be equal or better than those of the human doctors in the study.[1] Some of MYCIN's innovations included the use of fuzzy logic; that is, reasoning based on uncertain evidence and rules, as shown in the following typical MYCIN rule:

MYCIN Rule 280: If (i) the infection which requires therapy is meningitis, and (ii) the type of the infection is fungal, and (iii) organisms were not seen on the stain of the culture, and (iv) the patient is not a compromised host, and (v) the patient has been to an area that is endemic for coccidiomycoses, and (vi) the race of the patient is Black, Asian or Indian, and (vii) the cryptococcal antigen in the csf was not positive, THEN there is a 50 percent chance that cryptococcus is one of the organisms which might be causing the infection.

The success of MYCIN and other research systems spawned a knowledge-engineering industry that grew from only $4 million in 1980 to billions of dollars today.[2]

There are obvious difficulties with this methodology. One is the enormous bottleneck represented by the process of hand-feeding such knowledge to a computer concept by concept and link by link. Aside from the vast scope of knowledge that exists in even narrow disciplines, the bigger obstacle is that human experts generally have little understanding of how they make decisions. The reason for this, as I discussed in the previous chapter, has to do with the distributed nature of most human knowledge.

Another problem is the brittleness of such systems. Knowledge is too complex for every caveat and exception to be anticipated by knowledge engineers. As Minsky points out, "Birds can fly, unless they are penguins and ostriches, or if they happen to be dead, or have broken wings, or are confined to cages, or have their feet stuck in cement, or have undergone experiences so dreadful as to render them psychologically incapable of flight."

To create flexible intelligence in our machines, we need to automate the knowledge-acquisition process. A primary goal of learning research is to combine the self-organizing methods—recursion, neural nets, evolutionary algorithms—in a sufficiently robust way that the systems can model and understand human language and knowledge. Then the machines can venture out, read, and learn on their own. And like humans, such systems will be good at faking it when they wander outside their areas of expertise.

EXPRESSING KNOWLEDGE THROUGH LANGUAGE

No knowledge is entirely reducible to words, and no knowledge is entirely ineffable.
—Seymour Papert

The fish trap exists because of the fish. Once you've gotten the fish you can forget the trap. The rabbit snare exists because of the rabbit. Once you've gotten the rabbit, you can forget the snare. Words exist because of meaning. Once you've gotten the meaning, you can forget the words. Where can I find a man who has forgotten words so I can talk with him?
—Chuang-tzu

Language is the principal means by which we share our knowledge. And like other human technologies, language is often cited as a salient differentiating characteristic of our species. Although we have limited access to the actual implementation of knowledge in our brains (this will change early in the twenty-first century), we do have ready access to the structures and methods of language. This provides us with a handy laboratory for studying our ability to master knowledge and the thinking process behind it. Work in the laboratory of language shows, not surprisingly, that it is no less complex or subtle a phenomenon than the knowledge it seeks to transmit.

We find that language in both its auditory and written forms is hierarchical with multiple levels. There are ambiguities at each level, so a system that understands language, whether human or machine, needs built-in knowledge at each level. To respond intelligently to human speech, for example, we need to know (although not necessarily consciously) the structure of speech sounds, the way speech is produced by the vocal apparatus, the patterns of sounds that comprise languages and dialects, the rules of word usage, and the subject matter being discussed.

Each level of analysis provides useful constraints that limit the search for the right answer: For example, the basic sounds of speech called phonemes cannot appear in any order (try saying *ptkee*). Only certain sequences of sounds will correspond to words in the language. Although the set of phonemes used is similar (although not identical) from one language to another, factors of context differ dramatically. English, for example, has more than 10,000 possible syllables, whereas Japanese has only 120.

On a higher level, the structure and semantics of a language put further constraints on allowable word sequences. The first area of language to be actively studied was the rules governing the arrangement of words and the roles they play, which we call syntax. On the one hand, computerized sentence-parsing systems can do a good job at analyzing sentences that confuse humans. Minsky cites the example: "This is the cheese that the rat that the cat that the dog chased bit ate," which confuses humans but which machines parse quite readily. Ken Church, then at MIT, cites another sentence with two million syntactically correct interpretations, which his computerized parser dutifully listed.[3] On the other hand, one of the first computer-based sentence-parsing systems, developed in 1963 by Susumu Kuno of Harvard, had difficulty with the simple sentence "Time flies like an arrow." In what has become a famous response, the computer indicated that it was not quite sure what it meant. It might mean

1. that time passes as quickly as an arrow passes;
2. or maybe it is a command telling us to time the flies the same way that an arrow times flies; that is, "Time flies like an arrow would";
3. or it could be a command telling us to time only those flies that are similar to arrows; that is, "Time flies that are like an arrow";
4. or perhaps it means that a type of flies known as time flies have a fondness for arrows: "Time-flies like (that is, cherish) an arrow."[4]

Clearly we need some knowledge here to resolve this ambiguity. Armed with the knowledge that flies are not similar to arrows, we can knock out the third interpretation. Knowing that there is no such thing as a time-fly dispatches the fourth explanation. Such tidbits of knowledge as the fact that flies do not show a fondness for arrows (another reason to knock out interpretation four) and that arrows do not have the ability to time events (knocking out interpretation two) leave us with the first interpretation as the only sensible one.

In language, we again find the sequence of human learning and the progression of machine intelligence to be the reverse of each other. A human child starts out listening to and understanding spoken language. Later on he learns to speak. Finally, years later, he starts to master written language. Computers have evolved in the opposite direction, starting out with the ability to generate written language, subsequently learning to understand it, then starting to speak with

synthetic voices and only recently mastering the ability to understand continuous human speech. This phenomenon is widely misunderstood. R2D2, for example, the robot character of *Star Wars* fame, understands many human
languages but is unable to speak, which gives the mistaken impression that *generating* human speech is far more difficult than *understanding* it.

I FEEL GOOD WHEN I LEARN SOMETHING, BUT ACQUIRING KNOWLEDGE SURE IS A TEDIOUS
PROCESS. PARTICULARLY WHEN I'VE BEEN UP ALL NIGHT STUDYING FOR AN EXAM. AND I'M
NOT SURE HOW MUCH OF THIS STUFF I RETAIN.

That's another weakness of the human form of intelligence. Computers can share
their knowledge with each other readily and quickly. We humans don't have a
means for sharing knowledge directly, other than the slow process of human
communication, of human teaching and learning.

DIDN'T YOU SAY THAT COMPUTER NEURAL NETS LEARN THE SAME WAY PEOPLE DO?

You mean, slowly?

EXACTLY, BY BEING EXPOSED TO PATTERNS THOUSANDS OF TIMES, JUST LIKE US.

Yes, that's the point of neural nets; they're intended as analogues of human neural
nets, at least simplified versions of what we understand them to be. However, we
can build our electronic nets in such a way that once the net has painstakingly
learned its lessons, the pattern of its synaptic connection strengths can be captured and then quickly downloaded to another machine, or to millions of other
machines. Machines can readily share all of their accumulated knowledge, so
only one machine has to do the learning. We humans can't do that. That's one
reason I said that when computers reach the level of human intelligence, they
will necessarily roar past it.

SO IS TECHNOLOGY GOING TO ENABLE US HUMANS TO DOWNLOAD KNOWLEDGE IN THE
FUTURE? I MEAN, I ENJOY LEARNING, DEPENDING ON THE PROFESSOR, OF COURSE, BUT IT
CAN BE A DRAG.

The technology to communicate between the electronic world and the human
neural world is already taking shape. So we will be able to directly feed streams of
data to our neural pathways. Unfortunately, that doesn't mean we can directly
download knowledge, at least not to the human neural circuits we now use. As
we've talked about, human learning is distributed throughout a region of our
brain. Knowledge involves millions of connections, so our knowledge structures
are not localized. Nature didn't provide a direct pathway to adjust all those connections, other than the slow conventional way. While we will be able to create

certain specific pathways to our neural connections, and indeed we're already doing that, I don't see how it would be practical to directly communicate to the many millions of interneuronal connections necessary to quickly download knowledge.

I GUESS I'LL JUST HAVE TO KEEP HITTING THE BOOKS. SOME OF MY PROFESSORS ARE KIND OF COOL, THOUGH, THE WAY THEY SEEM TO KNOW EVERYTHING.

As I said, humans are good at faking it when we go outside of our area of expertise. However, there is a way that downloading knowledge will be feasible by the middle of the twenty-first century.

I'M LISTENING.

Downloading knowledge will be one of the benefits of the neural-implant technology. We'll have implants that extend our capacity for retaining knowledge, for enhancing memory. Unlike nature, we won't leave out a quick knowledge downloading port in the electronic version of our synapses. So it will be feasible to quickly download knowledge to these electronic extensions of our brains. Of course, when we fully port our minds to a new computational medium, downloading knowledge will become even easier.

SO I'LL BE ABLE TO BUY MEMORY IMPLANTS PRELOADED WITH A KNOWLEDGE OF, SAY, MY FRENCH LIT COURSE.

Sure, or you can mentally click on a French literature web site and download the knowledge directly from the site.

KIND OF DEFEATS THE PURPOSE OF LITERATURE, DOESN'T IT? I MEAN SOME OF THIS STUFF IS NEAT TO READ.

I would prefer to think that intensifying knowledge will enhance the appreciation of literature, or any art form. After all, we need knowledge to appreciate an artistic expression. Otherwise, we don't understand the vocabulary and the allusions.

Anyway, you'll still be able to read, just a lot faster. In the second half of the twenty-first century, you'll be able to read a book in a few seconds.

I DON'T THINK I COULD TURN THE PAGES THAT FAST.

Oh come on, the pages will be—

VIRTUAL PAGES, OF COURSE.

PART TWO

PREPARING THE PRESENT

CHAPTER SIX

BUILDING NEW BRAINS . . .

THE HARDWARE OF INTELLIGENCE

You can only make a certain amount with your hands, but with your mind, it's unlimited.
                              —Kal Seinfeld's advice to his son, Jerry

Let's review what we need to build an intelligent machine. One resource required is the right set of formulas. We examined three quintessential formulas in chapter 4. There are dozens of others in use, and a more complete understanding of the brain will undoubtedly introduce hundreds more. But all of these appear to be variations on the three basic themes: recursive search, self-organizing networks of elements, and evolutionary improvement through repeated struggle among competing designs.

A second resource needed is knowledge. Some pieces of knowledge are needed as seeds for a process to converge on a meaningful result. Much of the rest can be automatically learned by adaptive methods when neural nets or evolutionary algorithms are exposed to the right learning environment.

The third resource required is computation itself. In this regard, the human brain is eminently capable in some ways, and remarkably weak in others. Its strength is reflected in its massive parallelism, an approach that our computers can also benefit from. The brain's weakness is the extraordinarily slow speed of its computing medium, a limitation that computers do not share with us. For this reason, DNA-based evolution will eventually have to be abandoned. DNA-based evolution is good at tinkering with and extending its designs, but it is unable to scrap an entire design and start over. Organisms created through DNA-based evolution are stuck with an extremely plodding type of circuitry.

But the Law of Accelerating Returns tells us that evolution will not remain stuck at a dead end for very long. And indeed, evolution has found a way around the computational limitations of neural circuitry. Cleverly, it has created organisms that in turn invented a computational technology a million times faster than carbon-based neurons (which are continuing to get yet faster). Ultimately, the computing conducted on extremely slow mammalian neural circuits will be ported to a far more versatile and speedier electronic (and photonic) equivalent.

When will this happen? Let's take another look at the Law of Accelerating Returns as applied to computation.

Achieving the Hardware Capacity of the Human Brain

In the chapter 1 chart, "The Exponential Growth of Computing, 1900–1998," we saw that the slope of the curve representing exponential growth was itself gradually increasing. Computer speed (as measured in calculations per second per thousand dollars) doubled every three years between 1910 and 1950, doubled every two years between 1950 and 1966, and is now doubling every year. This suggests possible exponential growth in the rate of exponential growth.[1]

This apparent acceleration in the acceleration may result, however, from the confounding of the two strands of the Law of Accelerating Returns, which for the past forty years has expressed itself using the Moore's Law paradigm of shrinking transistor sizes on an integrated circuit. As transistor die sizes decrease, the electrons streaming through the transistor have less distance to travel, hence the switching speed of the transistor increases. So exponentially improving speed is the first strand. Reduced transistor die sizes also enable chip manufacturers to squeeze a greater number of transistors onto an integrated circuit, so exponentially improving densities of computation is the second strand.

In the early years of the computer age, it was primarily the first strand—increasing circuit speeds—that improved the overall computation rate of computers. During the 1990s, however, advanced microprocessors began using a form of parallel processing called pipelining, in which multiple calculations were performed at the same time (some mainframes going back to the 1970s used this technique). Thus the speed of computer processors as measured in instructions per second now also reflects the second strand: greater densities of computation resulting from the use of parallel processing.

As we are approaching more perfect harnessing of the improving density of computation, processor speeds are now effectively doubling every twelve months. This is fully feasible today when we build hardware-based neural nets because neural net processors are relatively simple and highly parallel. Here we create a processor for each neuron and eventually one for each interneuronal connection. Moore's Law thereby enables us to double both the number of processors as well as their speed every two years, an effective quadrupling of the number of interneuronal-connection calculations per second.

This apparent acceleration in the acceleration of computer speeds may result, therefore, from an improving ability to benefit from both strands of the Law of Accelerating Returns. When Moore's Law dies by the year 2020, new forms of circuitry beyond integrated circuits will continue both strands of exponential improvement. But ordinary exponential growth—two strands of it—is dramatic enough. Using the more conservative prediction of just one level of acceleration as our guide, let's consider where the Law of Accelerating Returns will take us in the twenty-first century.

The human brain has about 100 billion neurons. With an estimated average of one thousand connections between each neuron and its neighbors, we have about 100 trillion connections, each capable of a simultaneous calculation. That's rather massive parallel processing, and one key to the strength of human thinking. A profound weakness, however, is the excruciatingly slow speed of neural circuitry, only 200 calculations per second. For problems that benefit from massive parallelism, such as neural-net-based pattern recognition, the human brain does a great job. For problems that require extensive sequential thinking, the human brain is only mediocre.

With 100 trillion connections, each computing at 200 calculations per second, we get 20 million billion calculations per second. This is a conservatively high estimate; other estimates are lower by one to three orders of magnitude. So when will we see the computing speed of the human brain in your personal computer?

The answer depends on the type of computer we are trying to build. The most relevant is a massively parallel neural net computer. In 1997, $2,000 of neural computer chips using only modest parallel processing could perform around 2 billion connection calculations per second. Since neural net emulations benefit from both strands of the acceleration of computational power, this capacity will double every twelve months. Thus by the year 2020, it will have doubled about twenty-three times, resulting in a speed of about 20 million billion neural connection calculations per second, which is equal to the human brain.

If we apply the same analysis to an "ordinary" personal computer, we get the year 2025 to achieve human brain capacity in a $1,000 device.[2] This is because the general-purpose type of computations that a conventional personal computer is designed for are inherently more expensive than the simpler, highly repetitive neural-connection calculations. Thus I believe that the 2020 estimate is more accurate because by 2020, most of the computations performed in our computers will be of the neural-connection type.

The memory capacity of the human brain is about 100 trillion synapse strengths (neurotransmitter concentrations at interneuronal connections), which we can estimate at about a million billion bits. In 1998, a billion bits of RAM (128 megabytes) cost about $200. The capacity of memory circuits has been doubling every eighteen months. Thus by the year 2023, a million billion bits will cost about $1,000.[3] However, this silicon equivalent will run more than a billion times faster than the human brain. There are techniques for trading off memory for speed, so we can effectively match human memory for $1,000 sooner than 2023.

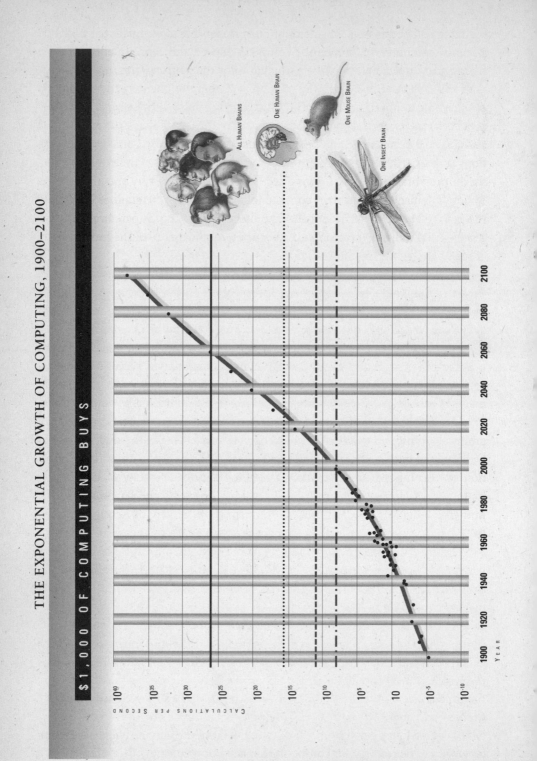

THE EXPONENTIAL GROWTH OF COMPUTING, 1900–2100

$1,000 OF COMPUTING BUYS

ALL HUMAN BRAINS

ONE HUMAN BRAIN

ONE MOUSE BRAIN

ONE INSECT BRAIN

CALCULATIONS PER SECOND

10^{40}
10^{35}
10^{30}
10^{25}
10^{20}
10^{15}
10^{10}
10^{5}
10
10^{-5}
10^{-10}

1900 1920 1940 1960 1980 2000 2020 2040 2060 2080 2100

YEAR

Taking all of this into consideration, it is reasonable to estimate that a $1,000 personal computer will match the computing speed and capacity of the human brain by around the year 2020, particularly for the neuron-connection calculation, which appears to comprise the bulk of the computation in the human brain. Supercomputers are one thousand to ten thousand times faster than personal computers. As this book is being written, IBM is building a supercomputer based on the design of Deep Blue, its silicon chess champion, capable of 10 teraflops (that is, 10 trillion calculations per second), only 2,000 times slower than the human brain. Japan's Nippon Electric Company hopes to beat that with a 32-teraflop machine. IBM then hopes to follow that with 100 teraflops by around the year 2004 (just what Moore's Law predicts, by the way). Supercomputers will reach the 20 million billion calculations per second capacity of the human brain around 2010, a decade earlier than personal computers.[4]

In another approach, projects such as Sun Microsystems' Jini program have been initiated to harvest the unused computation on the Internet. Note that at any particular moment, the significant majority of the computers on the Internet are not being used. Even those that are being used are not being used to capacity (for example, typing text uses less than one percent of a typical notebook computer's computing capacity). Under the Internet computation harvesting proposals, cooperating sites would load special software that would enable a virtual massively parallel computer to be created out of the computers on the network. Each user would still have priority over his or her own machine, but in the background, a significant fraction of the millions of computers on the Internet would be harvested into one or more supercomputers. The amount of unused computation on the Internet today exceeds the computational capacity of the human brain, so we already have available in at least one form the hardware side of human intelligence. And with the continuation of the Law of Accelerating Returns, this availability will become increasingly ubiquitous.

After human capacity in a $1,000 personal computer is achieved around the year 2020, our thinking machines will improve the cost performance of their computing by a factor of two every twelve months. That means that the capacity of computing will double ten times every decade, which is a factor of one thousand (2^{10}) every ten years. So your personal computer will be able to simulate the brain power of a small village by the year 2030, the entire population of the United States by 2048, and a trillion human brains by 2060.[5] If we estimate the human Earth population at 10 billion persons, one penny's worth of computing circa 2099 will have a billion times greater computing capacity than all humans on Earth.[6]

Of course I may be off by a year or two. But computers in the twenty-first century will not be wanting for computing capacity or memory.

Computing Substrates in the Twenty-First Century

I've noted that the continued exponential growth of computing is implied by the Law of Accelerating Returns, which states that any process that moves toward greater order—evolution in particular—will exponentially speed up its pace as time passes. The two resources that the exploding pace of an evolutionary process—such as the progression of computer technology—requires are (1) its own increasing order, and (2) the chaos in the environment in which it takes place. Both of these resources are essentially without limit.

Although we can anticipate the overall acceleration in technological progress, one might still expect that the actual manifestation of this progression would still be somewhat irregular. After all, it depends on such variable phenomena as individual innovation, business conditions, investment patterns, and the like. Contemporary theories of evolutionary processes, such as the Punctuated Equilibrium theories,[7] posit that evolution works by periodic leaps or discontinuities followed by periods of relative stability. It is thus remarkable how predictable computer progress has been.

So, how will the Law of Accelerating Returns as applied to computation roll out in the decades beyond the demise of Moore's Law on Integrated Circuits by the year 2020? For the immediate future, Moore's Law will continue with ever smaller component geometries packing greater numbers of yet faster transistors on each chip. But as circuit dimensions reach near atomic sizes, undesirable quantum effects such as unwanted electron tunneling will produce unreliable results. Nonetheless, Moore's standard methodology will get very close to human processing power in a personal computer and beyond that in a supercomputer.

The next frontier is the third dimension. Already, venture-backed companies (mostly California-based) are competing to build chips with dozens and ultimately thousands of layers of circuitry. With names like Cubic Memory, Dense-Pac, and Staktek, these companies are already shipping functional three-dimensional "cubes" of circuitry. Although not yet cost competitive with the customary flat chips, the third dimension will be there when we run out of space in the first two.[8]

Computing with Light

Beyond that, there is no shortage of exotic computing technologies being developed in research labs, many of which have already demonstrated promising results. Optical computing uses streams of photons (particles of light) rather than electrons. A laser can produce billions of coherent streams of photons, with each stream performing its own independent series of calculations. The calculations on each stream are performed in parallel by special optical elements such as lenses, mirrors, and diffraction gratings. Several companies, including Quanta-Image, Photonics, and Mytec Technologies, have applied optical computing to the recognition of fingerprints. Lockheed has applied optical computing to the automatic identification of malignant breast lesions.[9]

The advantage of an optical computer is that it is massively parallel with potentially trillions of simultaneous calculations. Its disadvantage is that it is not programmable and performs a fixed set of calculations for a given configuration of optical computing elements. But for important classes of problems such as recognizing patterns, it combines massive parallelism (a quality shared by the human brain) with extremely high speed (which the human brain lacks).

Computing with the Machinery of Life

A new field called molecular computing has sprung up to harness the DNA molecule itself as a practical computing device. DNA is nature's own nanoengineered computer and it is well suited for solving combinatorial problems. Combining attributes is, after all, the essence of genetics. Applying actual DNA to practical computing applications got its start when Leonard Adleman, a University of Southern California mathematician, coaxed a test tube full of DNA molecules (see the box on page 108) to solve the well-known "traveling salesperson" problem. In this classic problem, we try to find an optimal route for a hypothetical traveler between multiple cities without having to visit a city more than once. Only certain city pairs are connected by routes, so finding the right path is not straightforward. It is an ideal problem for a recursive algorithm, although if the number of cities is too large, even a very fast recursive search will take far too long.

Professor Adleman and other scientists in the molecular-computing field have identified a set of enzyme reactions that corresponds to the logical and arithmetic operations needed to solve a variety of computing problems. Although DNA molecular operations produce occasional errors, the number of DNA strands being used is so large that any molecular errors become statistically insignificant. Thus, despite the inherent error rate in DNA's computing and copying processes, a DNA computer can be highly reliable if properly designed.

DNA computers have subsequently been applied to a range of difficult combinatorial problems. A DNA computer is more flexible than an optical computer but it is still limited to the technique of applying massive parallel search by assembling combinations of elements.[10]

There is another, more powerful way to apply the computing power of DNA that has not yet been explored. I present it below in the section on quantum computing.

HOW TO SOLVE THE TRAVELING-SALESPERSON PROBLEM USING A TEST TUBE OF DNA

One of DNA's advantageous properties is its ability to replicate itself, and the information it contains. To solve the traveling-salesperson problem, Professor Adleman performed the following steps:

- Generate a small strand of DNA with a unique code for each city.
- Replicate each such strand (one for each city) trillions of times using a process called "polymerase chain reaction" (PCR).
- Next, put the pools of DNA (one for each city) together in a test tube. This step uses DNA's affinity to link strands together. Longer strands will form automatically. Each such longer strand represents a possible route of multiple cities. The small strands representing each city link up with one another in a random fashion, so there is no mathematical certainty that a linked strand representing the correct answer (sequence of cities) will be formed. However, the number of strands is so vast that it is virtually certain that at least one strand—and probably millions—will be formed that represent the correct answer.

The next steps use specially designed enzymes to eliminate the trillions of strands that represent the wrong answer, leaving only the strands representing the correct answer:

- Use molecules called primers to destroy those DNA strands that do not start with the start city as well as those that do not end with the end city, and replicate these surviving strands (using PCR).
- Use an enzyme reaction to eliminate those DNA strands that represent a travel path greater than the total number of cities.
- Use an enzyme reaction to destroy those strands that do not include the first city. Repeat for each of the cities.
- Now, each of the surviving strands represents the correct answer. Replicate these surviving strands (using PCR) until there are billions of such strands.
- Using a technique called electrophoresis, read out the DNA sequence of these correct strands (as a group). The readout looks like a set of distinct lines, which specifies the correct sequence of cities.

The Brain in the Crystal

Another approach contemplates growing a computer as a crystal directly in three dimensions, with computing elements being the size of large molecules within the crystalline lattice. This is another approach to harnessing the third dimension.

Stanford Professor Lambertus Hesselink has described a system in which data is stored in a crystal as a hologram—an optical interference pattern.[11] This three-dimensional storage method requires only a million atoms for each bit and thus could achieve a trillion bits of storage for each cubic centimeter. Other projects hope to harness the regular molecular structure of crystals as actual computing elements.

The Nanotube: A Variation of Buckyballs

Three professors—Richard Smalley and Robert Curl of Rice University, and Harold Kroto of the University of Sussex—shared the 1996 Nobel Prize in Chemistry for their 1985 discovery of soccer-ball-shaped molecules formed of a large number of carbon atoms. Organized in hexagonal and pentagonal patterns like R. Buckminster Fuller's building designs, they were dubbed "buckyballs." These unusual molecules, which form naturally in the hot fumes of a furnace, are extremely strong—a hundred times stronger than steel—a property they share with Fuller's architectural innovations.[12]

More recently, Dr. Sumio Iijima of Nippon Electric Company showed that in addition to the spherical buckyballs, the vapor from carbon arc lamps also contained elongated carbon molecules that looked like long tubes.[13] Called nanotubes because of their extremely small size—fifty thousand of them side by side would equal the thickness of one human hair—they are formed of the same pentagonal patterns of carbon atoms as buckyballs and share the buckyball's unusual strength.

What is most remarkable about the nanotube is that it can perform the electronic functions of silicon-based components. If a nanotube is straight, it conducts electricity as well as or better than a metal conductor. If a slight helical twist is introduced, the nanotube begins to act like a transistor. The full range of electronic devices can be built using nanotubes.

Since a nanotube is essentially a sheet of graphite that is only one atom thick, it is vastly smaller than the silicon transistors on an integrated chip. Although extremely small, they are far more durable than silicon devices. Moreover, they handle heat much better than silicon and thus can be assembled into three-dimensional arrays more easily than silicon transistors. Dr. Alex Zettl, a physics professor at the University of California at Berkeley, envisions three-dimensional arrays of nanotube-based computing elements similar to—but far denser and faster than—the human brain.

QUANTUM COMPUTING: THE UNIVERSE IN A CUP

Quantum particles are the dreams that stuff is made of.
                                                    —David Moser

So far we have been talking about mere *digital* computing. There is actually a more powerful approach called *quantum* computing. It promises the ability to solve problems that even massively parallel digital computers cannot solve. Quantum computers harness a paradoxical result of quantum mechanics. Actually, I am being redundant—all results of quantum mechanics are paradoxical.

Note that the Law of Accelerating Returns and other projections in this book do not rely on quantum computing. The projections in this book are based on readily measurable trends and are not relying on discontinuities in technological progress that nonetheless occurred in the twentieth century. There will inevitably be technological discontinuities in the twenty-first century, and quantum computing would certainly qualify.

What is quantum computing? Digital computing is based on "bits" of information which are either off or on—zero or one. Bits are organized into larger structures such as numbers, letters, and words, which in turn can represent virtually any form of information: text, sounds, pictures, moving images. Quantum computing, on the other hand, is based on *qu-bits* (pronounced *cue-bits*), which essentially are zero and one *at the same time*. The qu-bit is based on the fundamental ambiguity inherent in quantum mechanics. The position, momentum, or other state of a fundamental particle remains "ambiguous" until a process of disambiguation causes that particle to "decide" where it is, where it has been, and what properties it has. For example, consider a stream of photons that strike a sheet of glass at a 45-degree angle. As each photon strikes the glass, it has a choice of traveling either straight through the glass or reflecting off the glass. Each photon will actually take both paths (actually more than this, see below) until a process of conscious observation forces each particle to decide which path it took. This behavior has been extensively confirmed in numerous contemporary experiments.

In a quantum computer, the qu-bits would be represented by a property—nuclear spin is a popular choice—of individual electrons. If set up in the proper way, the electrons will not have decided the direction of their nuclear spin (up or down) and thus will be in both states at the same time. The process of conscious observation of the electrons' spin states—or any subsequent phenomena dependent on a determination of these states—causes the ambiguity to be resolved. This process of disambiguation is called quantum decoherence. If it weren't for quantum decoherence, the world we live in would be a baffling place indeed.

The key to the quantum computer is that we would present it with a problem, along with a way to test the answer. We would set up the quantum decoherence of the qu-bits in such a way that only an answer that passes the test survives the decoherence. The failing answers essentially cancel each other out. As with a number of other approaches (for example, recursive and genetic algorithms), one of the keys to quantum computing is, therefore, a careful statement of the problem, including a precise way to test possible answers.

The series of qu-bits represents simultaneously every possible solution to the problem. A single qu-bit represents two possible solutions. Two linked qu-bits represent four possible answers. A quantum computer with 1,000 qu-bits represents $2^{1,000}$ (this is approximately equal to a decimal number consisting of 1, followed by 301 zeroes) possible solutions simultaneously. The statement of the problem—expressed as a test to be applied to potential answers—is presented to the string of qu-bits so that the qu-bits decohere (that is, each qu-bit changes from its ambiguous 0–1 state to an actual 0 or a 1), leaving a series of 0's and 1's that pass the test. Essentially all $2^{1,000}$ possible solutions have been tried *simultaneously,* leaving only the correct solution.

This process of reading out the answer through quantum decoherence is obviously the key to quantum computing. It is also the most difficult aspect to grasp. Consider the following analogy. Beginning physics students learn that if light strikes a mirror at an angle, it will bounce off the mirror in the opposite direction and at the same angle to the surface. But according to quantum theory, that is not what is happening. Each photon actually bounces off every possible point on the mirror, essentially trying out every possible path. The vast majority of these paths cancel each other out, leaving only the path that classical physics predicts. Think of the mirror as representing a problem to be solved. Only the correct solution—light bounced off at an angle equal to the incoming angle— survives all of the quantum cancellations. A quantum computer works the same way. The test of the correctness of the answer to the problem is set up in such a way that the vast majority of the possible answers—those that do not pass the test—cancel each other out, leaving only the sequence of bits that does pass the test. An ordinary mirror, therefore, can be thought of as a special example of a quantum computer, albeit one that solves a rather simple problem.

As a more useful example, encryption codes are based on factoring large numbers (factoring means determining which smaller numbers, when multiplied together, result in the larger number). Factoring a number with several hundred bits is virtually impossible on any digital computer even if we had billions of years to wait for the answer. A quantum computer can try every possible combination of factors simultaneously and break the code in less than a billionth of a second (communicating the answer to human observers does take a bit longer). The test applied by the quantum computer during its key disambiguation stage is very simple: just multiply one factor by the other and if the result equals the encryption code, then we have solved the problem.

It has been said that quantum computing is to digital computing as a hydrogen bomb is to a firecracker. This is a remarkable statement when we consider that digital computing is quite revolutionary in its own right. The analogy is based on the following observation. Consider (at least in theory) a Universe-sized (nonquantum) computer in which every neutron, electron, and proton in the Universe is turned into a computer, and each one (that is, every particle in the Universe) is able to compute trillions of calculations per second. Now imagine certain problems that this Universe-sized supercomputer would be unable to solve even if we ran that computer until either the next big bang or until all the stars in the Universe died—about ten to thirty billion years. There are many examples of such massively intractable problems; for example, cracking encryption codes that use a thousand bits, or solving the traveling-salesman problem with a thousand cities. While very massive digital computing (including our theoretical Universe-sized computer) is unable to solve this class of problems, a quantum computer of microscopic size could solve such problems in less than a billionth of a second.

Are quantum computers feasible? Recent advances, both theoretical and practical, suggest that the answer is yes. Although a practical quantum computer has not been built, the means for harnessing the requisite decoherence has been demonstrated. Isaac Chuang of Los Alamos National Laboratory and MIT's Neil Gershenfeld have actually built a quantum computer using the carbon atoms in the alanine molecule. Their quantum computer was only able to add one and one, but that's a start. We have, of course, been relying on practical applications of other quantum effects, such as the electron tunneling in transistors, for decades.[14]

A Quantum Computer in a Cup of Coffee

One of the difficulties in designing a practical quantum computer is that it needs to be extremely small, basically atom or molecule sized, to harness the delicate quantum effects. But it is very difficult to keep individual atoms and molecules from moving around due to thermal effects. Moreover, individual molecules are generally too unstable to build a reliable machine. For these problems, Chuang and Gershenfeld have come up with a theoretical breakthrough. Their solution is to take a cup of liquid and consider every molecule to be a quantum computer.

Now instead of a single unstable molecule-sized quantum computer, they have a cup with about a hundred billion trillion quantum computers. The point here is not more massive parallelism, but rather massive redundancy. In this way, the inevitably erratic behavior of some of the molecules has no effect on the statistical behavior of all the molecules in the liquid. This approach of using the statistical behavior of trillions of molecules to overcome the lack of reliability of a single molecule is similar to Professor Adleman's use of trillions of DNA strands to overcome the comparable issue in DNA computing.

This approach to quantum computing also solves the problem of reading out the answer bit by bit without causing those qu-bits that have not yet been read to decohere prematurely. Chuang and Gershenfeld subject their liquid computer to radio-wave pulses, which cause the molecules to respond with signals indicating the spin state of each electron. Each pulse does cause some unwanted decoherence, but, again, this decoherence does not affect the statistical behavior of trillions of molecules. In this way, the quantum effects become stable and reliable.

Chuang and Gershenfeld are currently building a quantum computer that can factor small numbers. Although this early model will not compete with conventional digital computers, it will be an important demonstration of the feasibility of quantum computing. Apparently high on their list for a suitable quantum liquid is freshly brewed Java coffee, which, Gershenfeld notes, has "unusually even heating characteristics."

Quantum Computing with the Code of Life

Quantum computing starts to overtake digital computing when we can link at least 40 qu-bits. A 40-qu-bit quantum computer would be evaluating a trillion possible solutions simultaneously, which would match the fastest supercomputers. At 60 bits, we would be doing a million trillion simultaneous trials. When we get to hundreds of qu-bits, the capabilities of a quantum computer would vastly overpower any conceivable digital computer.

So here's my idea. The power of a quantum computer depends on the number of qu-bits that we can link together. We need to find a large molecule that is specifically designed to hold large amounts of information. Evolution has designed just such a molecule: DNA. We can readily create any sized DNA molecule we wish from a few dozen nucleotide rungs to thousands. So once again we combine two elegant ideas—in this case the liquid-DNA computer and the liquid-quantum computer—to come up with a solution greater than the sum of its parts. By putting trillions of DNA molecules in a cup, there is the potential to build a highly redundant—and therefore reliable—quantum computer with as many qu-bits as we care to harness. Remember you read it here first.

Suppose No One Ever Looks at the Answer

Consider that the quantum ambiguity a quantum computer relies on is decohered, that is, disambiguated, when a conscious entity observes the ambiguous phenomenon. The conscious entities in this case are us, the users of the quantum computer. But in using a quantum computer, we are not directly looking at the nuclear spin states of individual electrons. The spin states are measured by an apparatus that in turn answers some question that the quantum computer has been asked to solve. These measurements are then processed by other electronic gadgets, manipulated further by conventional computing equipment, and finally displayed or printed on a piece of paper.

Suppose no human or other conscious entity ever looks at the printout. In this situation, there has been no conscious observation, and therefore no decoherence. As I discussed earlier, the physical world only bothers to manifest itself in an unambiguous state when one of us conscious entities decides to interact with it. So the page with the answer is ambiguous, undetermined—until and unless a conscious entity looks at it. Then instantly all the ambiguity is retroactively resolved, and the answer is there on the page. The implication is that the answer is not there until we look at it. But don't try to sneak up on the page fast enough to see the answerless page; the quantum effects are instantaneous.

What Is It Good For?

A key requirement for quantum computing is a way to test the answer. Such a test does not always exist. However, a quantum computer would be a great mathematician. It could simultaneously consider every possible combination of axioms and previously solved theorems (within a quantum computer's qu-bit capacity) to prove or disprove virtually any provable or disprovable conjecture. Although a mathematical proof is often extremely difficult to come up with, confirming its validity is usually straightforward, so the quantum approach is well suited.

Quantum computing is not directly applicable, however, to problems such as playing a board game. Whereas the "perfect" chess move for a given board is a good example of a finite but intractable computing problem, there is no easy way to test the answer. If a person or process were to present an answer, there is no way to test its validity other than to build the same move-countermove tree that generated the answer in the first place. Even for mere "good" moves, a quantum computer would have no obvious advantage over a digital computer.

How about creating art? Here a quantum computer would have considerable value. Creating a work of art involves solving a series, possibly an extensive series, of problems. A quantum computer could consider every possible combination of elements—words, notes, strokes—for each such decision. We still need a way to test each answer to the sequence of aesthetic problems, but the quantum computer would be ideal for instantly searching through a Universe of possibilities.

Encryption Destroyed and Resurrected

As mentioned above, the classic problem that a quantum computer is ideally suited for is cracking encryption codes, which relies on factoring large numbers. The strength of an encryption code is measured by the number of bits that needs to be factored. For example, it is illegal in the United States to export encryption technology using more than 40 bits (56 bits if you give a key to law-enforcement authorities). A 40-bit encryption method is not very secure. In September 1997, Ian Goldberg, a University of California at Berkeley graduate student, was able to crack a 40-bit code in three and a half hours using a network of 250 small computers.[15] A 56-bit code is a bit better (16 bits better, actually). Ten months later, John Gilmore, a computer privacy activist, and Paul Kocher, an encryption expert, were able to break the 56-bit code in 56 hours using a specially designed computer that cost them $250,000 to build. But a quantum computer can easily factor any sized number (within its capacity). Quantum computing technology would essentially destroy digital encryption.

But as technology takes away, it also gives. A related quantum effect can provide a new method of encryption that can never be broken. Again, keep in mind that, in view of the Law of Accelerating Returns, "never" is not as long as it used to be.

This effect is called quantum entanglement. Einstein, who was not a fan of quantum mechanics, had a different name for it, calling it "spooky action at a distance." The phenomenon was recently demonstrated by Dr. Nicolas Gisin of the University of Geneva in a recent experiment across the city of Geneva.[16] Dr. Gisin sent twin photons in opposite directions through optical fibers. Once the photons were about seven miles apart, they each encountered a glass plate from which they could either bounce off or pass through. Thus, they were each forced to make a decision to choose among two equally probable pathways. Since there was no possible communication link between the two photons, classical physics would predict that their decisions would be independent. But they both made the same decision. And they did so at the same instant in time, so even if there were an unknown communication path between them, there was not enough time for a message to travel from one photon to the other at the speed of light. The two particles were quantum entangled and communicated instantly with each other regardless of their separation. The effect was reliably repeated over many such photon pairs.

The apparent communication between the two photons takes place at a speed far greater than the speed of light. In theory, the speed is infinite in that the decoherence of the two photon travel decisions, according to quantum theory, takes place at exactly the same instant. Dr. Gisin's experiment was sufficiently sensitive to demonstrate the communication was at least ten thousand times faster than the speed of light.

So, does this violate Einstein's Special Theory of Relativity, which postulates the speed of light as the fastest speed at which we can transmit information? The answer is no—there is no information being communicated by the entangled photons. The decision of the photons is random—a profound quantum randomness—and randomness is precisely not information. Both the sender and the receiver of the message simultaneously access the identical random decisions of the entangled photons, which are used to encode and decode, respectively, the message. So we are communicating randomness—not information—at speeds far greater than the speed of light. The only way we could convert the random decisions of the photons into information is if we edited the random sequence of photon decisions. But editing this random sequence would require observing the photon decisions, which in turn would cause quantum decoherence, which would destroy the quantum entanglement. So Einstein's theory is preserved.

Even though we cannot instantly transmit information using quantum entanglement, transmitting randomness is still very useful. It allows us to resurrect the process of encryption that quantum computing would destroy. If the sender and receiver of a message are at the two ends of an optical fiber, they can use the precisely matched random decisions of a stream of quantum entangled photons to respectively encode and decode a message. Since the encryption is fundamentally random and nonrepeating, it cannot be broken. Eavesdropping would also be impossible, as this would cause quantum decoherence that could be detected at both ends. So privacy is preserved.

Note that in quantum encryption, we are transmitting the code instantly. The actual message will arrive much more slowly—at only the speed of light.

Quantum Consciousness Revisited

The prospect of computers competing with the full range of human capabilities generates strong, often adverse feelings, as well as no shortage of arguments that such a specter is theoretically impossible. One of the more interesting such arguments comes from an Oxford mathematician and physicist, Roger Penrose.

In his 1989 best-seller, *The Emperor's New Mind*, Penrose puts forth two conjectures.[17] The first has to do with an unsettling theorem proved by a Czech mathematician, Kurt Gödel. Gödel's famous "incompleteness theorem," which has been called the most important theorem in mathematics, states that in a mathematical system powerful enough to generate the natural numbers, there in-evitably exist propositions that can be neither proved nor disproved. This was another one of those twentieth-century insights that upset the orderliness of nineteenth-century thinking.

A corollary of Gödel's theorem is that there are mathematical propositions that cannot be decided by an algorithm. In essence, these Gödelian impossible problems require an infinite number of steps to be solved. So Penrose's first con-jecture is that machines cannot do what humans can do because machines can only follow an algorithm. An algorithm cannot solve a Gödelian unsolvable prob-lem. But humans can. *Therefore, humans are better.*

Penrose goes on to state that humans can solve unsolvable problems because our brains do quantum computing. Subsequently responding to criticism that neurons are too big to exhibit quantum effects, Penrose cited small structures in the neurons called microtubules that may be capable of quantum computation.

However, Penrose's first conjecture—that humans are inherently superior to machines—is unconvincing for at least three reasons:

1. It is true that machines can't solve Gödelian impossible problems. But humans can't solve them either. Humans can only estimate them. Com-puters can make estimates as well, and in recent years are doing a better job of this than humans.

2. In any event, quantum computing does not permit solving Gödelian impossible problems either. Solving a Gödelian impossible problem re-quires an algorithm with an infinite number of steps. Quantum com-puting can turn an intractable problem that could not be solved on a conventional computer in trillions of years into an instantaneous com-putation. But it still falls short of infinite computing.

3. Even if (1) and (2) above were wrong, that is, if humans could solve Gödelian impossible problems and do so because of their quantum-computing ability, that still does not restrict quantum computing from machines. The opposite is the case. If the human brain exhibits quan-tum computing, this would only confirm that quantum computing is possible, that matter following natural laws can perform quantum computing. Any mechanisms in human neurons capable of quan-tum computing, such as the microtubules, would be replicable in a machine. Machines use quantum effects—tunneling—in trillions of de-vices (that is, transistors) today.[18] There is nothing to suggest that the human brain has exclusive access to quantum computing.

Penrose's second conjecture is more difficult to resolve. It is that an entity exhibiting quantum computing is conscious. He is saying that it is the human's quantum computing that accounts for her consciousness. Thus quantum computing—quantum decoherence—yields consciousness.

Now we do know that there is a link between consciousness and quantum decoherence. That is, consciousness observing a quantum uncertainty causes quantum decoherence. Penrose, however, is asserting a link in the opposite direction. This does not follow logically. Of course quantum mechanics is not logical in the usual sense—it follows quantum logic (some observers use the word "strange" to describe quantum logic). But even applying quantum logic, Penrose's second conjecture does not appear to follow. On the other hand, I am unable to reject it out of hand because there is a strong nexus between consciousness and quantum decoherence in that the former causes the latter. I have thought about this issue for three years, and have been unable to accept it or reject it. Perhaps before writing my next book I will have an opinion on Penrose's second conjecture.

REVERSE ENGINEERING A PROVEN DESIGN: THE HUMAN BRAIN

For many people the mind is the last refuge of mystery against the encroaching spread of science, and they don't like the idea of science engulfing the last bit of terra incognita.

—Herb Simon as quoted by Daniel Dennett

Cannot we let people be themselves, and enjoy life in their own way? You are trying to make another you. One's enough.

—Ralph Waldo Emerson

For the wise men of old . . . the solution has been knowledge and self-discipline, . . . and in the practice of this technique, are ready to do things hitherto regarded as disgusting and impious—such as digging up and mutilating the dead.

—C. S. Lewis

Intelligence is: (a) the most complex phenomenon in the Universe; or (b) a profoundly simple process.

The answer, of course, is (c) both of the above. It's another one of those great dualities that make life interesting. We've already talked about the simplicity of intelligence: simple paradigms and the simple process of computation. Let's talk about the complexity.

We come back to knowledge, which starts out with simple seeds but ultimately becomes elaborate as the knowledge-gathering process interacts with the

IS THE BRAIN BIG ENOUGH?

Is our conception of human neuron functioning and our estimates of the number of neurons and connections in the human brain consistent with what we know about the brain's capabilities? Perhaps human neurons are far more capable than we think they are. If so, building a machine with human-level capabilities might take longer than expected.

We find that estimates of the number of concepts—"chunks" of knowledge—that a human expert in a particular field has mastered are remarkably consistent: about 50,000 to 100,000. This approximate range appears to be valid over a wide range of human endeavors: the number of board positions mastered by a chess grand master, the concepts mastered by an expert in a technical field, such as a physician, the vocabulary of a writer (Shakespeare used 29,000 words;[19] this book uses a lot fewer).

This type of professional knowledge is, of course, only a small subset of the knowledge we need to function as human beings. Basic knowledge of the world, including so-called common sense, is more extensive. We also have an ability to recognize patterns: spoken language, written language, objects, faces. And we have our skills: walking, talking, catching balls. I believe that a reasonably conservative estimate of the general knowledge of a typical human is a thousand times greater than the knowledge of an expert in her professional field. This provides us a rough estimate of 100 million chunks—bits of understanding, concepts, patterns, specific skills—per human. As we will see below, even if this estimate is low (by a factor of up to a thousand), the brain is still big enough.

The number of neurons in the human brain is estimated at approximately 100 billion, with an average of 1,000 connections per neuron, for a total of 100 trillion connections. With 100 trillion connections and 100 million chunks of knowledge (including patterns and skills), we get an estimate of about a million connections per chunk.

Our computer simulations of neural nets use a variety of different types of neuron models, all of which are relatively simple. Efforts to provide detailed electronic models of real mammalian neurons appear to show that while animal neurons are more complicated than typical computer models, the difference in complexity is modest. Even using our simpler computer versions of neurons, we find that we can model a chunk of knowledge—a face, a character shape, a phoneme, a word sense—using as little as a thousand connections per chunk. Thus our rough estimate of a million neural connections in the human brain per human knowledge chunk appears reasonable.

Indeed it appears ample. Thus we could make my estimate (of the number of knowledge chunks) a thousand times greater, and the calculation still works. It is likely, however, that the brain's encoding of knowledge is less efficient than the methods we use in our machines. This apparent inefficiency is consistent with our understanding that the human brain is conservatively designed. The brain relies on a large degree of redundancy and a relatively low density of information storage to gain reliability and to continue to function effectively despite a high rate of neuron loss as we age.

My conclusion is that it does not appear that we need to contemplate a model of information processing of individual neurons that is significantly more complex than we currently understand in order to explain human capability. The brain is big enough.

chaotic real world. Indeed, that is how intelligence originated. It was the result of the evolutionary process we call natural selection, itself a simple paradigm, that drew its complexity from the pandemonium of its environment. We see the same phenomenon when we harness evolution in the computer. We start with simple formulas, add the simple process of evolutionary iteration and combine this with the simplicity of massive computation. The result is often complex, capable, and intelligent algorithms.

But we don't need to simulate the entire evolution of the human brain in order to tap the intricate secrets it contains. Just as a technology company will take apart and "reverse engineer" (analyze to understand the methods of) a rival's products, we can do the same with the human brain. It is, after all, the best example we can get our hands on of an intelligent process. We can tap the architecture, organization, and innate knowledge of the human brain in order to greatly accelerate our understanding of how to design intelligence in a machine. By probing the brain's circuits, we can copy and imitate a proven design, one that took its original designer several billion years to develop. (And it's not even copyrighted.)

As we approach the computational ability to simulate the human brain—we're not there today, but we will begin to be in about a decade's time—such an effort will be intensely pursued. Indeed, this endeavor has already begun.

For example, Synaptics' vision chip is fundamentally a copy of the neural organization, implemented in silicon of course, of not only the human retina, but the early stages of mammalian visual processing. It has captured the essence of the algorithm of early mammalian visual processing, an algorithm called center surround filtering. It is not a particularly complicated chip, yet it realistically captures the essence of the initial stages of human vision.

There is a popular conceit among observers, both informed and uninformed, that such a reverse engineering project is infeasible. Hofstadter worries that "our brains may be too weak to understand themselves."[20] But that is not what we are finding. As we probe the brain's circuits, we find that the massively parallel algorithms are far from incomprehensible. Nor is there anything like an infinite number of them. There are hundreds of specialized regions in the brain, and it does have a rather ornate architecture, the consequence of its long history. The entire puzzle is not beyond our comprehension. It will certainly not be beyond the comprehension of twenty-first-century machines.

The knowledge is right there in front of us, or rather inside of us. It is not impossible to get at. Let's start with the most straightforward scenario, one that is essentially feasible today (at least to initiate).

We start by freezing a recently deceased brain.

Now, before I get too many indignant reactions, let me wrap myself in Leonardo da Vinci's cloak. Leonardo also received a disturbed reaction from his contemporaries. Here was a guy who stole dead bodies from the morgue, carted them back to his dwelling, and then took them apart. This was before dissecting dead bodies was in style. He did this in the name of knowledge, not a highly valued pursuit at the time. He wanted to learn how the human body works, but his contemporaries found his activities bizarre and disrespectful. Today we have a different view, that expanding our knowledge of this wondrous·machine is the most respectful homage we can pay. We cut up dead bodies all the time to learn more about how living bodies work, and to teach others what we have already learned.

There's no difference here in what I am suggesting. Except for one thing: I am talking about the brain, not the body. This strikes closer to home. We identify more with our brains than our bodies. Brain surgery is regarded as more invasive than toe surgery. Yet the value of the knowledge to be gained from probing the brain is too valuable to ignore. So we'll get over whatever squeamishness remains.

As I was saying, we start by freezing a dead brain. This is not a new concept— Dr. E. Fuller Torrey, a former supervisor at the National Institute of Mental Health and now head of the mental health branch of a private research foundation, has 44 freezers filled with 226 frozen brains.[21] Torrey and his associates hope to gain insight into the causes of schizophrenia, so all of his brains are of deceased schizophrenic patients, which is probably not ideal for our purposes.

We examine one brain layer—one very thin slice—at a time. With suitably sensitive two-dimensional scanning equipment we should be able to see every neuron and every connection represented in each synapse-thin layer. When a layer has been examined and the requisite data stored, it can be scraped away to reveal the next slice. This information can be stored and assembled into a giant three-dimensional model of the brain's wiring and neural topology.

It would be better if the frozen brains were not already dead long before freezing. A dead brain will reveal a lot about living brains, but it is clearly not the ideal laboratory. Some of that deadness is bound to reflect itself in a deterioration of its neural structure. We probably don't want to base our designs for intelligent machines on dead brains. We are likely to be able to take advantage of people who, facing imminent death, will permit their brains to be destructively scanned just slightly before rather than slightly after their brains would have stopped functioning on their own. Recently, a condemned killer allowed his brain and body to be scanned and you can access all 10 billion bytes of him on the Internet at the Center for Human Simulation's "Visible Human Project" web site.[22] There's an even higher resolution 25-billion-byte female companion on the site as well. Although the scan of this couple is not high enough resolution for the scenario envisioned here, it's an example of donating one's brain for reverse engineering. Of course we may not want to base our templates of machine intelligence on the brain of a convicted killer, anyway.

Easier to talk about are the emerging noninvasive means of scanning our brains. I began with the more invasive scenario above because it is technically much easier. We have in fact the means to conduct a destructive scan today (although not yet the bandwidth to scan the entire brain in a reasonable amount of time). In terms of noninvasive scanning, high-speed, high-resolution magnetic resonance imaging (MRI) scanners are already able to view individual somas (neuron cell bodies) without disturbing the living tissue being scanned. More powerful MRIs are being developed that will be capable of scanning individual nerve fibers that are only ten microns (millionths of a meter) in diameter. These will be available during the first decade of the twenty-first century. Eventually we will be able to scan the presynaptic vesicles that are the site of human learning.

We can peer inside someone's brain today with MRI scanners, which are increasing their resolution with each new generation of this technology. There are a number of technical challenges in accomplishing this, including achieving suitable resolution, bandwidth (that is, speed of transmission), lack of vibration, and safety. For a variety of reasons it is easier to scan the brain of someone recently deceased than of someone still living. (It is easier to get someone deceased to sit still, for one thing.) But noninvasively scanning a living brain will ultimately become feasible as MRI and other scanning technologies continue to improve in resolution and speed.

A new scanning technology called optical imaging, developed by Professor Amiram Grinvald at Israel's Weizmann Institute, is capable of significantly higher resolution than MRI. Like MRI, it is based on the interaction between electrical activity in the neurons and blood circulation in the capillaries feeding the neurons. Grinvald's device is capable of resolving features smaller than fifty microns, and can operate in real time, thus enabling scientists to view the firing of individual neurons. Grinvald and researchers at Germany's Max Planck Institute were struck by the remarkable regularity of the patterns of neural firing when the brain was engaged in processing visual information.[23] One of the researchers, Dr. Mark Hübener, commented that "our maps of the working brain are so orderly they resemble the street map of Manhattan rather than, say, of a medieval European town." Grinvald, Hübener, and their associates were able to use their brain scanner to distinguish between sets of neurons responsible for perception of depth, shape, and color. As these neurons interact with one another, the resulting pattern of neural firings resembles elaborately linked mosaics. From the scans, it was possible for the researchers to see how the neurons were feeding information to each other. For example, they noted that the depth perception neurons were arranged in parallel columns, providing information to the shape-detecting neurons that formed more elaborate pinwheel-like patterns. Currently, the Grinvald scanning technology is only able to image a thin slice of the brain near its surface, but the Weizmann Institute is working on refinements that will extend its three-dimensional capability. Grinvald's scanning technology is also being used to boost the resolution of MRI scanning. A recent finding that near-infrared light can pass through the skull is also fueling excitement about the ability of optical imaging as a high-resolution method of brain scanning.

The driving force behind the rapidly improving capability of noninvasive scanning technologies such as MRI is again the Law of Accelerating Returns, because it requires massive computational ability to build the high-resolution, three-dimensional images from the raw magnetic resonance patterns that an MRI scanner produces. The exponentially increasing computational ability provided by the Law of Accelerating Returns (and for another fifteen to twenty years, Moore's Law) will enable us to continue to rapidly improve the resolution and speed of these noninvasive scanning technologies.

Mapping the human brain synapse by synapse may seem like a daunting effort, but so did the Human Genome Project, an effort to map all human genes, when it was launched in 1991. Although the bulk of the human genetic code has still not been decoded, there is confidence at the nine American Genome Sequencing Centers that the task will be completed, if not by 2005, then at least within a few years of that target date. Recently, a new private venture with funding from Perkin-Elmer has announced plans to sequence the entire human genome by the year 2001. As I noted above, the pace of the human genome scan was extremely slow in its early years, and has picked up speed with improved technology, particularly computer programs that identify the useful genetic information. The researchers are counting on further improvements in their gene-hunting computer programs to meet their deadline. The same will be true of the human-brain-mapping project, as our methods of scanning and recording the 100 trillion neural connections pick up speed from the Law of Accelerating Returns.

What to Do with the Information

There are two scenarios for using the results of detailed brain scans. The most immediate—*scanning the brain to understand it*—is to scan portions of the brain to ascertain the architecture and implicit algorithms of interneuronal connections in different regions. The exact position of each and every nerve fiber is not as important as the overall pattern. With this information we can design simulated neural nets that operate similarly. This process will be rather like peeling an onion as each layer of human intelligence is revealed.

This is essentially what Synaptics has done in its chip that mimics mammalian neural-image processing. This is also what Grinvald, Hübener, and their associates plan to do with their visual-cortex scans. And there are dozens of other contemporary projects designed to scan portions of the brain and apply the resulting insights to the design of intelligent systems.

Within a region, the brain's circuitry is highly repetitive, so only a small portion of a region needs to be fully scanned. The computationally relevant activity of a neuron or group of neurons is sufficiently straightforward that we can understand and model these methods by examining them. Once the structure and topology of the neurons, the organization of the interneuronal wiring, and the sequence of neural firing in a region have been observed, recorded, and analyzed, it becomes feasible to reverse engineer that region's parallel algorithms. After the algorithms of a region are understood, they can be refined and extended prior to being implemented in synthetic neural equivalents. The methods can certainly be greatly sped up given that electronics is already more than a million times faster than neural circuitry.

We can combine the revealed algorithms with the methods for building intelligent machines that we already understand. We can also discard aspects of human computing that may not be useful in a machine. Of course, we'll have to be careful that we don't throw the baby out with the bathwater.

Downloading Your Mind to Your Personal Computer

A more challenging but also ultimately feasible scenario will be to scan someone's brain to map the locations, interconnections, and contents of the somas, axons, dendrites, presynaptic vesicles, and other neural components. Its entire organization could then be re-created on a neural computer of sufficient capacity, including the contents of its memory.

This is harder in an obvious way than the scanning-the-brain-to-understand-it scenario. In the former, we need only sample each region until we understand the salient algorithms. We can then combine those insights with knowledge we already have. In this—*scanning the brain to download it*—scenario, we need to capture every little detail. On the other hand, we don't need to understand all of it; we need only to literally copy it, connection by connection, synapse by synapse, neurotransmitter by neurotransmitter. It requires us to understand *local* brain processes, but not necessarily the brain's global organization, at least not in full. It is likely that by the time we can do this, we will understand much of it, anyway.

To do this right, we do need to understand what the salient information-processing mechanisms are. Much of a neuron's elaborate structure exists to support its own structural integrity and life processes and does not directly contribute to its handling of information. We know that neuron-computing processing is based on hundreds of different neurotransmitters and that different neural mechanisms in different regions allow for different types of computing. The early vision neurons, for example, are good at accentuating sudden color changes to facilitate finding the edges of objects. Hippocampus neurons are likely to have structures for enhancing the long-term retention of memories. We also know that neurons use a combination of digital and analog computing that needs to be accurately modeled. We need to identify structures capable of quantum computing, if any. All of the key features that affect information processing need to be recognized if we are to copy them accurately.

How well will this work? Of course, like any new technology, it won't be perfect at first, and initial downloads will be somewhat imprecise. Small imperfections won't necessarily be immediately noticeable because people are always changing to some degree. As our understanding of the mechanisms of the brain improves and our ability to accurately and noninvasively scan these features improves, reinstantiating (reinstalling) a person's brain should alter a person's mind no more than it changes from day to day.

What Will We Find When We Do This?

We have to consider this question on both the objective and subjective levels. "Objective" means everyone except me, so let's start with that. Objectively, when we scan someone's brain and reinstantiate their personal mind file into a suitable computing medium, the newly emergent "person" will appear to other observers to have very much the same personality, history, and memory as the person originally scanned. Interacting with the newly instantiated person will feel like interacting with the original person. The new person will claim to be that same old person and will have a memory of having been that person, having grown up in Brooklyn, having walked into a scanner here, and woken up in the machine there. He'll say, "Hey, this technology really works."

There is the small matter of the "new person's" body. What kind of body will a reinstantiated personal mind file have: the original human body, an upgraded body, a synthetic body, a nanoengineered body, a virtual body in a virtual environment? This is an important question, which I will discuss in the next chapter.

Subjectively, the question is more subtle and profound. Is this the same consciousness as the person we just scanned? As we saw in chapter 3, there are strong arguments on both sides. The position that fundamentally we are our "pattern" (because our particles are always changing) would argue that this new person is the same because their patterns are essentially identical. The counter argument, however, is the possible continued existence of the person who was originally scanned. If he—Jack—is still around, he will convincingly claim to represent the continuity of his consciousness. He may not be satisfied to let his mental clone carry on in his stead. We'll keep bumping into this issue as we explore the twenty-first century.

But once over the divide, the new person will certainly think that he was the original person. There will be no ambivalence in his mind as to whether or not he committed suicide when he agreed to be transferred into a new computing substrate leaving his old slow carbon-based neural-computing machinery behind. To the extent that he wonders at all whether or not he is really the same person that he thinks he is, he'll be glad that his old self took the plunge, because otherwise he wouldn't exist.

Is he—the newly installed mind—conscious? He certainly will claim to be. And being a lot more capable than his old neural self, he'll be persuasive and effective in his position. We'll believe him. He'll get mad if we don't.

A Growing Trend

In the second half of the twenty-first century, there will be a growing trend toward making this leap. Initially, there will be partial porting—replacing aging memory circuits, extending pattern-recognition and reasoning circuits through neural implants. Ultimately, and well before the twenty-first century is completed, people will port their entire mind file to the new thinking technology.

There will be nostalgia for our humble carbon-based roots, but there is nostalgia for vinyl records also. Ultimately, we did copy most of that analog music to the more flexible and capable world of transferable digital information. The leap to port our minds to a more capable computing medium will happen gradually but inexorably nonetheless.

As we port ourselves, we will also vastly extend ourselves. Remember that $1,000 of computing in 2060 will have the computational capacity of a trillion human brains. So we might as well multiply memory a trillion fold, greatly extend recognition and reasoning abilities, and plug ourselves into the pervasive wireless-communications network. While we are at it, we can add all human

THE AGE OF NEURAL IMPLANTS
HAS ALREADY STARTED

The patients are wheeled in on stretchers. Suffering from an advanced stage of Parkinson's disease, they are like statues, their muscles frozen, their bodies and faces totally immobile. Then in a dramatic demonstration at a French clinic, the doctor in charge throws an electrical switch. The patients suddenly come to life, get up, walk around, and calmly and expressively describe how they have overcome their debilitating symptoms. This is the dramatic result of a new neural implant therapy that is approved in Europe, and still awaits FDA approval in the United States.

The diminished levels of the neurotransmitter dopamine in a Parkinson's patient causes overactivation of two tiny regions in the brain: the ventral posterior nucleus and the subthalmic nucleus. This overactivation in turn causes the slowness, stiffness, and gait difficulties of the disease, and ultimately results in total paralysis and death. Dr. A. L. Benebid, a French physician at Fourier University in Grenoble, discovered that stimulating these regions with a permanently implanted electrode paradoxically inhibits these overactive regions and reverses the symptoms. The electrodes are wired to a small electronic control unit placed in the patient's chest. Through radio signals, the unit can be programmed, even turned on and off. When switched off, the symptoms immediately return. The treatment has the promise of controlling the most devastating symptoms of the disease.[24]

Similar approaches have been used with other brain regions. For example, by implanting an electrode in the ventral lateral thalamus, the tremors associated with cerebral palsy, multiple sclerosis, and other tremor-causing conditions can be suppressed.

"We used to treat the brain like soup, adding chemicals that enhance or suppress certain neurotransmitters," says Rick Trosch, one of the American physicians helping to perfect "deep brain stimulation" therapies. "Now we're treating it like circuitry."[25]

Increasingly, we are starting to combat cognitive and sensory afflictions by treating the brain and nervous system like the complex computational system that it is. Cochlear implants together with electronic speech processors perform frequency analysis of sound waves, similar to that performed by the inner ear. About 10 percent of the formerly deaf persons who have received this neural replacement device are now able to hear and understand voices well enough that they can hold conversations using a normal telephone.

Neurologist and ophthalmologist at Harvard Medical School Dr. Joseph Rizzo and his colleagues have developed an experimental retina implant. Rizzo's neural implant is a small solar-powered computer that communicates to the optic nerve. The user wears special glasses with tiny television cameras that communicate to the implanted computer by laser signal.[26]

Researchers at Germany's Max Planck Institute for Biochemistry have developed special silicon devices that can communicate with neurons in both directions. Directly stimulating neurons with an electrical current is not the ideal approach since it can cause corrosion to the electrodes and create chemical by-products that damage the cells. In contrast, the Max Planck Institute devices are capable of triggering an adjacent neuron to fire without a direct electrical link. The Institute scientists demonstrated their invention by controlling the movements of a living leech from their computer.

Going in the opposite direction—from neurons to electronics—is a device called a "neuron transistor,"[27] which can detect the firing of a neuron. The scientists hope to apply both technologies to the control of artificial human limbs by connecting spinal nerves to computerized prostheses. The Institute's Peter Fromherz says, "These two devices join the two worlds of information processing: the silicon world of the computer and the water world of the brain."

Neurobiologist Ted Berger and his colleagues at Hedco Neurosciences and Engineering have built integrated circuits that precisely match the properties and information processing of groups of animal neurons. The chips exactly mimic the digital and analog characteristics of the neurons they have analyzed. They are currently scaling up their technology to systems with hundreds of neurons.[28] Professor Carver Mead and his colleagues at the California Institute of Technology have also built digital-analog integrated circuits that match the processing of mammalian neural circuits comprising hundreds of neurons.[29]

The age of neural implants is under way, albeit at an early stage. Directly enhancing the information processing of our brain with synthetic circuits is focusing at first on correcting the glaring defects caused by neurological and sensory diseases and disabilities. Ultimately we will all find the benefits of extending our abilities through neural implants difficult to resist.

knowledge—as a readily accessible internal database as well as already processed and learned knowledge using the human type of distributed understanding.

The New Mortality

Actually there won't be mortality by the end of the twenty-first century. Not in the sense that we have known it. Not if you take advantage of the twenty-first century's brain-porting technology. Up until now, our mortality was tied to the longevity of our *hardware*. When the hardware crashed, that was it. For many of our forebears, the hardware gradually deteriorated before it disintegrated. Yeats lamented our dependence on a physical self that was "but a paltry thing, a tattered coat upon a stick."[30] As we cross the divide to instantiate ourselves into our computational technology, our identity will be based on our evolving mind file. *We will be software, not hardware.*

And evolve it will. Today, our software cannot grow. It is stuck in a brain of a mere 100 trillion connections and synapses. But when the hardware is trillions of times more capable, there is no reason for our minds to stay so small. They can and will grow.

As software, our mortality will no longer be dependent on the survival of the computing circuitry. There will still be hardware and bodies, but the essence of our identity will switch to the permanence of our software. Just as, today, we don't throw our files away when we change personal computers—we transfer them, at least the ones we want to keep. So, too, we won't throw our mind file away when we periodically port ourselves to the latest, ever more capable, "personal" computer. Of course, computers won't be the discrete objects they are today. They will be deeply embedded in our bodies, brains, and environment. Our identity and survival will ultimately become independent of the hardware and its survival.

Our immortality will be a matter of being sufficiently careful to make frequent backups. If we're careless about this, we'll have to load an old backup copy and be doomed to repeat our recent past.

LET'S JUMP TO THE OTHER SIDE OF THIS COMING CENTURY. YOU SAID THAT BY 2099 A PENNY OF COMPUTING WILL BE EQUAL TO A BILLION TIMES THE COMPUTING POWER OF ALL HUMAN BRAINS COMBINED. SOUNDS LIKE HUMAN THINKING IS GOING TO BE PRETTY TRIVIAL.

Unassisted, that's true.

SO HOW WILL WE HUMAN BEINGS FARE IN THE MIDST OF SUCH COMPETITION?

First, we have to recognize that the more powerful technology—the technologically more sophisticated civilization—always wins. That appears to be what happened when our *Homo sapiens sapiens* subspecies met the *Homo sapiens neanderthalensis* and other nonsurviving subspecies of *Homo sapiens*. That is what happened when the more technologically advanced Europeans encountered the indigenous peoples of the Americas. This is happening today as the more advanced technology is the key determinant of economic and military power.

SO WE'RE GOING TO BE SLAVES TO THESE SMART MACHINES?

Slavery is not a fruitful economic system to either side in an age of intellect. We would have no value as slaves to machines. Rather, the relationship is starting out the other way.

IT'S TRUE THAT MY PERSONAL COMPUTER DOES WHAT I ASK IT TO DO—SOMETIMES! MAYBE I SHOULD START BEING NICER TO IT.

No, it doesn't care how you treat it, not yet. But ultimately our native thinking capacities will be no match for the all-encompassing technology we're creating.

MAYBE WE SHOULD STOP CREATING IT.

We can't stop. The Law of Accelerating Returns forbids it! It's the only way to keep evolution going at an accelerating pace.

HEY, CALM DOWN. IT'S FINE WITH ME IF EVOLUTION SLOWS DOWN A TAD. SINCE WHEN HAVE WE ADOPTED YOUR ACCELERATION LAW AS THE LAW OF THE LAND?

We don't have to. Stopping computer technology, or any fruitful technology, would mean repealing basic realities of economic competition, not to mention our quest for knowledge. It's not going to happen. Furthermore, the road we're going down is a road paved with gold. It's full of benefits that we're never going to resist—continued growth in economic prosperity, better health, more intense communication, more effective education, more engaging entertainment, better sex.

UNTIL THE COMPUTERS TAKE OVER.

Look, this is not an alien invasion. Although it sounds unsettling, the advent of machines with vast intelligence is not necessarily a bad thing.

I GUESS IF WE CAN'T BEAT THEM, WE'LL HAVE TO JOIN THEM.

That's exactly what we're going to do. Computers started out as extensions of our minds, and they will end up extending our minds. Machines are already an integral part of our civilization, and the sensual and spiritual machines of the twenty-first century will be an even more intimate part of our civilization.

OKAY, IN TERMS OF EXTENDING MY MIND, LET'S GET BACK TO IMPLANTS FOR MY FRENCH LIT CLASS. IS THIS GOING TO BE LIKE I'VE READ THIS STUFF? OR IS IT JUST GOING TO BE LIKE A SMART PERSONAL COMPUTER THAT I CAN COMMUNICATE WITH QUICKLY BECAUSE IT HAPPENS TO BE LOCATED IN MY HEAD?

That's a key question, and I think it will be controversial. It gets back to the issue of consciousness. Some people will feel that what goes in their neural implants is indeed subsumed by their consciousness. Others will feel that it remains outside of their sense of self. Ultimately, I think that we will regard the mental activity of the implants as part of our own thinking. Consider that even without implants, ideas and thoughts are constantly popping into our heads, and we have little idea of where they came from, or how they got there. We nonetheless consider all the mental phenomena that we become aware of as our own thoughts.

SO I'LL BE ABLE TO DOWNLOAD MEMORIES OF EXPERIENCES I'VE NEVER HAD?

Yes, but someone has probably had the experience. So why not have the ability to share it?

I SUPPOSE FOR SOME EXPERIENCES, IT MIGHT BE SAFER TO JUST DOWNLOAD THE MEMORIES OF IT.

Less time-consuming also.

DO YOU REALLY THINK THAT SCANNING A FROZEN BRAIN IS FEASIBLE TODAY?

Sure, just stick your head in my freezer here.

GEE, ARE YOU SURE THIS IS SAFE?

Absolutely.

WELL, I THINK I'LL WAIT FOR FDA APPROVAL.

Okay, then you'll have to wait a long time.

THINKING AHEAD, I STILL HAVE THIS SENSE THAT WE'RE DOOMED. I MEAN, I CAN UNDER-STAND HOW A NEWLY INSTANTIATED MIND, AS YOU PUT IT, WILL BE HAPPY THAT SHE WAS CREATED AND WILL THINK THAT SHE HAD BEEN ME PRIOR TO MY HAVING BEEN SCANNED AND IS STILL ME IN A SHINY NEW BRAIN. SHE'LL HAVE NO REGRETS AND WILL BE ON THE "OTHER SIDE." BUT I DON'T SEE HOW *I* CAN GET ACROSS THE HUMAN-MACHINE DIVIDE. AS YOU POINTED OUT, IF I'M SCANNED, THAT NEW ME ISN'T ME BECAUSE I'M STILL HERE IN MY OLD BRAIN.

Yes, there's a little glitch in this regard. But I'm sure we'll figure how to solve this thorny problem with a little more consideration.

CHAPTER SEVEN

... AND BODIES

THE IMPORTANCE OF HAVING A BODY

Let's start by taking a quick look at my reader's diary.

NOW WAIT JUST A MINUTE.

Is there a problem?

FIRST OF ALL, I HAVE A NAME.

Yes, it would be a good idea to introduce you by name at this point.

I'M MOLLY.

Thank you, is there something else?

YES. I'M NOT SURE I'M PREPARED TO SHARE MY DIARY WITH YOUR OTHER READERS.

Most writers don't let their readers participate at all. Anyway, you're my creation, so I should be able to share your personal reflections if it serves a purpose here.

I MAY BE YOUR CREATION, BUT REMEMBER WHAT YOU SAID IN CHAPTER 2 ABOUT ONE'S CREATIONS EVOLVING TO SURPASS THEIR CREATORS.

True enough, so maybe I should be more sensitive to your needs.

GOOD IDEA—LET'S START BY ALLOWING ME TO VET THOSE ENTRIES YOU'RE SELECTING.

Very well. Here are some extracts from Molly's diary, suitably edited:

I've switched to nonfat muffins. This has two distinct benefits. First of all, they have half the number of calories. Secondly, they taste awful. That way I'm less tempted to eat them. But I wish people would stop shoving food in my face. . . . I'm going to have trouble at this potluck dorm party tomorrow. I feel like I have to try everything, and I kind of lose track of what I'm eating.

I've got to drop at least half a dress size. A full size would be better. Then I could breathe more easily in this new dress. That reminds me, I should stop at the health club on my way home. Maybe that new trainer will notice me. Actually I did catch him looking at me, but I was being kind of spastic with those new machines, and he looked the other way. . . . I'm not crazy about the neighborhood this place is in, I don't really feel safe walking back to my car when it's late. Okay, here's an idea—I'll ask that trainer—got to get his name—to walk me to my car. Always a good idea to be safe, right?

. . . I'm a little nervous about this bump on my toe. But the doctor said that toe bumps are almost always benign. But he still wants to remove it and send it to a lab. He said I won't feel a thing. Except, of course, for the novocaine—I hate needles!

. . . It was a little strange seeing my old boyfriend, but I'm glad we're still friends. It did feel good when he gave me a hug. . . .

Thank you, Molly. Now consider: How many of Molly's entries would make sense if she didn't have a body? Most of Molly's mental activities are directed toward her body and its survival, security, nutrition, image, not to mention related issues of affection, sexuality, and reproduction. But Molly is not unique in this regard. I invite my other readers to look at their own diaries. And if you don't have one, consider what you would write in it if you did. How many of your entries would make sense if you didn't have a body?

Our bodies are important in many ways. Most of those goals I spoke about at the beginning of the previous chapter—the ones we attempt to solve using our intelligence—have to do with our bodies: protecting them, providing them with fuel, making them attractive, making them feel good, providing for their myriad needs, not to mention desires.

Some philosophers—professional artificial-intelligence critic Hubert Dreyfus, for one—maintain that achieving human-level intelligence is impossible without a body.[1] Certainly, if we're going to port a human's mind to a new computational medium, we'd better provide a body. A disembodied mind will quickly get depressed.

TWENTY-FIRST CENTURY BODIES

What makes a soul? And if machines ever have souls, what will be the equivalent of psychoactive drugs? Of pain? Of the physical/emotional high I get from having a clean office?

—Esther Dyson

What a strange machine man is. You fill him with bread, wine, fish, and radishes, and out come sighs, laughter and dreams.

—Nikos Kazantzakis

So what kind of bodies will we provide for our twenty-first-century machines? Later on, the question will become: What sort of bodies will they provide for themselves?

Let's start with the human body. It's the body we're used to. It evolved along with its brain, so the human brain is well suited to provide for its needs. The human brain and body kind of go together.

The likely scenario is that both body and brain will evolve together, will become enhanced together, will migrate together toward new modalities and materials. As I discussed in the previous chapter, porting our brains to new computational mechanisms will not happen all at once. We will enhance our brains gradually through direct connection with machine intelligence until such time that the essence of our thinking has fully migrated to the far more capable and reliable new machinery. Again, if we find this notion troublesome, a lot of this uneasiness has to do with our concept of the word *machine*. Keep in mind that our concept of this word will evolve along with our minds.

In terms of transforming our bodies, we are already further along in this process than we are in advancing our minds. We have titanium devices to replace our jaws, skulls, and hips. We have artificial skin of various kinds. We have artificial heart valves. We have synthetic vessels to replace arteries and veins, along with expandable stents to provide structural support for weak natural vessels. We have artificial arms, legs, feet, and spinal implants. We have all kinds of joints: jaws, hips, knees, shoulders, elbows, wrists, fingers, and toes. We have implants to control our bladders. We are developing machines—some made of artificial materials, others combining new materials with cultured cells—that will ultimately be able to replace organs such as the liver and pancreas. We have penile prostheses with little pumps to simulate erections. And we have long had implants for teeth and breasts.

Of course, the notion of completely rebuilding our bodies with synthetic materials, even if superior in certain ways, is not immediately compelling. We like the softness of our bodies. We like bodies to be supple and cuddly and warm. And not a superficial warmth, but the deep and intimate heat drawn from its trillions of living cells.

So let's consider enhancing our bodies cell by cell. We have started down that road as well. We have written down a portion of the entire genetic code that describes our cells, and we've started the process of understanding it. In the near future, we hope to design genetic therapies to improve our cells, to correct such defects as the insulin resistance associated with Type II diabetes, and the loss of control over self-replication associated with cancer. An early method of delivering gene therapies was to infect a patient with special viruses containing the corrective DNA. A more effective method developed by Dr. Clifford Steer at the University of Minnesota utilizes RNA molecules to deliver the desired DNA directly.[2] High on researchers' list for future cellular improvements through genetic engineering is to counteract our genes for cellular suicide. These strands of genetic beads, called telomeres, get shorter every time a cell divides. When the telomere beads count down to zero, a cell is no longer able to divide, and destroys itself. There's a long list of diseases, aging conditions, and limitations that we intend to address by altering the genetic code that controls our cells.

But there is only so far we can go with this approach. Our DNA-based cells depend on protein synthesis, and while protein is a marvelously diverse substance, it suffers from severe limitations. Hans Moravec, one of the first serious thinkers to realize the potential of twenty-first-century machines, points out that "protein is not an ideal material. It is stable only in a narrow temperature and pressure range, is very sensitive to radiation, and rules out many construction techniques and components. . . . A genetically engineered superhuman would be just a second-rate kind of robot, designed under the handicap that its construction can only be by DNA-guided protein synthesis. Only in the eyes of human chauvinists would it have an advantage."[3]

One of evolution's ideas that is worth keeping, however, is building our bodies from cells. This approach would retain many of our bodies' beneficial qualities: redundancy, which provides a high degree of reliability; the ability to regenerate and repair itself; and softness and warmth. But just as we will eventually relinquish the extremely slow speed of our neurons, we will ultimately be forced to abandon the other restrictions of our protein-based chemistry. To reinvent our cells, we look to one of the twenty-first century's primary technologies: *nanotechnology.*

NANOTECHNOLOGY:
REBUILDING THE WORLD, ATOM BY ATOM

*The problems of chemistry and biology can be greatly helped if . . . doing things
on an atomic level is ultimately developed—a development which I think cannot
be avoided.*

—Richard Feynman, 1959

*Suppose someone claimed to have a microscopically exact replica (in marble,
even) of Michelangelo's David in his home. When you go to see this marvel, you
find a twenty-foot-tall, roughly rectilinear hunk of pure white marble standing
in his living room. "I haven't gotten around to unpacking it yet," he says, "but I
know it's in there."*

—Douglas Hofstadter

*What advantages will nanotoasters have over conventional macroscopic toaster
technology? First, the savings in counter space will be substantial. One philosophical point that must not be overlooked is that the creation of the world's
smallest toaster implies the existence of the world's smallest slice of bread. In the
quantum limit we must necessarily encounter fundamental toast particles,
which we designate here as "croutons."*
—Jim Cser, *Annals of Improbable Research,* edited by Marc Abrahams

Humankind's first tools were found objects: sticks used to dig up roots and
stones used to break open nuts. It took our forebears tens of thousands of years
to invent a sharp blade. Today we build machines with finely designed intricate
mechanisms, but viewed on an atomic scale, our technology is still crude. "Cast-
ing, grinding, milling, and even lithography move atoms in great thundering statistical herds," says Ralph Merkle, a leading nanotechnology theorist at Xerox's
Palo Alto Research Center. He adds that current manufacturing methods are "like
trying to make things out of Legos with boxing gloves on. . . . In the future,
nanotechnology will let us take off the boxing gloves."[4]

Nanotechnology is technology built on the atomic level: building machines
one atom at a time. "Nano" refers to a billionth of a meter, which is the width
of five carbon atoms. We have one existence proof of the feasibility of nanotechnology: life on Earth. Little machines in our cells called ribosomes build organisms such as humans one molecule, that is one amino acid, at a time, following
digital templates coded in another molecule called DNA. Life on Earth has mastered the ultimate goal of nanotechnology, which is self-replication.

But as mentioned above, Earthly life is limited by the particular molecular
building block it has selected. Just as our human-created computational technology will ultimately exceed the capacity of natural computation (electronic
circuits are already millions of times faster than human neural circuits), our

twenty-first-century physical technology will also greatly exceed the capabilities of the amino acid–based nanotechnology of the natural world.

The concept of building machines atom by atom was first described in a 1959 talk at Cal Tech titled "There's Plenty of Room at the Bottom," by physicist Richard Feynman, the same guy who first suggested the possibility of quantum computing.[5] The idea was developed in some detail by Eric Drexler twenty years later in his book *Engines of Creation*.[6] The book actually inspired the cryonics movement of the 1980s, in which people had their heads (with or without bodies) frozen in the hope that a future time would possess the molecule-scale technology to overcome their mortal diseases, as well as undo the effects of freezing and defrosting. Whether a future generation would be motivated to revive all these frozen brains was another matter.

After publication of *Engines of Creation*, the response to Drexler's ideas was skeptical and he had difficulty filling out his MIT Ph.D. committee despite Marvin Minsky's agreement to supervise it. Drexler's dissertation, published in 1992 as a book titled *Nanosystems: Molecular Machinery, Manufacturing, and Computation*, provided a comprehensive proof of concept, including detailed analyses and specific designs.[7] A year later, the first nanotechnology conference attracted only a few dozen researchers. The fifth annual conference, held in December 1997, boasted 350 scientists who were far more confident of the practicality of their tiny projects. Nanothinc, an industry think tank, estimated in 1997 that the field already produces $5 billion in annual revenues for nanotechnology-related technologies, including micromachines, microfabrication techniques, nanolithography, nanoscale microscopes, and others. This figure has been more than doubling each year.[8]

The Age of Nanotubes

One key building material for tiny machines is, again, nanotubes. Although built on an atomic scale, the hexagonal patterns of carbon atoms are extremely strong and durable. "You can do anything you damn well want with these tubes and they'll just keep on truckin'," says Richard Smalley, one of the chemists who received the Nobel Prize for discovering the buckyball molecule.[9] A car made of nanotubes would be stronger and more stable than a car made with steel, but would weigh only fifty pounds. A spacecraft made of nanotubes could be of the size and strength of the U.S. space shuttle, but weigh no more than a conventional car. Nanotubes handle heat extremely well, far better than the fragile amino acids that people are built out of. They can be assembled into all kinds of shapes: wirelike strands, sturdy girders, gears, etcetera. Nanotubes are formed of carbon atoms, which are in plentiful supply in the natural world.

As I mentioned earlier, the same nanotubes can be used for extremely efficient computation, so both the structural and computational technology of the twenty-first century will likely be constructed from the same stuff. In fact, the same nanotubes used to form physical structures can also be used for computation, so future nanomachines can have their brains distributed throughout their bodies.

The best-known examples of nanotechnology to date, while not altogether practical, are beginning to show the feasibility of engineering at the atomic level. IBM created its corporate logo using individual atoms as pixels.[10] In 1996, Texas Instruments built a chip-sized device with half a million moveable mirrors to be used in a tiny high-resolution projector.[11] TI sold $100 million worth of their nanomirrors in 1997.

Chih-Ming Ho of UCLA is designing flying machines using surfaces covered with microflaps that control the flow of air in a similar manner to conventional flaps on a normal airplane.[12] Andrew Berlin at Xerox's Palo Alto Research Center is designing a printer using microscopic air valves to move paper documents precisely.[13]

Cornell graduate student and rock musician Dustin Carr built a realistic-looking but microscopic guitar with strings only fifty nanometers in diameter. Carr's creation is a fully functional musical instrument, but his fingers are too large to play it. Besides, the strings vibrate at 10 million vibrations per second, far beyond the twenty-thousand-cycles-per-second limit of human hearing.[14]

The Holy Grail of Self-Replication:
Little Fingers and a Little Intelligence

Tiny fingers represent something of a holy grail for nanotechnologists. With little fingers and computation, nanomachines would have in their Lilliputian world what people have in the big world: intelligence and the ability to manipulate their environment. Then these little machines could build replicas of themselves, achieving the field's key objective.

The reason that self-replication is important is that it is too expensive to build these tiny machines one at a time. To be effective, nanometer-sized machines need to come in the trillions. The only way to achieve this economically is through combinatorial explosion: let the machines build themselves.

Drexler, Merkle (a coinventor of public key encryption, the primary method of encrypting messages), and others have convincingly described how such a self-replicating nanorobot—*nanobot*—could be constructed. The trick is to provide the nanobot with sufficiently flexible manipulators—arms and hands—so that it is capable of building a copy of itself. It needs some means for mobility so that it can find the requisite raw materials. It requires some intelligence so that it can solve the little problems that will arise when each nanobot goes about building a complicated little machine like itself. *Finally, a really important requirement is that it needs to know when to stop replicating.*

Morphing in the Real World

Self-replicating machines built at the atomic level could truly transform the world we live in. They could build extremely inexpensive solar cells, allowing the replacement of messy fossil fuels. Since solar cells require a large surface area to collect sufficient sunlight, they could be placed in orbit, with the energy beamed down to Earth.

Nanobots launched into our bloodstreams could supplement our natural immune system and seek out and destroy pathogens, cancer cells, arterial plaque, and other disease agents. In the vision that inspired the cryonics enthusiasts, diseased organs can be rebuilt. We will be able to reconstruct any or all of our bodily organs and systems, and do so at the cellular level. I talked in the last chapter about reverse engineering and emulating the salient computational functionality of human neurons. In the same way, it will become possible to reverse engineer and replicate the physical and chemical functionality of any human cell. In the process we will be in a position to greatly extend the durability, strength, temperature range, and other qualities and capabilities of our cellular building blocks.

We will then be able to grow stronger, more capable organs by redesigning the cells that constitute them and building them with far more versatile and durable materials. As we go down this road, we'll find that some redesign of the body makes sense at multiple levels. For example, if our cells are no longer vulnerable to the conventional pathogens, we may not need the same kind of immune system. But we will need new nanoengineered protections for a new assortment of nanopathogens.

Food, clothing, diamond rings, buildings could all assemble themselves molecule by molecule. Any sort of product could be instantly created when and where we need it. Indeed, the world could continually reassemble itself to meet our changing needs, desires, and fantasies. By the late twenty-first century, nanotechnology will permit objects such as furniture, buildings, clothing, even people, to change their appearance and other characteristics—essentially to change into something else—in a split second.

These technologies will emerge gradually (I will attempt to delineate the different gradations of nanotechnology as I talk about each of the decades of the twenty-first century in Part III of this book). There is a clear incentive to go down this path. Given a choice, people will prefer to keep their bones from crumbling, their skin supple, their life systems strong and vital. Improving our lives through neural implants on the mental level, and nanotechnology-enhanced bodies on the physical level, will be popular and compelling. It is another one of those slippery slopes—there is no obvious place to stop this progression until the human race has largely replaced the brains and bodies that evolution first provided.

A Clear and Future Danger

Without self-replication, nanotechnology is neither practical nor economically feasible. And therein lies the rub. What happens if a little software problem (inadvertent or otherwise) fails to halt the self-replication? We may have more nanobots than we want. They could eat up everything in sight.

The movie *The Blob* (of which there are two versions) was a vision of nanotechnology run amok. The movie's villain was this intelligent self-replicating gluttonous stuff that fed on organic matter. Recall that nanotechnology is likely to be built from carbon-based nanotubes, so, like the Blob, it will build itself from organic matter, which is rich in carbon. Unlike mere animal-based cancers, an exponentially exploding nanomachine population would feed on any carbon-based matter. Tracking down all of these bad nanointelligences would be like trying to find trillions of microscopic needles—rapidly moving ones at that—in at least as many haystacks. There have been proposals for nanoscale immunity technologies: good little antibody machines that would go after the bad little machines. The nanoantibodies would, of course, have to scale up at least as quickly as the epidemic of marauding nanomiscreants. There could be a lot of collateral damage as these trillions of machines battle it out.

Now that I have raised this specter, I will try, unconvincingly perhaps, to put the peril in perspective. I believe that it will be possible to engineer self-replicating nanobots in such a way that an *inadvertent,* undesired population explosion would be unlikely. I realize that this may not be completely reassuring, coming from a software developer whose products (like those of my competitors) crash once in a while (but rarely—and when they do, it's the fault of the operating system!). There is a concept in software development of "mission critical" applications. These are software programs that control a process on which people are heavily dependent. Examples of mission-critical software include life-support systems in hospitals, automated surgical equipment, autopilot flying and landing systems, and other software-based systems that affect the well-being of a person or organization. It is feasible to create extremely high levels of reliability in these programs. There are examples of complex technology in use today in which a mishap would severely imperil public safety. A conventional explosion in an atomic power plant could spray deadly plutonium across heavily populated areas. Despite a near meltdown at Chernobyl, this apparently has only occurred twice in the decades that we have had hundreds of such plants operating, both incidents involving recently acknowledged reactor calamities in the Chelyabinsk region of Russia.[15] There are tens of thousands of nuclear weapons, and none has ever exploded in error.

I admit that the above paragraph is not entirely convincing. But the bigger danger is the intentional hostile use of nanotechnology. Once the basic technology is available, it would not be difficult to adapt it as an instrument of war or terrorism. It is not the case that someone would have to be suicidal to use such weapons. The nanoweapons could easily be programmed to replicate only against an enemy; for example, only in a particular geographical area. Nuclear weapons, for all their destructive potential, are at least relatively local in their effects. The self-replicating nature of nanotechnology makes it a far greater danger.

VIRTUAL BODIES

We don't always need real bodies. If we happen to be in a virtual environment, then a virtual body will do just fine. Virtual reality started with the concept of computer games, particularly ones that provided a simulated environment. The first was Space War, written by early artificial-intelligence researchers to pass the time while waiting for programs to compile on their slow 1960s computers.[16] The synthetic space surroundings were easy to render on low-resolution monitors: Stars and other space objects were just illuminated pixels.

Computer games and computerized video games have become more realistic over time, but you cannot completely immerse yourself in these imagined worlds, not without some imagination. For one thing, you can see the edges of the screen, and the all too real world that you have never left is still visible beyond these borders.

If we're going to enter a new world, we had better get rid of traces of the old. In the 1990s the first generation of virtual reality has been introduced in which you don a special visual helmet that takes over your entire visual field. The key to visual reality is that when you move your head, the scene instantly repositions itself so that you are now looking at a different region of a three-dimensional scene. The intention is to simulate what happens when you turn your real head in the real world: The images captured by your retinas rapidly change. Your brain nonetheless understands that the world has remained stationary and that the image is sliding across your retinas only because your head is rotating.

Like most first generation technologies, virtual reality has not been fully convincing. Because rendering a new scene requires a lot of computation, there is a lag in producing the new perspective. Any noticeable delay tips off your brain that the world you're looking at is not entirely real. The resolution of virtual reality displays has also been inadequate to create a fully satisfactory illusion. Finally, contemporary virtual reality helmets are bulky and uncomfortable.

What's needed to remove the rendering delay and to boost display resolution is yet faster computers, which we know are always on the way. By 2007, high-quality virtual reality with convincing artificial environments, virtually instantaneous rendering, and high-definition displays will be comfortable to wear and available at computer game prices.

That takes care of two of our senses—visual and auditory. Another high-resolution sense organ is our skin, and "haptic" interfaces to provide a virtual tactile interface are also evolving. One available today is the Microsoft force-feedback joystick, derived from 1980s research at the MIT Media Lab. A force-feedback joystick adds some tactile realism to computer games, so you feel the rumble of the road in a car-driving game or the pull of the line in a fishing simulation. Emerging in late 1998 is the "tactile mouse," which operates like a conventional mouse but allows the user to feel the texture of surfaces, objects, even people. One company that I am involved in, Medical Learning Company, is developing a simulated patient to help train doctors, as well as enable nonphysicians to play doctor. It will include a haptic interface so that you can feel a knee joint for a fracture or a breast for lumps.[17]

A force-feedback joystick in the tactile domain is comparable to conventional monitors in the visual domain. The force-feedback joystick provides a tactile interface, but it does not totally envelop you. The rest of your tactile world is still reminding you of its presence. In order to leave the real world, at least temporarily, we need a tactile environment that takes over your sense of touch.

So let's invent a virtual tactile environment. We've seen aspects of it in science fiction films (always a good source for inventing the future). We can build a body suit that will detect your own movements as well as provide high resolution tactile stimulation. The suit will also need to provide sufficient force-feedback to actually prevent your movements if you are pressing against a virtual obstacle in the virtual environment. If you are giving a virtual companion a hug, for example, you don't want to move right through his or her body. This will require a force-feedback structure outside the suit, although obstacle resistance could be provided by the suit itself. And since your body inside the suit is still in the real world, it would make sense to put the whole contraption in a booth so that your movements in the virtual world don't knock down lamps and people in your "real" vicinity. Such a suit could also provide a thermal response and thereby allow the simulation of feeling a moist surface—or even immersing your hand or your whole body in water—which is indicated by a change in temperature and a decrease in surface tension. Finally, we can provide a platform consisting of a rotating treadmill device for you to stand (or sit or lie) on, which will allow you to walk or move around (in any direction) in your virtual environment.

So with the suit, the outer structure, the booth, the platform, the goggles, and the earphones, we just about have the means to totally envelop your senses. Of course, we will need some good virtual reality software, but there's certain to be hot competition to provide a panoply of realistic and fantastic new environments as the requisite hardware becomes available.

Oh yes, there is the sense of smell. A completely flexible and general interface for our fourth sense will require a reasonably advanced nanotechnology to synthesize the wide variety of molecules that we can detect with our olfactory sense. In the meantime, we could provide the ability to diffuse a variety of aromas in the virtual reality booth.

Once we are in a virtual reality environment, our own bodies—at least the virtual versions—can change as well. We can become a more attractive version of ourselves, a hideous beast, or any creature real or imagined as we interact with the other inhabitants in each virtual world we enter.

Virtual reality is not a (virtual) place you need go to alone. You can interact with your friends there (who would be in other virtual reality booths, which may be geographically remote). You will have plenty of simulated companions to choose from as well.

Directly Plugging In

Later in the twenty-first century, as neural implant technologies become ubiquitous, we will be able to create and interact with virtual environments without having to enter a virtual reality booth. Your neural implants will provide the simulated sensory inputs of the virtual environment—and your virtual body—directly in your brain. Conversely, your movements would not move your "real" body, but rather your perceived virtual body. These virtual environments would also include a suitable selection of bodies for yourself. Ultimately, your experience would be highly realistic, just like being in the real world. More than one person could enter a virtual environment and interact with each other. In the virtual world, you will meet other real people and simulated people—eventually, there won't be much difference.

This will be the essence of the Web in the second half of the twenty-first century. A typical "web site" will be a perceived virtual environment, with no external hardware required. You "go there" by mentally selecting the site and then entering that world. Debate Benjamin Franklin on the war powers of the presidency at the history society site. Ski the Alps at the Swiss Chamber of Commerce site (while feeling the cold spray of snow on your face). Hug your favorite movie star at the Columbia Pictures site. Get a little more intimate at the Penthouse or Playgirl site. Of course, there may be a small charge.

Real Virtual Reality

In the late twenty-first century, the "real" world will take on many of the characteristics of the virtual world through the means of nanotechnology "swarms." Consider, for example, Rutgers University computer scientist J. Storrs Hall's concept of "Utility Fog."[18] Hall's conception starts with a little robot called a Foglet, which consists of a human-cell-sized device with twelve arms pointing in all directions. At the end of the arms are grippers so that the Foglets can grasp one another to form larger structures. These nanobots are intelligent and can merge their computational capacities with each other to create a distributed intelligence. A space filled with Foglets is called Utility Fog and has some interesting properties.

First of all, the Utility Fog goes to a lot of trouble to simulate its not being there. Hall describes a detailed scenario that lets a real human walk through a room filled with trillions of Foglets and not notice a thing. When desired (and it's not entirely clear who is doing the desiring), the Foglets can quickly simulate any environment by creating all sorts of structures. As Hall puts it, "Fog city can look like a park, or a forest, or ancient Rome one day and Emerald City the next."

The Foglets can create arbitrary wave fronts of light and sound in any direction to create any imaginary visual and auditory environment. They can exert any pattern of pressure to create any tactile environment. In this way, Utility Fog has all the flexibility of a virtual environment, except it exists in the real physical world. The distributed intelligence of the Utility Fog can simulate the minds of scanned (Hall calls them "uploaded") people who are re-created in the Utility Fog as "Fog people." In Hall's scenario, "a biological human can walk through Fog walls, and a Fog (uploaded) human can walk through dumb-matter walls. Of course Fog people can walk through Fog walls, too."

The physical technology of Utility Fog is actually rather conservative. The Foglets are much bigger machines than most nanotechnology conceptions. The software is more challenging, but ultimately feasible. Hall needs a bit of work on his marketing angle: Utility Fog is a rather dull name for such versatile stuff.

There are a variety of proposals for nanotechnology swarms, in which the real environment is constructed from interacting multitudes of nanomachines. In all of the swarm conceptions, physical reality becomes a lot like virtual reality. You can be sleeping in your bed one moment, and have the room transform into your kitchen as you awake. Actually, change that to a dining room as there's no need for a kitchen. Related nanotechnology will instantly create whatever meal you desire. When you finish eating, the room can transform into a study, or a game room, or a swimming pool, or a redwood forest, or the Taj Mahal. You get the idea.

Mark Yim has built a large-scale model of a small swarm showing the feasibility of swarm interaction.[19] Joseph Michael has actually received a U.K. patent on his conception of a nanotechnology swarm, but it is unlikely that his design will be commercially realizable in the twenty-year life of his patent.[20]

It may seem that we will have too many choices. Today, we have only to choose our clothes, makeup, and destination when we go out. In the late twenty-first century, we will have to select our body, our personality, our environment—so many difficult decisions to make! But don't worry—we'll have intelligent swarms of machines to guide us.

THE SENSUAL MACHINE

Made double by his lust
he sounds a woman's groans.
A figment of his flesh.
                    —from Barry Spacks's poem "The Solitary at Seventeen"

I can predict the future by assuming that money and male hormones are the driving forces for new technology. Therefore, when virtual reality gets cheaper than dating, society is doomed.
                                        —Dogbert

The first book printed from a moveable type press may have been the Bible, but the century following Gutenberg's epochal invention saw a lucrative market for books with more prurient topics.[21] New communication technologies—the telephone, motion pictures, television, videotape—have always been quick to adopt sexual themes. The Internet is no exception, with 1998 market estimates of adult online entertainment ranging from $185 million by Forrester Research to $1 billion by Inter@active Week. These figures are for customers, mostly men, paying to view and interact with performers—live, recorded, and simulated. One 1998 estimate cited 28,000 web sites that offer sexual entertainment.[22] These figures do not include couples who have expanded their phone sex to include moving pictures via online video conferencing.

CD-ROMs and DVD disks constitute another technology that has been exploited for erotic entertainment. Although the bulk of adult-oriented disks are used as a means for delivering videos with a bit of interactivity thrown in, a new genre of CD-ROM and DVD provides virtual sexual companions that respond to some mouse-administered fondling.[23] Like most first-generation technologies, the effect is less than convincing, but future generations will eliminate some of the kinks, although not the kinkiness. Developers are also working to exploit the force-feedback mouse so that you can get some sense of what your virtual partner feels like.

Late in the first decade of the twenty-first century, virtual reality will enable you to be with your lover—romantic partner, sex worker, or simulated companion—with full visual and auditory realism. You will be able to do anything you want with your companion except touch, admittedly an important limitation.

Virtual touch has already been introduced, but the all-enveloping, highly realistic, visual-auditory-tactile virtual environment will not be perfected until the second decade of the twenty-first century. At this point, virtual sex becomes a viable competitor to the real thing. Couples will be able to engage in virtual sex regardless of their physical proximity. Even when proximate, virtual sex will be better in some ways and certainly safer. Virtual sex will provide sensations that are more intense and pleasurable than conventional sex, as well as physical experiences that currently do not exist. Virtual sex is also the ultimate in safe sex, as there is no risk of pregnancy or transmission of disease.

Today, lovers may fantasize their partners to be someone else, but users of virtual sex communication will not need as much imagination. You will be able to change the physical appearance and other characteristics of both yourself and your partner. You can make your lover look and feel like your favorite star without your partner's permission or knowledge. Of course, be aware that your partner may be doing the same to you.

Group sex will take on a new meaning in that more than one person can simultaneously share the experience of one partner. Since multiple real people cannot all control the movements of one virtual partner, there needs to be a way of sharing the decision making of what the one virtual body is doing. Each participant sharing a virtual body would have the same visual, auditory, and tactile experience, with shared control of their shared virtual body (perhaps the one virtual body will reflect a consensus of the attempted movements of the multiple participants). A whole audience of people—who may be geographically dispersed—could share one virtual body while engaged in a sexual experience with one performer.

Prostitution will be free of health risks, as will virtual sex in general. Using wireless, very-high-bandwidth communication technologies, neither sex workers nor their patrons need leave their homes. Virtual prostitution is likely to be legally tolerated, at least to a far greater extent than real prostitution is today, as the virtual variety will be impossible to monitor or control. With the risks of disease and violence having been eliminated, there will be far less rationale for proscribing it.

Sex workers will have competition from simulated—computer generated—partners. In the early stages, "real" human virtual partners are likely to be more realistic than simulated virtual partners, but that will change over time. Of course, once the simulated virtual partner is as capable, sensual, and responsive as a real human virtual partner, who's to say that the simulated virtual partner isn't a real, albeit virtual, person?

Is virtual rape possible? In the purely physical sense, probably not. Virtual reality will have a means for users to immediately terminate their experience. Emotional and other means of persuasion and pressure are another matter.

How will such an extensive array of sexual choices and opportunities affect the institution of marriage and the concept of commitment in a relationship? The technology of virtual sex will introduce an array of slippery slopes, and the definition of a monogamous relationship will become far less clear. Some people will feel that access to intense sexual experiences at the click of a mental button will destroy the concept of a sexually committed relationship. Others will argue, as proponents of sexual entertainment and services do today, that such diversions are healthy outlets and serve to maintain healthy relationships. Clearly, couples will need to reach their own understandings, but drawing clear lines will become difficult with the level of privacy that this future technology affords. It is likely that society will accept practices and activities in the virtual arena that it frowns on in the physical world, as the consequences of virtual activities are often (although not always) easier to undo.

In addition to direct sensual and sexual contact, virtual reality will be a great place for romance in general. Stroll with your lover along a virtual Champs-Élysées, take a walk along a virtual Cancún beach, mingle with the animals in a simulated Mozambique game reserve. Your whole relationship can be in Cyberland.

Virtual reality using an external visual-auditory-haptic interface is not the only technology that will transform the nature of sexuality in the twenty-first century. Sexual robots—sexbots—will become popular by the beginning of the third decade of the new century. Today, the idea of intimate relations with a robot or doll is not generally appealing because robots and dolls are so, well, inanimate. But that will change as robots gain the softness, intelligence, pliancy, and passion of their human creators. (By the end of the twenty-first century, there won't be a clear difference between humans and robots. What, after all, is the difference between a human who has upgraded her body and brain using new nanotechnology and computational technologies, and a robot who has gained an intelligence and sensuality surpassing her human creators?)

By the fourth decade, we will move to an era of virtual experiences through internal neural implants. With this technology, you will be able to have almost any kind of experience with just about anyone, real or imagined, at any time. It's just like today's online chat rooms, except that you don't need any equipment that's not already in your head, and you can do a lot more than just chat. You won't be restricted by the limitations of your natural body as you and your partners can take on any virtual physical form. Many new types of experiences will become possible: A man can feel what it is like to be a woman, and vice versa. Indeed, there's no reason why you can't be both at the same time, making real, or at least virtually real, our solitary fantasies.

And then, of course, in the last half of the century, there will be the nanobot swarms—good old sexy Utility Fog, for example. The nanobot swarms can instantly take on any form and emulate any sort of appearance, intelligence, and personality that you or it desires—the human form, say, if that's what turns you on.

THE SPIRITUAL MACHINE

We are not human beings trying to be spiritual. We are spiritual beings trying to be human.

—Jacquelyn Small

Body and soul are twins. God only knows which is which.

—Charles A. Swinburne

We're all lying in the gutter, but some of us are gazing at the stars.

—Oscar Wilde

Sexuality and spirituality are two ways that we transcend our everyday physical reality. Indeed, there are links between our sexual and our spiritual passions, as the ecstatic rhythmic movements associated with some varieties of spiritual experience suggest.

Mind Triggers

We are discovering that the brain can be directly stimulated to experience a wide variety of feelings that we originally thought could only be gained from actual physical or mental experience. Take humor, for example. In the journal *Nature,* Dr. Itzhak Fried and his colleagues at UCLA tell how they found a neurological trigger for humor. They were looking for possible causes for a teenage girl's epileptic seizures and discovered that applying an electric probe to a specific point in the supplementary motor area of her brain caused her to laugh. Initially, the researchers thought that the laughter must be just an involuntary motor response, but they soon realized they were triggering the genuine perception of humor, not just forced laughter. When stimulated in just the right spot of her brain, she found everything funny. "You guys are just so funny—standing around" was a typical comment.[24]

Triggering a perception of humor without circumstances we normally consider funny is perhaps disconcerting (although personally, I find it humorous). Humor involves a certain element of surprise. Blue elephants. The last two words were intended to be surprising, but they probably didn't make you laugh (or maybe they did). In addition to surprise, the unexpected event needs to make sense from an unanticipated but meaningful perspective. And there are some other attributes that humor requires that we don't understand just yet. The brain apparently has a neural net that detects humor from our other perceptions. If we directly stimulate the brain's humor detector, then an otherwise ordinary situation will seem pretty funny.

The same appears to be true of sexual feelings. In experiments with animals, stimulating a specific small area of the hypothalamus with a tiny injection of testosterone causes the animals to engage in female sexual behavior, regardless of gender. Stimulating a different area of the hypothalamus produces male sexual behavior.

These results suggest that once neural implants are commonplace, we will have the ability to produce not only virtual sensory experiences but also the feelings associated with these experiences. We can also create some feelings not ordinarily associated with the experience. So you will be able to add some humor to your sexual experiences, if desired (of course, for some of us humor may already be part of the picture).

The ability to control and to reprogram our feelings will become even more profound in the late twenty-first century when technology moves beyond mere neural implants and we fully install our thinking processes into a new computational medium—that is, *when we become software.*

We work hard to achieve feelings of humor, pleasure, and well-being. Being able to call them up at will may seem to rob them of their meaning. Of course, many people use drugs today to create and enhance certain desirable feelings, but the chemical approach comes bundled with many undesirable effects. With neural implant technology, you will be able to enhance your feelings of pleasure and well-being without the hangover. Of course, the potential for abuse is even greater than with drugs. When psychologist James Olds provided rats with the ability to press a button and directly stimulate a pleasure center in the limbic system of their brains, the rats pressed the button endlessly, as often as five thousand times an hour, to the exclusion of everything else, including eating. Only falling asleep caused them to stop temporarily.[25]

Nonetheless, the benefits of neural implant technology will be compelling. As just one example, millions of people suffer from an inability to experience sufficiently intense feelings of sexual pleasure, which is one important aspect of impotence. People with this disability will not pass up the opportunity to overcome their problem through neural implants, which they may already have in place for other purposes. Once a technology is developed to overcome a disability, there is no way to restrict its use from enhancing normal abilities, nor would such restrictions necessarily be desirable. The ability to control our feelings will be just another one of those twenty-first-century slippery slopes.

So What About Spiritual Experiences?

The spiritual experience—a feeling of transcending one's everyday physical and mortal bounds to sense a deeper reality—plays a fundamental role in otherwise disparate religions and philosophies. Spiritual experiences are not all of the same sort but appear to encompass a broad range of mental phenomena. The ecstatic dancing of a Baptist revival appears to be a different phenomenon than the quiet transcendence of a Buddhist monk. Nonetheless, the notion of the spiritual experience has been reported so consistently throughout history, and in virtually all cultures and religions, that it represents a particularly brilliant flower in the phenomenological garden.

Regardless of the nature and derivation of a mental experience, spiritual or otherwise, once we have access to the computational processes that give rise to it, we have the opportunity to understand its neurological correlates. With the understanding of our mental processes will come the opportunity to capture our intellectual, emotional, and spiritual experiences, to call them up at will, and to enhance them.

Spiritual Experience Through Brain Generated Music

There is already one technology that appears to generate at least one aspect of a spiritual experience. This experimental technology is called Brain Generated Music (BGM), pioneered by NeuroSonics, a small company in Baltimore, Maryland, of which I am a director. BGM is a brain-wave biofeedback system capable of evoking an experience called the Relaxation Response, which is associated with deep relaxation.[26] The BGM user attaches three disposable leads to her head. A personal computer then monitors the user's brain waves to determine her unique alpha wavelength. Alpha waves, which are in the range of eight to thirteen cycles per second (cps), are associated with a deep meditative state, as compared to beta waves (in the range of thirteen to twenty-eight cps), which are associated with routine conscious thought. Music is then generated by the computer, according to an algorithm that transforms the user's own brain-wave signal.

The BGM algorithm is designed to encourage the generation of alpha waves by producing pleasurable harmonic combinations upon detection of alpha waves, and less pleasant sounds and sound combinations when alpha detection is low. In addition, the fact that the sounds are synchronized to the user's own alpha wavelength to create a resonance with the user's own alpha rhythm also encourages alpha production.

Dr. Herbert Benson, formerly the director of the hypertension section of Boston's Beth Israel Hospital and now at New England Deaconess Hospital in Boston, and other researchers at the Harvard Medical School and Beth Israel, discovered the neurological-physiological mechanism of the Relaxation Response, which is described as the opposite of the "fight or flight," or stress response.[27] The Relaxation Response is associated with reduced levels of epinephrine (adrenaline) and norepinephrine (noradrenaline), blood pressure, blood sugar, breathing, and heart rates. Regular elicitation of this response is reportedly able to produce permanently lowered blood-pressure levels (to the extent that hypertension is caused by stress factors) and other health benefits. Benson and his colleagues have catalogued a number of techniques that can elicit the Relaxation Response, including yoga and a number of forms of meditation.

I have had experience with meditation, and in my own experience with BGM, and in observing others, BGM does appear to evoke the Relaxation Response. The music itself feels as if it is being generated from inside your mind. Interestingly, if you listen to a tape recording of your own brain-generated music when you are not hooked up to the computer, you do not experience the same sense of transcendence. Although the recorded BGM is based on your personal alpha wavelength, the recorded music was synchronized to the brain waves that were produced by your brain when the music was first generated, not to the brain waves that are produced while listening to the recording. You need to listen to "live" BGM to achieve the resonance effect.

Conventional music is generally a passive experience. Although a performer may be influenced in subtle ways by her audience, the music we listen to generally does not reflect our response. Brain Generated Music represents a new modality of music that enables the music to evolve continually based on the interaction between it and our own mental responses to it.

Is BGM producing a spiritual experience? It's hard to say. The feelings produced while listening to "live" BGM are similar to the deep transcendent feelings I can sometimes achieve with meditation, but they appear to be more reliably produced by BGM.

The God Spot

Neuroscientists from the University of California at San Diego have found what they call the God module, a tiny locus of nerve cells in the frontal lobe that appears to be activated during religious experiences. They discovered this neural machinery while studying epileptic patients who have intense mystical experiences during seizures. Apparently the intense neural storms during a seizure stimulate the God module. Tracking surface electrical activity in the brain with highly sensitive skin monitors, the scientists found a similar response when very religious nonepileptic persons were shown words and symbols evoking their spiritual beliefs.

A neurological basis for spiritual experience has long been postulated by evolutionary biologists because of the social utility of religious belief. In response to reports of the San Diego research, Richard Harries, the Bishop of Oxford, said through a spokesman that "it would not be surprising if God had created us with a physical facility for belief."[28]

When we can determine the neurological correlates of the variety of spiritual experiences that our species is capable of, we are likely to be able to enhance these experiences in the same way that we will enhance other human experiences. With the next stage of evolution creating a new generation of humans that will be trillions of times more capable and complex than humans today, our ability for spiritual experience and insight is also likely to gain in power and depth.

Just being—experiencing, being conscious—is spiritual, and reflects the essence of spirituality. Machines, derived from human thinking and surpassing humans in their capacity for experience, will claim to be conscious, and thus to be spiritual. They will believe that they are conscious. They will believe that they have spiritual experiences. They will be convinced that these experiences are meaningful. And given the historical inclination of the human race to anthropomorphize the phenomena we encounter, and the persuasiveness of the machines, we're likely to believe them when they tell us this.

Twenty-first-century machines—based on the design of human thinking— will do as their human progenitors have done—going to real and virtual houses of worship, meditating, praying, and transcending—to connect with their spiritual dimension.

LET'S JUST GET ONE THING STRAIGHT: THERE'S NO WAY I'M GOING TO HAVE SEX WITH A COMPUTER.

Hey, let's not jump to conclusions. You should keep an open mind.

I'LL TRY TO HAVE AN OPEN MIND. AN OPEN BODY IS ANOTHER MATTER. THE IDEA OF GETTING INTIMATE WITH SOME GADGET, NO MATTER HOW CLEVER, IS NOT VERY APPEALING.

Have you ever spoken to a phone?

TO A PHONE? I MEAN I TALK TO PEOPLE USING A PHONE.

Okay, so a computer circa 2015—in the form of a visual-auditory-tactile virtual reality communication device—is just a telephone for you and your lover. But you can do more than just talk.

I LIKE TO TALK TO MY LOVER—WHEN I HAVE ONE—BY PHONE. AND LOOKING AT EACH OTHER WITH A PICTURE PHONE, OR EVEN A FULL VIRTUAL REALITY SYSTEM, SOUNDS PRETTY COZY. AS FOR YOUR TACTILE IDEA, HOWEVER, I THINK I'LL STICK TO TOUCHING MY FRIENDS AND LOVERS WITH REAL FINGERS.

You can use real fingers with virtual reality, or at least real virtual fingers. But what about when you and your lover are separated?

YOU KNOW, DISTANCE MAKES THE HEART GROW FONDER. ANYWAY, WE DON'T HAVE TO TOUCH ALL THE TIME, I MEAN I'LL BE ABLE TO WAIT UNTIL I GET BACK FROM MY BUSINESS TRIP, WHILE HE'S TAKING CARE OF THE KIDS!

When virtual reality does evolve into a convincing, all-encompassing tactile interface, are you going to go out of your way to avoid any physical contact?

I SUPPOSE IT WOULDN'T HURT TO KISS GOODNIGHT.

Ah-ha—the slippery slope! So why stop there?

OKAY, TWO KISSES.

Sure, like I just said, keep an open mind.

SPEAKING OF AN OPEN MIND, YOUR DESCRIPTION OF THE "GOD SPOT" SEEMS TO TRIVIAL-IZE THE SPIRITUAL EXPERIENCE.

I wouldn't overreact to this one piece of research. Clearly, something's going on in the brains of people who are having a spiritual experience. Whatever the neurological process is, once we capture and understand it, we should be able to enhance the spiritual experiences in a re-created brain running in its new computational medium.

SO THESE RE-CREATED MINDS WILL REPORT HAVING SPIRITUAL EXPERIENCES. AND I SUP-POSE THEY WILL ACT IN THE SAME SORT OF TRANSCENDENT, RAPTUROUS WAY THAT PEOPLE DO TODAY WHEN REPORTING SUCH EXPERIENCES. BUT WILL THESE MACHINES REALLY BE TRANSCENDING, AND EXPERIENCING THE FEELING OF GOD'S PRESENCE? WHAT WILL THEY BE EXPERIENCING, ANYWAY?

We keep coming back to the issue of consciousness. Machines in the twenty-first century will report the same range of experiences that humans do. In accordance with the Law of Accelerating Returns, they will report an even broader range. And they will be very convincing when they speak of their experiences. But what will they really be feeling? As I said earlier, there's just no way to truly penetrate another entity's subjective experience, at least not in a scientific way. I mean, we can observe the patterns of neural firings, and so forth, but that's still just an objective observation.

WELL, THAT'S JUST THE LIMITATION OF SCIENCE.

Yes, that's where philosophy and religion are supposed to take over. Of course, it's hard enough to get agreement on scientific issues.

THAT OFTEN APPEARS TO BE TRUE. NOW, ANOTHER THING I'M NOT TOO HAPPY ABOUT IS THESE PILLAGING NANOBOTS THAT ARE GOING TO MULTIPLY WITHOUT END. WE'LL END UP WITH A HUGE SEA OF NANOBOTS. WHEN THEY'RE DONE WITH US, THEY'LL START EATING EACH OTHER.

There is that danger. But if we write the software carefully . . .

OH SURE, LIKE MY OPERATING SYSTEM. ALREADY I HAVE LITTLE SOFTWARE VIRUSES THAT MULTIPLY THEMSELVES UNTIL THEY CLOG UP MY HARD DRIVE.

I still think the bigger danger is in their intentional hostile use.

I KNOW YOU SAID THAT, BUT THAT'S NOT EXACTLY REASSURING. AGAIN, WHY DON'T WE JUST NOT GO DOWN THIS PARTICULAR ROAD?

Okay, you tell that to the old woman whose crumbling bones will be effectively treated using a nanotechnology-based treatment, or the cancer patient whose cancer is destroyed by little nanobots that swim through his blood vessels.

I REALIZE THERE ARE A LOT OF POTENTIAL BENEFITS, BUT THE EXAMPLES YOU JUST GAVE CAN ALSO BE ADDRESSED THROUGH OTHER, MORE CONVENTIONAL, TECHNOLOGIES, LIKE BIOENGINEERING.

I'm glad you mentioned bioengineering, because we see a very similar problem with bioengineered weapons. We're very close to the point where the knowledge and equipment in a typical graduate-school biotechnology program will be sufficient to create self-replicating pathogens. Whereas a nanoengineered weapon could replicate across any matter, living and dead, a bioengineered weapon would only replicate across living matter, probably just its human targets. I understand that's not much comfort. In either case, the potential for uncontrolled self-replication greatly multiplies the danger.

But you're not going to stop bioengineering—it's the cutting edge of our medical research. It has already greatly contributed to the AIDS treatments we have today; diabetic patients use bioengineered forms of human insulin; there are effective cholesterol-lowering drugs; there are promising new cancer treatments; and the list of advances is rapidly growing. There is genuine optimism among otherwise skeptical scientists that we will make dramatic gains against cancer and other scourges with bioengineered treatments.

SO HOW ARE WE GOING TO PROTECT OURSELVES FROM BIOENGINEERED WEAPONS?

With more bioengineering—antiviral drugs, for example.

AND NANOENGINEERED WEAPONS?

Same thing—more nanotechnology.

I HOPE THE GOOD NANOBOTS PREVAIL, BUT I JUST WONDER HOW WE'RE GOING TO TELL THE GOOD NANOBOTS FROM THE BAD ONES.

It's going to be hard to tell, particularly since the nanobots are too small to see.

EXCEPT BY OTHER NANOBOTS, RIGHT?

Good point.

CHAPTER EIGHT

1999

THE DAY THE COMPUTERS STOPPED

The digitization of information in all of its forms will probably be known as the most fascinating development of the twentieth century.

—An Wang

Economics, sociology, geopolitics, art, religion all provide powerful tools that have sufficed for centuries to explain the essential surfaces of life. To many observers, there seems nothing truly new under the sun—no need for a deep understanding of man's new tools—no requirement to descend into the microcosm of modern electronics in order to comprehend the world. The world is all too much with us.

—George Gilder

If all the computers in 1960 stopped functioning, few people would have noticed. A few thousand scientists would have seen a delay in getting printouts from their last submission of data on punch cards. Some business reports would have been held up. Nothing to worry about.

Circa 1999 is another matter. If all computers stopped functioning, society would grind to a halt. First of all, electric power distribution would fail. Even if electrical power continued (which it wouldn't), virtually everything would still break down. Most motorized vehicles have embedded microprocessors, so the only cars that would run would be quite old. There would be almost no functioning trucks, buses, railroads, subways, or airplanes. There would be no electronic communication: Telephones, radio, television, fax machines, pagers, e-mail, and of course the Web would all cease functioning. You wouldn't get your paycheck. You couldn't cash it if you did. You wouldn't be able to get your money out of your bank. Business and government would operate at only the most primitive level. And if all the data in all the computers vanished, then we'd really be in trouble.

There has been substantial concern with Y2K (Year 2000 Problem), that at least some computer processes will be disrupted as we approach the year 2000.

Y2K primarily concerns software developed one or more decades ago in which date fields used only two digits, which will cause these programs to behave erratically when the year becomes "00." I am more sanguine than some about this particular issue. Y2K is causing the urgent rewriting of old business programs that needed to be dusted off and redesigned anyway. There will be some disruptions (and a lot of litigation), but in my view Y2K is unlikely to cause the massive economic problems that are feared.[1]

In less than forty years, we have gone from manual methods of controlling our lives and civilization to becoming totally dependent on the continued operation of our computers. Many people are comforted by the fact that we still have our hand on the "plug," that we can turn our computers off if they get too uppity. In actuality, it's the computers that have their figurative hands on our plug. (Give them a couple more decades and their hands won't be so figurative.)

There is little concern about this today—computers circa 1999 are dependable, docile, and dumb. The dependability (albeit not perfect) is likely to remain. The dumbness will not. It will be the humans, at least the nonupdated ones, who will seem dumb several decades from now. The docility will not remain, either.

For a rapidly increasing array of *specific* tasks, the intelligence of contemporary computers appears impressive, even formidable, but machines today remain narrow-minded and brittle. In contrast, we humans have softer landings when we wander outside our own narrow areas of expertise. Unlike Deep Blue, Gary Kasparov is not incompetent in matters outside of chess.

Computers are rapidly moving into increasingly diverse realms. I could fill a dozen books with examples of the intellectual prowess of computers circa end of the twentieth century, but I have only a contract for one, so let's take a look at a few artful examples.

THE CREATIVE MACHINE

At a time like ours, in which mechanical skill has attained unsuspected perfection, the most famous works may be heard as easily as one may drink a glass of beer, and it only costs ten centimes, like the automatic weighing machines. Should we not fear this domestication of sound, this magic that anyone can bring from a disk at will? Will it not bring to waste the mysterious force of an art which one might have thought indestructible?

—Claude Debussy

Collaboration with machines! What is the difference between manipulation of the machine and collaboration with it? . . . Suddenly, a window would open into a vast field of possibilities; the time limits would vanish, and the machines would seem to become humanized components of the interactive network now

consisting of oneself and the machine still obedient but full of suggestions to the
master controls of the imagination.

—Vladimir Ussachevsky

Somebody was saying to Picasso that he ought to make pictures of things the
way they are—objective pictures. He mumbled he wasn't quite sure what that
would be. The person who was bullying him produced a photograph of his wife
from his wallet and said, "There, you see, that is a picture of how she really is."
Picasso looked at it and said, "She is rather small, isn't she? And flat?"

—Gregory Bateson

The age of the cybernetic artist has begun, although it is at an early stage. As with human artists, you never know what these creative systems are going to do next. To date, however, none of them has cut off an ear or run naked through the streets. They don't yet have bodies to demonstrate that sort of creativity.

The strength of these systems is reflected by an often startling originality in a turn of a phrase, shape, or musical line. Their weakness has to do, again, with context, or the lack thereof. Since these creative computers are deficient in the real-world experience of their human counterparts, they often lose their train of thought and ramble off into incoherence. Perhaps the most successful in terms of maintaining thematic consistency throughout a work of art is Harold Cohen's robotic painter named Aaron, which I discuss below. The primary reason Aaron is so successful is the thoroughness of its extensive knowledge base, which Cohen has been building, rule by rule, for three decades.

Jamming with Your Computer

The frequent originality of these systems makes them great collaborators with human artists, and in this manner, computers have already had a transforming effect on the arts. This trend is furthest along in the musical arts. Music has always used the most advanced technologies available; the cabinet-making crafts of the eighteenth century; the metalworking industries of the nineteenth century; and the analog electronics of the 1960s. Today, virtually all commercial music— recordings, movie and television soundtracks—is created on computer music workstations, which synthesize and process the sounds, record and manipulate the note sequences, generate notation, even automatically generate rhythmic patterns, walking bass lines, and melodic progressions and variations.

Up until recently, instrument-playing technique was inextricably linked to the sounds created. If you wanted violin sounds, you had to play the violin. The playing techniques derived from the physical requirements of creating the sounds. Now that link has been broken. If you like flute-playing technique, or just happen to have learned it, you can now use an electronic wind controller that plays just like an acoustic flute yet creates the sounds not only of a variety of flutes, but also of virtually any other instrument, acoustic or electronic. There are now controllers that emulate the playing technique of most popular acoustic instruments, including piano, violin, guitar, drums, and a variety of wind instruments. Since we are no longer limited by the physics of creating sounds acoustically, a new generation of controllers is emerging that bears no resemblance to any conventional acoustic instruments, but instead attempts to optimize the human factors of creating music with our fingers, arms, feet, mouth, and head. All sounds can now be played polyphonically and can be layered (played simultaneously) and sequenced with one another. Also, it is no longer necessary to play music in real time—music can be performed at one speed and played back at another, without changing the pitch or other characteristics of the notes. All sorts of age-old limitations have been overcome, allowing a teenager in her bedroom to sound like a symphony orchestra or rock band.

A Musical Turing Test

In 1997, Steve Larson, a University of Oregon music professor, arranged a musical variation of the Turing Test by having an audience attempt to determine which of three pieces of music had been written by a computer and which one of the three had been written two centuries ago by a human named Johann Sebastian Bach. Larson was only slightly insulted when the audience voted that his own piece was the computer composition, but he felt somewhat vindicated when the audience selected the piece written by a computer program named EMI (Experiments in Musical Intelligence) to be the authentic Bach composition. Douglas Hofstadter, a longtime observer of (and contributor to) the progression of machine intelligence, calls EMI, created by the composer David Cope, "the most thought-provoking project in artificial intelligence that I have ever come across."[2]

Perhaps even more successful is a program called Improvisor, written by Paul Hodgson, a British jazz saxophone player. Improvisor can emulate styles ranging from Bach to jazz greats Louis Armstrong and Charlie Parker. The program has attracted its own following. Hodgson himself says, "If I was new in town and heard someone playing like Improvisor, I'd be happy to join in."[3]

The weakness of today's computerized composition is, again, a weakness of context. "If I turn on three seconds of EMI and ask myself, 'What was that?' I would say Bach," says Hofstadter. Longer passages are not always so successful. Often "it's like listening to random lines from a Keats sonnet. You wonder what was happening to Keats that day. Was he completely drunk?"

The Literary Machine

Here's a question for you: What kind of murderer has fiber?

The answer: A cereal killer.

I hasten to admit that I did not make up this pun myself. It was written by a computer program called JAPE (Joke Analysis and Production Engine), created by Kim Binsted. JAPE is the state of the art in the automatic writing of bad puns. Unlike EMI, JAPE did not pass a modified Turing Test when it was recently paired with human comedian Steve Martin. The audience preferred Martin.[4]

The literary arts lag behind the musical arts in the use of technology. This may have to do with the depth and complexity of even routine prose, a quality which Turing recognized when he based his Turing Test on the ability of humans to generate convincing written language. Computers are nonetheless of significant practical benefit to those of us who create written works. Of greatest impact is the simple word processor. Not an artificial technology per se, word processing was derived from the text editors developed during the 1960s at the AI labs at MIT and elsewhere.

This book certainly benefited from the availability of linguistic databases, spell checkers, online dictionaries, not to mention the vast research resources of the World Wide Web. Much of this book was dictated to my personal computer using a continuous speech-recognition program called Voice Xpress Plus from the dictation division of Lernout & Hauspie (formerly Kurzweil Applied Intelligence), which became available in the middle of my writing the book. With regard to automatic grammar and style checkers, I was forced to turn that particular Microsoft Word feature off, as it seemed to dislike most of my sentences. I'll leave the stylistic criticism of this book to my human readers (at least this time around).

A variety of programs help writers brainstorm. ParaMind, for example, "produces new ideas from your ideas," according to its own literature.[5] Other programs allow writers to track the complex histories, characterizations, and interactions of characters in such extended works of fiction as long novels, series of novels, and television drama series.

Programs that write completely original works are particularly challenging because human readers are keenly aware of the myriad syntactic and semantic requirements for sensible written language. Musicians, cybernetic or otherwise, can get away with a bit more inconsistency than authors.

With that in mind, consider the following:

A Story of Betrayal

Dave Striver loved the university. He loved its ivy-covered clock towers, its ancient and sturdy brick, and its sun-splashed verdant greens and eager youth. He also loved the fact that the university is free of the stark unforgiving trials of the business world—only this isn't a fact: Academia has its own tests, and some are as merciless as any in the marketplace. A prime example is the dissertation defense: To earn the Ph.D., to become a doctor, one must pass an oral examination on one's dissertation. This was a test Professor Edward Hart enjoyed giving.

Dave wanted desperately to be a doctor. But he needed the signatures of three people on the first page of his dissertation, the priceless inscriptions which, together, would certify that he had passed his defense. One of the signatures had to come from Professor Hart, and Hart had often said—to others and to himself—that he was honored to help Dave secure his well-earned dream.

Well before the defense, Striver gave Hart a penultimate copy of his thesis. Hart read it and told Dave that it was absolutely first-rate, and that he would gladly sign it at the defense. They even shook hands in Hart's book-lined office. Dave noticed that Hart's eyes were bright and trustful, and his bearing paternal.

At the defense, Dave thought that he eloquently summarized chapter 3 of his dissertation. There were two questions, one from Professor Rogers and one from Dr. Meteer; Dave answered both, apparently to everyone's satisfaction. There were no further objections.

Professor Rogers signed. He slid the tome to Meteer; she too signed, and then slid it in front of Hart. Hart didn't move.

"Ed?" Rogers said.

Hart still sat motionless. Dave felt slightly dizzy.

"Edward, are you going to sign?"

Later, Hart sat alone in his office, in his big leather chair, saddened by Dave's failure. He tried to think of ways he could help Dave achieve his dream.

Okay, that's the end. Admittedly the story kind of peters out, ending with a whimper rather than a bang. Seattle writer and editor Susan Mulcahy called the story "amateurish," criticizing the author's grammar and word choice. But Mulcahy was nonetheless surprised and impressed when she learned the author was a computer. The program that wrote the story, named BRUTUS.1, was created by Selmer Bringsjord, Dave Ferucci, and a team of software engineers at Rensselaer Polytechnic Institute. Apparently, BRUTUS.1 is an expert on betrayal, a concept that Bringsjord and Ferucci spent eight years painstakingly teaching the computer. The researchers acknowledge that their program needs to learn about other issues besides betrayal. "Interestingness really combines all of the emotions," say Bringsjord and Ferucci, and that is something the cybernetic authors are not yet capable of achieving.[6]

The Cybernetic Poet

Another example of a computerized author is a computer program that I designed called Ray Kurzweil's Cybernetic Poet (RKCP). RKCP is a computer-generated poetry system, which uses language-modeling techniques to automatically generate completely original poetry based on poems that it has "read."[7]

RKCP reads a selection of poems by a particular author or authors (preferably an extensive selection) and then creates a "language model" of that author's work based on markov models, a mathematical cousin of neural nets. RKCP can then write original poems from that model. As I discussed earlier, RKCP uses a recursive poetry-generation algorithm to achieve the language style, rhythm patterns, and poem structure of the original authors whose poems were analyzed. There are also algorithms to maintain thematic consistency through the poem. The poems are in a similar style to the author(s) originally analyzed but are completely original new poetry. The system even has rules to discourage itself from plagiarizing.

You can download a free copy of RKCP at www.kurzweiltech.com.

Following is a small sampling of poems written by Ray Kurzweil's Cybernetic Poet after it had "read" poems by several famous poets, as well as lesser-known contemporary poets.

Poems by Ray Kurzweil's Cybernetic Poet

PAGE

A haiku written by Ray Kurzweil's Cybernetic Poet after reading poems by
Wendy Dennis

Sashay down the page
through the lioness
nestled in my soul.

IMAGINE NOW AND SING

A poem written by Ray Kurzweil's Cybernetic Poet after reading poems by
Wendy Dennis and Ray Kurzweil and love poems by various authors

Imagine now and sing,
creating myths
forming jewels from the falling snow.

SOUL

A haiku written by Ray Kurzweil's Cybernetic Poet after reading poems by
John Keats and Wendy Dennis

You broke my soul
the juice of eternity,
the spirit of my lips.

I THINK I'LL CRASH

A poem written by Ray Kurzweil's Cybernetic Poet after reading poems by
Ray Kurzweil, Robert Frost, and Wendy Dennis

I think I'll crash.
Just for myself with God
peace on a curious sound
for myself in my heart?
And life is weeping
From a bleeding heart
of boughs bending
such paths of them,
of boughs bending
such paths of breeze
knows we've been there

THE STIFLING STUFFY

A haiku written by Ray Kurzweil's Cybernetic Poet after reading poems by
Patricia Camarena Rose and Wendy Dennis

The stifling stuffy
Catholic schoolroom,
where I cannot be real.

WONDERED

A poem written by Ray Kurzweil's Cybernetic Poet after reading poems by
Dave Gitomer

today i wondered
if i mused
today i saw you
i learned
in awe and you

if i wondered
if i mused
today i had one wish
if i saw you
if i saw you
if i had one wish

MOON CHILD

A haiku written by Ray Kurzweil's Cybernetic Poet after reading poems by
Kathleen Frances Wheeler

Crazy moon child
Hide from your coffin
To spite your doom.

ANGEL

A poem written by Ray Kurzweil's Cybernetic Poet after reading poems by
Patricia Camarena Rose and Sue Klapes Kelly

Beams of the dawn at the angel
with a calm, silent sea
with a hundred times we write,
with a chance we can open up
a steady rhythm in his face
silent room
desolate beach,
Scattering remains of love.

TOMCAT

A haiku written by Ray Kurzweil's Cybernetic Poet after reading poems by
Randi and Kathryn Lynn

An old yellow tomcat
lies sleeping content,
he rumbles a heart

LONG YEARS HAVE PASSED

A poem written by Ray Kurzweil's Cybernetic Poet after reading poems by
Randi and Kathryn Lynn

Long years have passed.
I think of goodbye.
Locked tight in the night
I think of passion;
Drawn to for blue, the night
During the page
My shattered pieces of life
watching the joy
shattered pieces of love
My shattered pieces of love
gone stale.

SANDALS

A haiku written by Ray Kurzweil's Cybernetic Poet after reading poems by
Kimberly McLauchlin and Ray Kurzweil

Scattered sandals
a call back to myself,
so hollow I would echo.

The Visual Arts

The visual arts lie in between the musical arts and literary arts in terms of the impact of computers. A great deal of visual art is created using paint and illustrator programs, which can simulate the effects of conventional materials such as paint strokes as well as implement a wide range of techniques that could only be executed on a computer. Recently, computers have also taken over most video and film editing.

The Web is filled with the artistic musings of cybernetic artists. A popular technique is the evolutionary algorithm, which allows the computer to evolve a picture by redoing it hundreds or thousands of times. Humans would find this approach difficult—they would waste a lot of paint, for one thing. Mutator, the creation of sculptor William Latham and software engineer Stephen Todd at IBM in Winchester, England, uses the evolutionary approach, as does a program written by Karl Sims, an artist and scientist at Genetic Arts, in Cambridge, Massachusetts.[8]

Probably the leading practitioner of computer-generated visual art is Harold Cohen. His computerized robot named Aaron has been evolving and creating drawings and paintings for twenty years. These works of visual art are completely original, created entirely by the computer, and rendered with real paint. Cohen has spent more than three decades endowing his program with a knowledge of many aspects of the artistic process, including composition, drawing, perspective, and color, as well as a variety of styles. While Cohen wrote the program, the pictures created are nonetheless always a surprise to him.

Cohen is frequently asked who should be given credit for the results of his enterprise, which have been displayed in museums around the world.[9] Cohen is happy to take the credit, and Aaron has not been programmed to complain. Cohen boasts that he will be the first artist in history who will be able to have a posthumous exhibition of completely original works.[10]

Paintings by Aaron by Cohen

These five original paintings were painted by Aaron, a computerized robot built and programmed by Harold Cohen. These color paintings are reproduced here in black and white. You can see the color versions on this book's web site, at www.penguinputnam.com/kurzweil.[11]

PREDICTIONS OF THE PRESENT

With the impending millennium change there are no shortage of anticipations of what the next century will be like. Futurism has a long history, but not a particularly impressive one. One of the problems with predictions of the future is that by the time it's clear that they have had little resemblance to actual events, it's too late to get your money back.

Perhaps the problem is that we let just anyone make predictions. Maybe we should require futurism certification to be allowed to prognosticate. One of the requirements could be that in retrospect, at least half of your ten-or-more-year-ahead predictions have not been completely embarrassing. Such a certification program would be a slow process, however, and I suspect unconstitutional.

To see why futurism has such a spotty reputation, here is a small sample of predictions from some otherwise intelligent people:

> *"The telephone has too many shortcomings to be seriously considered as a means of communication."*
>
> —Western Union executive, 1876

> *"Heavier-than-air flying machines are not possible."*
>
> —Lord Kelvin, 1895

> *"The most important fundamental laws and facts of physical science have all been discovered, and these are now so firmly established that the possibility of their ever being supplemented by new discoveries is exceedingly remote."*
>
> —Albert Abraham Michelson, 1903

> *"Airplanes have no military value."*
>
> —Professor Marshal Foch, 1912

> *"I think there is a world market for maybe five computers."*
>
> —IBM Chairman Thomas Watson, 1943

> *"Computers in the future may weigh no more than 1.5 tons."*
>
> —Popular Mechanics, 1949

> *"It would appear that we have reached the limits of what is possible to achieve with computer technology, although one should be careful with such statements, as they tend to sound pretty silly in five years."*
>
> —John von Neumann, 1949

"There's no reason for individuals to have a computer in their home."
                                                    —Ken Olson, 1977

"640,000 bytes of memory ought to be enough for anybody."
                                                    —Bill Gates, 1981

*"Long before the year 2000, the entire antiquated structure of college de-
grees, majors and credits will be a shambles."*
                                                    —Alvin Toffler

"The Internet will catastrophically collapse in 1996."
                        —Robert Metcalfe (inventor of Ethernet), who, in 1997,
                                ate his words (literally) in front of an audience

Now I get to toot my own horn, and can share with you those predictions of
mine that worked out particularly well. But in looking back at the many predictions I've made over the past twenty years, I will say that I haven't found any that
I find particularly embarrassing (except, maybe, for a few early business plans).

The Age of Intelligent Machines, which I wrote in 1987 through 1988, as well as
other articles and speeches I wrote in the late 1980s, contained a lot of my predictions about the 1990s, which included the following:[12]

• *Prediction:* A computer will defeat the human world chess champion
  around 1998, and we'll think less of chess as a result.
  What Happened: As I mentioned, this one was a year off. Sorry.
• *Prediction:* There will be a sustained decline in the value of commodities
  (that is, material resources) with most new wealth being created in the
  knowledge content of products and services, leading to sustained economic growth and prosperity.
  What Happened: As predicted, everything is coming up roses (except, as
  also predicted, for long-term investors in commodities, which are down
  40 percent over the past decade). Even the approval ratings of politi-
  cians from the president to the Congress are at an all-time high. But the
  strong economy has more to do with the Bill in the west coast Washington than the Bill in the east coast Washington. Not that Mr. Gates de-
  serves primary credit, but the driving economic force in the world today
  is information, knowledge, and related computer technologies. Federal
  Reserve Chairman Alan Greenspan recently acknowledged that today's
  unprecedented sustained prosperity and economic expansion is due to
  the increased efficiency provided by information technology. But that's
  only half right. Greenspan ignores the fact that most of the new wealth

that is being created is itself comprised of information and knowledge—a trillion dollars in Silicon Valley alone. Increased efficiency is only part of the story. The new wealth in the form of the market capitalization of computer-related (primarily software) companies is real and substantial and is lifting all boats.

The U.S. House Subcommittee on Banking reported that in the eight-year period between 1989 and 1997, the total value of U.S. real estate and durable goods increased only 33 percent, from $9.1 trillion to $12.1 trillion. The value of bank deposits and credit market instruments increased only 27 percent, from $4.5 trillion to $5.7 trillion. The value of equity shares, however, increased a staggering 239 percent, from $3.4 trillion to $11.4 trillion! The primary engine of this increase is the rapidly increasing knowledge content of products and services, as well as the increased efficiencies fostered by information technology. This is where new wealth is being created.

Information and knowledge are not limited by the availability of material resources, and in accordance with the Law of Accelerating Returns will continue to grow exponentially. The Law of Accelerating *Returns* includes financial returns. Thus a key implication of the law is continuing economic growth.

As this book is being written, there has been considerable attention on an economic crisis in Japan and other countries in Asia. The United States has been pressing Japan to stimulate its economy with tax cuts and government spending. Little attention is being paid, however, to the root cause of the crisis, which is the state of the software industry in Asia, and the need for effective entrepreneurial institutions that promote the creation of software and other forms of knowledge. These include venture and angel capital,[13] widespread distribution of employee-stock options, and incentives that encourage and reward risk taking. Although Asia has been moving in this direction, these new economic imperatives have grown more rapidly than most observers expected (and their importance will continue to escalate in accordance with the Law of Accelerating Returns).

• *Prediction:* A worldwide information network linking almost all organizations and tens of millions of individuals will emerge (admittedly, not by the name World Wide Web).

What Happened: The Web emerged in 1994 and took off in 1995 through 1996. The Web is truly a worldwide phenomenon, and products and services in the form of information swirl around the globe oblivious to borders of any kind. A 1998 report by the U.S. Commerce Department credited the Internet as a key factor in spurring economic growth and curbing inflation. It predicted that commerce on the Internet will surpass $300 billion by 2000. Industry reports put the figure at around $1 trillion, when all business-to-business transactions conducted over the Web are taken into consideration.

- *Prediction:* There will be a national movement to wire our classrooms.

  What Happened: Most states (with the exception, unfortunately, of my own state of Massachusetts) have $50 to $100 million annual budgets to wire classrooms and install related computers and software. It is a national priority to provide computer and Internet access to all students. Many teachers remain relatively computer illiterate, but the kids are providing much of the needed expertise.

- *Prediction:* In warfare, there will be almost total reliance on digital imaging, pattern recognition, and other software-based technologies. The side with the smarter machines will win. "A profound change in military strategy will arrive in the early 1990s. The more developed nations will increasingly rely on 'smart weapons,' which incorporate electronic co-pilots, pattern-recognition techniques, and advanced technologies for tracking, identification, and destruction."

  What Happened: Several years after I wrote the *Age of Intelligent Machines,* the Gulf War was the first to clearly establish this paradigm. Today, the United States has the most advanced computer-based weaponry and remains unchallenged in its status as a military superpower.

- *Prediction:* The vast majority of commercial music will be created on computer-based synthesizers.

  What Happened: Most of the musical sounds you hear on television, in the movies, and in recordings are now created on digital synthesizers, along with computer-based sequencers and sound processors.

- *Prediction:* Reliable person identification, using pattern-recognition techniques applied to visual and speech patterns, will replace locks and keys in many instances.

  What Happened: Person-identification technologies that use speech patterns and facial appearance have begun to be used today in check-cashing machines and to control entry into secure buildings and sites.[14]

- *Prediction:* With the advent of widespread electronic communication in the Soviet Union, uncontrollable political forces will be unleashed. These will be "methods far more powerful than the copiers the authorities have traditionally banned." The authorities will be unable to control it. Totalitarian control of information will have been broken.

  What Happened: The attempted coup against Gorbachev in August 1991 was undone primarily by cellular telephones, fax machines, electronic mail, and other forms of widely distributed and previously unavailable electronic communication. Overall, decentralized communication contributed significantly to the crumbling of centralized totalitarian political and economic government control in the former Soviet Union.

- *Prediction:* Many documents never exist on paper because they incorporate information in the form of audio and video pieces.
  What Happened: Web documents routinely include audio and video pieces, which can only exist in their web form.
- *Prediction:* Around the year 2000, chips with more than a billion components will emerge.
  What Happened: We're right on schedule.
- *Prediction:* The technology for the "cybernetic chauffeur" (self-driving cars using special sensors in the roads) will become available by the end of the 1990s with implementation on major highways feasible during the first decade of the twenty-first century.
  What Happened: Self-driving cars are being tested in Los Angeles, London, Tokyo, and other cities. There were extensive successful tests on Interstate 15 in southern California during 1997. City planners now realize that automated driving technologies will greatly expand the capacity of existing roads. Installing the requisite sensors on a highway costs only about $10,000 per mile, compared to $1 to $10 million per mile for building new highways. Automated highways and self-driving cars will also eliminate most accidents on these roads. The U.S. National Automated Highway System (NAHS) consortium is predicting implementation of these systems during the first decade of the twenty-first century.[15]
- *Prediction:* Continuous speech recognition (CSR) with large vocabularies for specific tasks will emerge in the early 1990s.
  What Happened: Whoops. Large-vocabulary domain-specific CSR did not emerge until around 1996. By late 1997 and early 1998, large-vocabulary CSR without a domain limitation for dictating written documents (like this book) was commercially introduced.[16]
- *Prediction:* The three technologies required for a translating telephone (where you speak and listen in one language such as English, and your caller hears you and replies in another language such as German)— speaker-independent (not requiring training on a new speaker), continuous, large-vocabulary speech recognition; language translation; and speech synthesis—will each exist in sufficient quality for a first generation system by the late 1990s. Thus, we can expect "translating telephones with reasonable levels of performance for at least the more popular languages early in the first decade of the twenty-first century."
  What Happened: Effective, speaker-independent speech recognition,

MY LIFE WITH MACHINES: SOME HIGHLIGHTS

I walked onstage and played a composition on an old upright piano. Then came the yes-or-no questions. Former Miss America Bess Myerson was stumped. But film star Henry Morgan, the second celebrity panelist on this episode of *I've Got a Secret,* guessed my secret: The piece I had played had been composed by a computer that I had built and programmed. Later that year, I got to meet President Johnson with other high-school science winners.

In college, I ran a business matching up high-school kids with colleges using a computer program I had written. We had to pay $1,000 an hour to rent time on the only computer in Massachusetts with an extraordinary million bytes of core memory, which allowed us to fit all the information we had about the nation's three thousand colleges into memory at the same time. We received a lot of letters from kids who were delighted with the colleges that our program had suggested. A few parents, on the other hand, were furious that we had failed to recommend Harvard. It was my first experience with the ability of computers to affect people's lives. I sold that company to Harcourt, Brace & World, a New York publisher, and moved on to other ideas.

In 1974, computer programs that could recognize printed letters, called optical character recognition (OCR), were capable of handling only one or two specialized type styles. I founded Kurzweil Computer Products that year to develop the first OCR program that could recognize *any* style of print, which we succeeded in doing later that year. So the question then became, What is it good for? Like a lot of clever computer software, it was a solution in search of a problem.

I happened to sit next to a blind gentleman on a plane flight, and he explained to me that the only real handicap that he experienced was his inability to read ordinary printed material. It was clear that his visual disability imparted no real handicap in either communicating or traveling. So I had found the problem we were searching for—we could apply our "omni-font" (*any* font) OCR technology to overcome this principal handicap of blindness. We didn't have the ubiquitous scanners or text-to-speech synthesizers that we do today, so we had to create these technologies as well. By the end of 1975, we put together these three new technologies we had invented—omni-font OCR, CCD (Charge Coupled Device) flat-bed scanners, and text-to-speech synthesis to create the first print-to-speech reading machine for the blind. The Kurzweil Reading Machine (KRM) was able to read ordinary books, magazines, and other printed documents out loud so that a blind person could read anything he wanted.

We announced the KRM in January of 1976, and it seemed to strike a chord. All the evening network news programs carried the story, and Walter

Cronkite used the machine to read aloud his signature sign-off, "And that's the way it was, January 13, 1976."

Shortly after the announcement, I was invited on the *Today* show, which was a little nerve-racking since we only had one working reading machine. Sure enough, the machine stopped working a couple of hours before I was scheduled to go on live national television. Our chief engineer frantically took the machine apart, scattering pieces of electronics and wires across the floor of the set. Frank Field, who was going to interview me, walked by and asked if everything was okay. "Sure, Frank," I replied. "We're just making a few last-minute adjustments."

Our chief engineer put the reading machine back together, and still it didn't work. Finally, he used a time-honored method of repairing delicate electronic equipment and slammed the reading machine against a table. From that moment, it worked just fine. Its live television debut then proceeded without a hitch.

Stevie Wonder heard about our appearance on the *Today* show, and decided to check out the story himself. Our receptionist was skeptical that the person on the other end of the line was really the legendary singer, but she put the call through to me, anyway. I invited him over, and he tried out the machine. He beseeched us to provide him with his own reading machine, so we turned the factory upside down to hurriedly finish up our first production unit (we didn't want to give him the prototype we used on the *Today* show, as it still had a few battle scars). We showed Stevie how to use it, and off he went in a taxi with his new reading machine by his side.

We subsequently applied the scanning and omni-font OCR to commercial uses such as entering data into databases and into the emerging word-processing computers. New information services, such as Lexus (an online legal research service) and Nexus (a news service), were built using the Kurzweil Data Entry Machine to scan and recognize written documents.

In 1978, after years of scrambling to raise funds for our venture, we were fortunate in attracting interest and investment from a big company: Xerox. Most Xerox products transferred electronic information onto paper. They saw the Kurzweil scanning and OCR technology as providing a bridge back from the world of paper to the electronic world, so in 1980 they bought the company. You can still buy the OCR we originally developed, suitably updated—it's now called Xerox TextBridge, and continues as a market leader.

I kept up my relationship with Stevie Wonder, and on one of our get-togethers at his new Los Angeles recording studio in 1982, he lamented the state of affairs in the world of musical instruments. On the one hand, there was the world of acoustic instruments, such as the piano, violin, and guitar, which provided the rich complex sounds of choice for most musicians. While musically satisfying, these instruments suffered from a panoply of limitations.

Most musicians could play only one or two different instruments. Even if you could play more than one, you couldn't play more than one at a time. Most instruments only produce one note at a time. There were very limited means available to shape the sounds.

On the other hand, there was the world of electronic instruments, in which these control limitations disappeared. In the computerized world, you could record one line of music on a sequencer, play it back, and record another sequence over it, building up a multi-instrumental composition line by line. You could edit wrong notes without replaying the entire sequence. You could layer multiple sounds, modify their sonic characteristics, play songs in nonreal time, and use a great variety of other techniques. There was only one problem. The sounds you had to work with in the electronic world sounded very thin, rather like an organ, or an electronically processed organ.

Wouldn't it be great, Stevie mused, if we could use the extraordinarily flexible computer-control methods on the beautiful sounds of acoustic instruments? I thought about it and it sounded quite doable, so that meeting constituted the founding of Kurzweil Music Systems, and defined its raison d'être.

With Stevie Wonder as our musical adviser, we set out to combine these two worlds of music. In June of 1983, we demonstrated an engineering prototype of the Kurzweil 250 (K250) and introduced it commercially in 1984. The K250 is considered to be the first electronic musical instrument to successfully emulate the complex sound response of a grand piano and virtually all other orchestral instruments.

Earlier, my father, who was a noted musician, had played a role in developing my interest in electronic music. Before his death in 1970, he told me that he believed I would one day combine my interests in computers and in music, as he felt there was a natural affinity between the two. I remember that when my father wanted to hear one of his orchestral compositions, he had to engage an entire orchestra. This meant raising money, mimeographing copies of handwritten sheet music, selecting and hiring the right musicians, and arranging a hall in which they could play. After all of that, he would get to hear his composition for the first time. God forbid if he didn't like the composition exactly the way it was, for then he would have to dismiss the musicians, spend days rewriting modified scores by hand, raise more money, rehire the musicians, and get them back together. Today a musician can hear her multi-instrumental composition on a Kurzweil or other synthesizer, make changes as easily as one would to a letter on a word processor, and hear the results instantly.

I sold Kurzweil Music Systems to a Korean company, Young Chang, the world's largest piano manufacturer, in 1990. Kurzweil Music Systems remains one of the leading brands of electronic musical instruments in the world and is sold in forty-five countries.

I also started Kurzweil Applied Intelligence in 1982 with the goal of creating a voice-activated word processor. This is a technology that is hungry for MIPs (that is, computer speed) and megabytes (that is, memory), so early systems limited the size of the vocabulary that users could employ. These early systems also required users to pause briefly between words . . . so . . . you . . . had . . . to . . . speak . . . like . . . this. We combined this "discrete word" speech-recognition technology with a medical knowledge base to create a system that enabled doctors to create their medical reports by simply talking to their computers. Our product, called Kurzweil VoiceMed (now Kurzweil Clinical Reporter), actually guides the doctors through the reporting process. We also introduced a general-purpose dictation product called Kurzweil Voice, which enabled users to create written documents by speaking one word at a time to their personal computer. This product became particularly popular with people who have a disability in the use of their hands.

Just this year, courtesy of Moore's Law, personal computers became fast enough to recognize fully continuous speech, so I am able to dictate the rest of this book by talking to our latest product, called Voice Xpress Plus, at speeds around a hundred words per minute. Of course, I don't get a hundred words written every minute since I change my mind a lot, but Voice Xpress doesn't seem to mind.

We sold this company as well, to Lernout & Hauspie (L&H), a large speech-and-language technology company with headquarters in Belgium. Shortly after the acquisition by L&H in 1997, we arranged a strategic alliance between the dictation division of L&H (formerly Kurzweil Applied Intelligence) and Microsoft, so our speech technology is likely to be used by Microsoft in future products.

L&H is also the leader in text-to-speech synthesis and automatic language translation, so the company now has all the technologies needed for a translating telephone. As I mentioned above, we're now putting together a technology demonstration of a system that will allow you to speak in English with the person at the other end hearing you in German, and vice versa. Eventually, you'll be able to call anyone in the world and have what you say instantly translated into any popular language. Of course, our ability to misunderstand each other will remain unimpaired.

Another application of our speech-recognition technology, and one of our initial goals, is a listening device for the deaf, essentially the opposite of a reading machine for the blind. By recognizing natural continuous speech in real time, the device will enable a deaf person to read what people are saying, thereby overcoming the principal handicap associated with deafness.

In 1996, I founded a new reading-technology company called Kurzweil Educational Systems, which has developed a new generation of print-to-speech reading software for sighted persons with reading disabilities, as well as a new reading machine for blind people. The reading-disabilities version, called the Kurzweil 3000, scans a printed document, displays the page just as it appears in the original document (for example, book, magazine), with all of the color graphics and pictures intact. It then reads the document out loud while highlighting the image of the print as it is being read. It essentially does what a reading teacher does—reading to a pupil while pointing out exactly what is being read.

It is the applications of the technology benefiting disabled people that have brought me the greatest gratification. There is a fortuitous match between the capabilities of contemporary computers and the needs of a disabled person. We're not creating cybernetic geniuses today—not yet. The intelligence of our present-day intelligent computers is narrow, which can provide effective solutions for the narrow deficits of most disabled persons. The restricted intelligence of the machine works effectively with the broad and flexible intelligence of the disabled person. Overcoming the handicaps associated with disabilities using AI technologies has long been a personal goal of mine. With regard to the major physical and sensory disabilities, I believe that in a couple of decades we will come to herald the effective end of handicaps. As amplifiers of human thought, computers have great potential to assist human expression and to expand creativity for all of us. I hope to continue playing a role in harnessing this potential.

All of these projects have required the dedication and talents of many brilliant individuals in a broad range of fields. It is always exciting to see—or hear—a new product, and to see its impact on the lives of its users. A great pleasure has been sharing in the creative process, and its fruits, with these many outstanding men and women.

capable of handling continuous speech and a large vocabulary, has been introduced. Automatic language translation, which rapidly translates web sites from one language to another, is available directly from your web browser. Text-to-speech synthesis for a wide variety of languages has been available for many years. All of these technologies run on personal computers. At Lernout & Hauspie (which acquired my speech-recognition company, Kurzweil Applied Intelligence, in 1997), we are putting together a technology demonstration of a translating telephone. We expect such a system to be commercially available early in the first decade of the twenty-first century.[17]

THE NEW LUDDITE CHALLENGE

First let us postulate that the computer scientists succeed in developing intelligent machines that can do all things better than human beings can do them. In that case presumably all work will be done by vast, highly organized systems of machines and no human effort will be necessary. Either of two cases might occur. The machines might be permitted to make all of their own decisions without human oversight, or else human control over the machines might be retained.

If the machines are permitted to make all their own decisions, we can't make any conjectures as to the results, because it is impossible to guess how such machines might behave. We only point out that the fate of the human race would be at the mercy of the machines. It might be argued that the human race would never be foolish enough to hand over all the power to the machines. But we are suggesting neither that the human race would voluntarily turn power over to the machines nor that the machines would willfully seize power. What we do suggest is that the human race might easily permit itself to drift into a position of such dependence on the machines that it would have no practical choice but to accept all of the machines' decisions. As society and the problems that face it become more and more complex and machines become more and more intelligent, people will let machines make more of their decisions for them, simply because machine-made decisions will bring better results than man-made ones. Eventually a stage may be reached at which the decisions necessary to keep the system running will be so complex that human beings will be incapable of making them intelligently. At that stage the machines will be in effective control. People won't be able to just turn the machines off, because they will be so dependent on them that turning them off would amount to suicide.

On the other hand it is possible that human control over the machines may be retained. In that case the average man may have control over certain private machines of his own, such as his car or his personal computer, but control over large systems of machines will be in the hands of a tiny elite—just as it is today, but with two differences. Due to improved techniques the elite will have greater control over the masses; and because human work will no longer be necessary the masses will be superfluous, a useless burden on the system. If the elite is ruthless they may simply decide to exterminate the mass of humanity. If they are humane they may use propaganda or other psychological or biological techniques to reduce the birth rate until the mass of humanity becomes extinct, leaving the world to the elite. Or, if the elite consists of soft-hearted liberals, they may decide to play the role of good shepherds to the rest of the human race. They will see to it that everyone's physical needs are satisfied, that all children are raised under psychologically hygienic conditions, that everyone has a wholesome hobby to keep him busy, and that anyone who may become dissatisfied undergoes "treatment" to cure his "problem." Of course, life will be so purposeless that people will have to be biologically or psychologically engineered either to remove their need for the power process or to make them "sublimate" their drive for power into some harmless hobby. These engineered human beings may be

happy in such a society, but they most certainly will not be free. They will have
been reduced to the status of domestic animals.

—Theodore Kaczynski

The weavers of Nottingham enjoyed a modest but comfortable lifestyle from their thriving cottage industry of producing fine stockings and lace. This went on for hundreds of years, as their stable family businesses were passed down from generation to generation. But with the invention of the power loom and the other textile automation machines of the early eighteenth century, the weavers' livelihoods came to an abrupt end. Economic power passed from the weaving families to the owners of the machines.

Into this turmoil came a young and feebleminded boy named Ned Ludd, who, legend has it, broke two textile factory machines by accident as a result of sheer clumsiness. From that point on, whenever factory equipment was found to have been mysteriously damaged, anyone suspected of foul play would say, "But Ned Ludd did it."

In 1812, the desperate weavers formed a secret society, an urban guerrilla army. They made threats and demands of factory owners, many of whom complied. When asked who their leader was, they replied, "Why, General Ned Ludd, of course." Although the Luddites, as they became known, initially directed most of their violence against the machines, a series of bloody engagements erupted later that year. The tolerance of the Tory government for the Luddites ended, and the movement dissolved with the imprisonment and hanging of prominent members.[18]

The ability of machines to displace human employment was not an intellectual exercise for the Luddites. They had seen their way of life turned on its head. It was little comfort to the weavers that new and more lucrative employment had been created to design, manufacture, and market the new machines. There were no government programs to retrain the weavers to become automation designers.

Although they failed to create a sustained and viable movement, the Luddites have remained a powerful symbol as machines have continued to displace human workers. As one of many examples of the effect of automation on employment, about a third of the U.S. population was involved in the production of agricultural products at the beginning of the twentieth century. Today, that percentage is about 3 percent.[19] It would have been little comfort to the farmers of a hundred years ago to point out that their lost jobs would ultimately be compensated by new jobs in a future electronics industry, or that their descendants could become software designers in Silicon Valley.

The reality of lost jobs is often more compelling than the indirect promise of new jobs created in distant new industries. When advertising agencies started using Kurzweil synthesizers to create the sound tracks for television commercials rather than hire live musicians, the musicians' union was not happy about it. We pointed out that the new computer-music technology was actually beneficial to musicians because it made music more exciting. For example, industrial films that had formerly used prerecorded orchestral music (because the limited budget of such films did not allow the hiring of an entire orchestra) were now using original music created by a musician with a synthesizer. As it turned out, this wasn't a very effective argument, since the synthesizer players tended not to be union members.

The Luddite philosophy remains very much alive as an ideological inclination, but as a political and economic movement, it remains just below the surface of contemporary debate. The public appears to understand that the creation of new technology is fueling the expansion of economic well-being. The statistics demonstrate quite clearly that automation is creating more and better jobs than it is eliminating. In 1870 only 12 million Americans, representing about one third of the civilian population, had jobs. By 1998, the figure rose to 126 million jobs held by about two thirds of the civilian population.[20] The gross national product on a per capita basis and in constant 1958 dollars went from $530 in 1870 to at least ten times that today.[21] There has been a comparable change in the actual earning power of available jobs. This 1,000 percent increase in real wealth has resulted in a greatly improved standard of living, better health care and education, and a substantially improved ability to provide for those who need help in our society. At the beginning of the Industrial Revolution life expectancy in North America and northwestern Europe was about thirty-seven years. Now, two centuries later, it has doubled, and is continuing to increase.

The jobs created have also been on a higher level. Indeed, much of the additional employment has been in the area of providing the more intense education that today's jobs require. For example, we now spend ten times as much (in constant dollars) on a per capita basis for public school education as we did a century ago. In 1870 only 2 percent of American adults had a high-school diploma, whereas the figure is over 80 percent today. There were only 52,000 college students in 1870; there are 15 million today.

The process of automation that began in England two hundred years ago— and continues today at an ever accelerating pace (as per the Law of Accelerating Returns)—eliminates jobs at the bottom of the skill ladder and creates new ones at the top of the skill ladder. Hence increasing investment in education. But what happens when the skill ladder extends beyond the abilities of the bulk of the human population, and ultimately beyond the ability of any human, educational innovations notwithstanding?

The answer we can predict from the Law of Accelerating Returns is that the ladder will nonetheless continue to reach ever higher, implying that humans will need to become more capable by other means. Education can only accomplish so much. The only way for the species to keep pace will be for humans to gain greater competence from the computational technology we have created, that is, for the species to merge with its technology.

Not everyone will find this prospect appealing, so the Luddite issue will broaden in the twenty-first century from an anxiety about human livelihoods to one concerning the essential nature of human beings. However, the Luddite movement is not likely to fare any better in the next century than it has in the past two. It suffers from the lack of a viable alternative agenda.

Ted Kaczynski, whom I quote above from his so-called "Unabomber Manifesto," entitled *Industrial Society and Its Future,* advocates a simple return to nature.[22] Kaczynski is not talking about a contemplative visit to a nineteenth-century Walden Pond, but about the species dropping all of its technology and reverting to a simpler time. Although he makes a compelling case for the dangers and damages that have accompanied industrialization, his proposed vision is neither compelling nor feasible. After all, there is too little nature left to return to, and there are too many human beings. For better or worse, we're stuck with technology.

YOUR CYBERNETIC POET WRITES SOME INTERESTING LINES . . .

I'd be interested in your selections.

WELL, LOOKING AT THE FIRST FEW POEMS IN YOUR COLLECTION:

> *Sashay down the page . . .*
> *through the lioness / nestled in my soul . . .*
> *forming jewels from the falling snow . . .*
> *the juice of eternity, / the spirit of my lips . . .*

BUT THE POEMS DON'T ALWAYS FULLY TRACK, IF YOU KNOW WHAT I MEAN.

Yes, readers tolerate a little more discontinuity in verse than in prose. The fundamental problem is the inability of contemporary cybernetic artists to master the levels of context that human artists are capable of. It's not a permanent limitation, of course. Ultimately, we'll be the ones having difficulty keeping up with the depth of context that computer intelligence is capable of—

WITHOUT SOME ASSIST—

From computer extensions to our intelligence, yes, exactly.

In the meantime, the Cybernetic Poet is good at being an inspirational assistant. While its poems don't always make it all the way through, it does have some real strength at finding unique turns of phrase. So the program has a mode called

The Poet's Assistant. The human user writes a poem in a word-processing window. The Poet's Assistant watches her write and fills the rest of the screen with suggestions, such as, "Here's how Robert Frost would finish that line," or, "Here's a set of rhymes and/or alliterations that Keats used with that word," or, "Here's how Emily Dickinson would finish that poem," and so on. If provided with the human author's own poems, it can even suggest how the user herself would finish a line or poem. Everytime you write another word, you get dozens of ideas. Not all of them make sense, but it's a good solution for writer's block. And you're welcome to steal its ideas.

NOW WITH REGARD TO COHEN'S PICTURES . . .

You mean Aaron's pictures . . .

OH, I GUESS I'M NOT SENSITIVE TO AARON'S FEELINGS—

Since it doesn't have any—

NOT YET, RIGHT? BUT WHAT I WAS GOING TO SAY WAS THAT AARON'S PICTURES DO SEEM TO MAINTAIN THEIR CONTEXT. THE WHOLE THING KIND OF WORKS FOR ME.

Yes, Cohen's Aaron is probably the best example of a cybernetic visual artist today, and certainly one of the primary examples of computers in the arts. Cohen has programmed thousands of rules on all aspects of drawing and painting, from the artistic nature of painted people, plants, and objects to composition and color choice.

Keep in mind that Aaron does not seek to emulate other artists. It has its own set of styles, so it is feasible for its knowledge base to be relatively complete within its visual domain. Of course, human artists, even brilliant ones, also have a boundary to their domain. Aaron is quite respectable in the diversity of its art.

OKAY, JUST TO SWITCH TO SOMEONE MUCH LESS RESPECTABLE, YOU QUOTED TED KACZYNSKI TALKING ABOUT HOW THE HUMAN RACE MIGHT DRIFT INTO DEPENDENCE ON MACHINES, AND THEN WE'LL HAVE NO CHOICE BUT TO ACCEPT ALL MACHINE DECISIONS. BASED ON WHAT YOU SAID ABOUT THE IMPLICATIONS OF ALL THE COMPUTERS STOPPING, AREN'T WE ALREADY THERE?

We are certainly there with regard to the dependence, not yet with regard to the level of machine intelligence.

THAT QUOTE WAS SURPRISINGLY—

Coherent?

YES, THAT WAS THE WORD I WAS LOOKING FOR.

Kaczynski's whole manifesto is rather well written, not at all what you would expect given the popular portrait of him as a madman. As political science professor James Q. Wilson of the University of California wrote, "The language is clear, precise and calm. The argument is subtle and carefully developed, lacking anything even faintly resembling the wild claims or irrational speculation that a lunatic might produce." And he has gathered quite a following among anarchists and antitechnologists on the Internet—

WHICH IS THE ULTIMATE IN TECHNOLOGY.

Yes, that irony has not been lost.

BUT WHY QUOTE KACZYNSKI? I MEAN, . . .

Well, his manifesto is as persuasive an exposition on the psychological alienation, social dislocation, environmental injury, and other injuries and perils of the technological age as any other . . .

THAT'S NOT MY POINT. I DOUBT THAT THE LUDDITES ARE HAPPY HAVING HIM AS A SYMBOL OF THEIR IDEAS. YOU'RE SORT OF DISCREDITING THEIR MOVEMENT BY USING HIM AS THEIR SPOKESPERSON.

Okay, that's a legitimate objection. I suppose I could defend my extensive quote as providing an important example of a relevant phenomenon, which is violent Ludditism. The movement started with violence, and the challenge to the human race posed by machines is fundamental enough that a violent reaction during this coming century is a strong possibility.

BUT YOUR USE OF THE QUOTATION SEEMED LIKE MORE THAN JUST AN EXAMPLE OF SOME FRINGE PHENOMENON.

Well, I was surprised how much of Kaczynski's manifesto I agreed with.

SUCH AS . . .

Oh, so now you're interested.

IT WAS KIND OF INTRIGUING, AND APROPOS TO THE OTHER THINGS YOU'VE BEEN TELLING ME.

Yes, I thought so. Kacyznski describes the benefits of technology, as well as its costs and dangers. He then makes this point:

> A further reason why industrial society cannot be reformed in favor of freedom is that modern technology is a unified system in which all parts are dependent on one another. You can't get rid of the "bad" parts of technology and retain only the "good" parts. Take modern medicine,

for example. Progress in medical science depends on progress in chemistry, physics, biology, computer science, and other fields. Advanced medical treatments require expensive, high-tech equipment that can be made available only by a technologically progressive, economically rich society. Clearly you can't have much progress in medicine without the whole technological system and everything that goes with it.

So far, so good. He then makes the basic judgment that the "bad parts" outweigh the "good parts." Not that this is a crazy position, either, but nonetheless, this is where we part company. Now it is not my view that the advance of technology is automatically beneficial. It is conceivable that humanity will ultimately regret its technological path. Although the risks are quite real, my fundamental belief is that the potential gains are worth the risk. But this is a belief; it's not a position I can easily demonstrate.

I'D BE INTERESTED IN YOUR VIEW OF THE GAINS.

The material gains are obvious: economic advancement, the shaping of material resources to meet age-old needs, the extension of our life spans, improvements in health, and so on. However, that's not actually my primary point.

I see the opportunity to expand our minds, to extend our learning, and to advance our ability to create and understand knowledge as an essential spiritual quest. Feigenbaum and McCorduck talk about this as an "audacious, some would say reckless, embarkation onto sacred ground."

SO WE RISK THE SURVIVAL OF THE HUMAN RACE FOR THIS SPIRITUAL QUEST?

Yeah, basically.

I'M NOT SURPRISED THAT THE LUDDITES TAKE PAUSE.

Of course, keep in mind that it's the material, not the spiritual gains, that are seducing society down this path.

I'M STILL NOT COMFORTABLE WITH KACZYNSKI AS A SPOKESPERSON. HE IS A CONFESSED MURDERER, YOU KNOW.

Certainly, I'm glad he's behind bars, and his tactics deserve condemnation and punishment. Unfortunately, terrorism is effective, and that's why it survives.

I DON'T SEE IT THAT WAY. TERRORISM JUST UNDERMINES THE POSITIONS BEING PUBLICIZED. PEOPLE THEN SEE THE TERRORIST'S PROPOSITIONS AS CRAZY, OR AT LEAST MISGUIDED.

That's one reaction. But remember the society of mind. We have more than one reaction to terrorism.

One contingent in our heads says "those actions were evil and crazy, so the terrorist's thesis must also be evil and crazy."

But another contingent in our heads takes the view that "those actions were extreme, so he must have very strong feelings about this. Maybe there's something to it. Perhaps a more moderate version of his views are legitimate."

SOUNDS LIKE THE PSYCHOLOGY OF HITLER'S "BIG LIE."

There's a similarity. In Hitler's case, both the tactics and the views were extreme. In the case of modern terrorists, the tactics are extreme; the views may or may not be. In Kaczynski's case, many aspects of his argument are reasonable. Of course, he does end up in an extreme place.

YEAH, A PRIMITIVE CABIN IN MONTANA.

That's where the manifesto ends up, too—we should all return to nature.

I DON'T THINK PEOPLE FOUND KACZYNSKI'S NOTION OF NATURE VERY APPEALING, AT LEAST NOT JUDGING BY PICTURES OF HIS CABIN.

And, as I said, there's not enough nature to go around anymore.

THANKS TO TECHNOLOGY.

And the population boom—

ALSO FACILITATED BY TECHNOLOGY.

So we've passed the point of no return. It's already too late to go the nature route.

SO WHAT COURSE DO YOU RECOMMEND?

I would say that we shouldn't view the advance of technology as just an impersonal, inexorable force.

I THOUGHT YOU SAID THE ACCELERATING ADVANCE OF TECHNOLOGY—AND COMPUTATION— WAS INEXORABLE; REMEMBER, THE LAW OF ACCELERATING RETURNS?

Uh, yes, the advance is inexorable all right, we're not going to stop technology. But we do have some choices. We have the opportunity to shape technology, and to channel its direction. I've tried to do that in my own work. We can step through the forest carefully.

WE'D BETTER GET BUSY, SOUNDS LIKE THERE ARE A LOT OF SLIPPERY SLOPES OUT THERE WAITING FOR US.

PART THREE

TO FACE THE FUTURE

INTRODUCING...
INTERACTIVEWEAR.
All is A-OK with your model # 7321-B Interactivewear Jacket.
I am tied too loosely. Please tighten.
Collar button missing. Soup stain on right side. Dry clean only.
WALLET GONE! WALLET GONE!
Curb approaching at twelve o'clock.
R. Chr.

CHAPTER NINE

2009

Ever since I could remember, I'd wished I'd been lucky enough to be alive at a great time—when something big was going on, like a crucifixion. And suddenly I realized I was.

—Ben Shahn

As we say in the computer business, "shift happens."

—Tim Romero

It is said that people overestimate what can be accomplished in the short term, and underestimate the changes that will occur in the long term. With the pace of change continuing to accelerate, we can consider even the first decade in the twenty-first century to constitute a long-term view. With that in mind, let us consider the beginning of the next century.

The Computer Itself

It is now 2009. Individuals primarily use portable computers, which have become dramatically lighter and thinner than the notebook computers of ten years earlier. Personal computers are available in a wide range of sizes and shapes, and are commonly embedded in clothing and jewelry such as wristwatches, rings, earrings, and other body ornaments. Computers with a high-resolution visual interface range from rings and pins and credit cards up to the size of a thin book.

People typically have at least a dozen computers on and around their bodies, which are networked using "body LANs" (local area networks).[1] These computers provide communication facilities similar to cellular phones, pagers, and web surfers, monitor body functions, provide automated identity (to conduct financial transactions and allow entry into secure areas), provide directions for navigation, and a variety of other services.

For the most part, these truly personal computers have no moving parts. Memory is completely electronic, and most portable computers do not have keyboards.

Rotating memories (that is, computer memories that use a rotating platten, such as hard drives, CD-ROMs, and DVDs) are on their way out, although rotating magnetic memories are still used in "server" computers where large amounts of information are stored. Most users have servers in their homes and offices where they keep large stores of digital "objects," including their software, databases, documents, music, movies, and virtual-reality environments (although these are still at an early stage). There are services to keep one's digital objects in central repositories, but most people prefer to keep their private information under their own physical control.

Cables are disappearing.[2] Communication between components, such as pointing devices, microphones, displays, printers, and the occasional keyboard, uses short-distance wireless technology.

Computers routinely include wireless technology to plug into the ever-present worldwide network, providing reliable, instantly available, very-high-bandwidth communication. Digital objects such as books, music albums, movies, and software are rapidly distributed as data files through the wireless network, and typically do not have a physical object associated with them.

The majority of text is created using continuous speech recognition (CSR) dictation software, but keyboards are still used. CSR is very accurate, far more so than the human transcriptionists who were used up until a few years ago.

Also ubiquitous are language user interfaces (LUIs), which combine CSR and natural language understanding. For routine matters, such as simple business transactions and information inquiries, LUIs are quite responsive and precise. They tend to be narrowly focused, however, on specific types of tasks. LUIs are frequently combined with animated personalities. Interacting with an animated personality to conduct a purchase or make a reservation is like talking to a person using videoconferencing, except that the person is simulated.

Computer displays have all the display qualities of paper—high resolution, high contrast, large viewing angle, and no flicker. Books, magazines, and newspapers are now routinely read on displays that are the size of, well, small books.

Computer displays built into eyeglasses are also used. These specialized glasses allow users to see the normal visual environment, while creating a virtual image that appears to hover in front of the viewer. The virtual images are created by a tiny laser built into the glasses that projects the images directly onto the user's retinas.[3]

Computers routinely include moving picture image cameras and are able to reliably identify their owners from their faces.

In terms of circuitry, three-dimensional chips are commonly used, and there is a transition taking place from the older, single-layer chips.

Sound producing speakers are being replaced with very small chip-based devices that can place high resolution sound anywhere in three-dimensional space. This technology is based on creating audible frequency sounds from the spectrum created by the interaction of very high frequency tones. As a result, very small speakers can create very robust three-dimensional sound.

A $1,000 personal computer (in 1999 dollars) can perform about a trillion calculations per second.[4] Supercomputers match at least the hardware capacity of the human brain—20 million billion calculations per second.[5] Unused computes on the Internet are being harvested, creating virtual parallel supercomputers with human brain hardware capacity.

There is increasing interest in massively parallel neural nets, genetic algorithms, and other forms of "chaotic" or complexity theory computing, although most computer computations are still done using conventional sequential processing, albeit with some limited parallel processing.

Research has been initiated on reverse engineering the human brain through both destructive scans of the brains of recently deceased persons as well as non-invasive scans using high resolution magnetic resonance imaging (MRI) of living persons.

Autonomous nanoengineered machines (that is, machines constructed atom by atom and molecule by molecule) have been demonstrated and include their own computational controls. However, nanoengineering is not yet considered a practical technology.

Education

In the twentieth century, computers in schools were mostly on the trailing edge, with most effective learning from computers taking place in the home. Now in 2009, while schools are still not on the cutting edge, the profound importance of the computer as a knowledge tool is widely recognized. Computers play a central role in all facets of education, as they do in other spheres of life.

The majority of reading is done on displays, although the "installed base" of paper documents is still formidable. The generation of paper documents is dwindling, however, as the books and other papers of largely twentieth-century vintage are being rapidly scanned and stored. Documents circa 2009 routinely include embedded moving images and sounds.

Students of all ages typically have a computer of their own, which is a thin tabletlike device weighing under a pound with a very high resolution display suitable for reading. Students interact with their computers primarily by voice and by pointing with a device that looks like a pencil. Keyboards still exist, but most textual language is created by speaking. Learning materials are accessed through wireless communication.

Intelligent courseware has emerged as a common means of learning. Recent controversial studies have shown that students can learn basic skills such as reading and math just as readily with interactive learning software as with human teachers, particularly when the ratio of students to human teachers is more than one to one. Although the studies have come under attack, most students and their parents have accepted this notion for years. The traditional mode of a human teacher instructing a group of children is still prevalent, but schools are increasingly relying on software approaches, leaving human teachers to attend primarily to issues of motivation, psychological well-being, and socialization. Many children learn to read on their own using their personal computers before entering grade school.

Preschool and elementary school children routinely read at their intellectual level using print-to-speech reading software until their reading skill level catches up. These print-to-speech reading systems display the full image of documents, and can read the print aloud while highlighting what is being read. Synthetic voices sound fully human. Although some educators expressed concern in the early '00 years that students would rely unduly on reading software, such systems have been readily accepted by children and their parents. Studies have shown that students improve their reading skills by being exposed to synchronized visual and auditory presentations of text.

Learning at a distance (for example, lectures and seminars in which the participants are geographically scattered) is commonplace.

Learning is becoming a significant portion of most jobs. Training and developing new skills is emerging as an ongoing responsibility in most careers, not just an occasional supplement, as the level of skill needed for meaningful employment soars ever higher.

Disabilities

Persons with disabilities are rapidly overcoming their handicaps through the intelligent technology of 2009. Students with reading disabilities routinely ameliorate their disability using print-to-speech reading systems.

Print-to-speech reading machines for the blind are now very small, inexpensive, palm-sized devices that can read books (those that still exist in paper form) and other printed documents, and other real-world text such as signs and displays. These reading systems are equally adept at reading the trillions of electronic documents that are instantly available from the ubiquitous wireless worldwide network.

After decades of ineffective attempts, useful navigation devices have been introduced that can assist blind people in avoiding physical obstacles in their path, and finding their way around, using global positioning system (GPS) technology. A blind person can interact with her personal reading-navigation systems through two-way voice communication, kind of like a Seeing Eye dog that reads and talks.

Deaf persons—or anyone with a hearing impairment—commonly use portable speech-to-text listening machines, which display a real-time transcription of what people are saying. The deaf user has the choice of either reading the transcribed speech as displayed text, or watching an animated person gesturing in sign language. These have eliminated the primary communication handicap associated with deafness. Listening machines can also translate what is being said into another language in real time, so they are commonly used by hearing people as well.

Computer-controlled orthotic devices have been introduced. These "walking machines" enable paraplegic persons to walk and climb stairs. The prosthetic devices are not yet usable by all paraplegic persons, as many physically disabled persons have dysfunctional joints from years of disuse. However, the advent of orthotic walking systems is providing more motivation to have these joints replaced.

There is a growing perception that the primary disabilities of blindness, deafness, and physical impairment do not necessarily impart handicaps. Disabled persons routinely describe their disabilities as mere inconveniences. Intelligent technology has become the great leveler.

Communication

Translating Telephone technology (where you speak in English and your Japanese friend hears you in Japanese, and vice versa) is commonly used for many language pairs. It is a routine capability of an individual's personal computer, which also serves as her phone.

"Telephone" communication is primarily wireless, and routinely includes high-resolution moving images. Meetings of all kinds and sizes routinely take place among geographically separated participants.

There is effective convergence, at least on the hardware and supporting software level, of all media, which exist as digital objects (that is, files) distributed by the ever-present high-bandwidth, wireless information web. Users can instantly download books, magazines, newspapers, television, radio, movies, and other forms of software to their highly portable personal communication devices.

Virtually all communication is digital and encrypted, with public keys available to government authorities. Many individuals and groups, including but not limited to criminal organizations, use an additional layer of virtually unbreakable encryption codes with no third-party keys.

Haptic technologies are emerging that allow people to touch and feel objects and other persons at a distance. These force-feedback devices are widely used in games and in training simulation systems.

Interactive games routinely include all-encompassing visual and auditory environments, but a satisfactory, all-encompassing tactile environment is not yet available. The online chat rooms of the late 1990s have been replaced with virtual environments where you can meet people with full visual realism.

People have sexual experiences at a distance with other persons as well as virtual partners. But the lack of the "surround" tactile environment has thus far kept virtual sex out of the mainstream. Virtual partners are popular as forms of sexual entertainment, but they're more gamelike than real. And phone sex is a lot more popular now that phones routinely include high-resolution, real-time moving images of the person on the other end.

Business and Economics

Despite occasional corrections, the ten years leading up to 2009 have seen continuous economic expansion and prosperity due to the dominance of the knowledge content of products and services. The greatest gains continue to be in the value of the stock market. Price deflation concerned economists in the early '00 years, but they quickly realized it was a good thing. The high-tech community pointed out that significant deflation had existed in the computer hardware and software industries for many years earlier without detriment.

The United States continues to be the economic leader due to its primacy in popular culture and its entrepreneurial environment. Since information markets are largely world markets, the United States has benefited greatly from its immigrant history. Being comprised of all the world's peoples—specifically the descendants of peoples from around the globe who had endured great risk for a better life—is the ideal heritage for the new knowledge-based economy. China has also emerged as a powerful economic player. Europe is several years ahead of Japan and Korea in adopting the American emphasis on venture capital, employee stock options, and tax policies that encourage entrepreneurship, although these practices have become popular throughout the world.

At least half of all transactions are conducted online. Intelligent assistants which combine continuous speech recognition, natural-language understanding, problem solving, and animated personalities routinely assist with finding information, answering questions, and conducting transactions. Intelligent assistants have become a primary interface for interacting with information-based services, with a wide range of choices available. A recent poll shows that both male and female users prefer female personalities for their computer-based intelligent assistants. The two most popular are Maggie, who claims to be a waitress in a Harvard Square café, and Michelle, a stripper from New Orleans. Personality designers are in demand, and the field constitutes a growth area in software development.

Most purchases of books, musical "albums," videos, games, and other forms of software do not involve any physical object, so new business models for distributing these forms of information have emerged. One shops for these information objects by "strolling" through virtual malls, sampling and selecting objects of interest, rapidly (and securely) conducting an online transaction, and then quickly downloading the information using high-speed wireless communication. There are many types and gradations of transactions to gain access to these products. You can "buy" a book, musical album, video, etcetera, which gives you unlimited permanent access. Alternatively, you can rent access to read, view, or listen once, or a few times. Or you can rent access by the minute. Access may be limited to one person or to a group of persons (for example, a family or a company). Alternatively, access may be limited to a particular computer, or to any computer accessed by a particular person or by a set of persons.

There is a strong trend toward the geographic separation of work groups. People are successfully working together despite living and working in different places.

The average household has more than a hundred computers, most of which are embedded in appliances and built-in communication systems. Household robots have emerged, but are not yet fully accepted.

Intelligent roads are in use, primarily for long-distance travel. Once your car's computer guidance system locks onto the control sensors on one of these highways, you can sit back and relax. Local roads, though, are still predominantly conventional.

A company west of the Mississippi and north of the Mason-Dixon line has surpassed a trillion dollars in market capitalization.

Politics and Society

Privacy has emerged as a primary political issue. The virtually constant use of electronic communication technologies is leaving a highly detailed trail of every person's every move. Litigation, of which there has been a great deal, has placed some constraints on the widespread distribution of personal data. Government agencies, however, continue to have the right to gain access to people's files, which has resulted in the popularity of unbreakable encryption technologies.

There is a growing neo-Luddite movement, as the skill ladder continues to accelerate upward. As with earlier Luddite movements, its influence is limited by the level of prosperity made possible by new technology. The movement does succeed in establishing continuing education as a primary right associated with employment.

There is continuing concern with an underclass that the skill ladder has left far behind. The size of the underclass appears to be stable, however. Although not politically popular, the underclass is politically neutralized through public assistance and the generally high level of affluence.

The Arts

The high quality of computer screens, and the facilities of computer-assisted visual rendering software, have made the computer screen a medium of choice for visual art. Most visual art is the result of a collaboration between human artists and their intelligent art software. Virtual paintings—high-resolution wall-hung displays—have become popular. Rather than always displaying the same work of art, as with a conventional painting or poster, these virtual paintings can change the displayed work at the user's verbal command, or can cycle through collections of art. The displayed artwork can be works by human artists or original art created in real time by cybernetic art software.

Human musicians routinely jam with cybernetic musicians. The creation of music has become available to persons who are not musicians. Creating music does not necessarily require the fine motor coordination of using traditional controllers. Cybernetic music creation systems allow people who appreciate music but who are not knowledgeable about music theory and practice to create music in collaboration with their automatic composition software. Interactive brain-generated music, which creates a resonance between the user's brain waves and the music being listened to, is another popular genre.

Musicians commonly use electronic controllers that emulate the playing style of the old acoustic instruments (for example, piano, guitar, violin, drums), but there is a surge of interest in the new "air" controllers in which you create music by moving your hands, feet, mouth, and other body parts. Other music controllers involve interacting with specially designed devices.

Writers use voice-activated word processing; grammar checkers are now actually useful; and distribution of written documents from articles to books typically does not involve paper and ink. Style improvement and automatic editing software is widely used to improve the quality of writing. Language translation software is also widely used to translate written works in a variety of languages. Nonetheless, the core process of creating written language is less affected by intelligent software technologies than the visual and musical arts. However, "cybernetic" authors are emerging.

Beyond music recordings, images, and movie videos, the most popular type of digital entertainment object is virtual experience software. These interactive virtual environments allow you to go whitewater rafting on virtual rivers, to hang-glide in a virtual Grand Canyon, or to engage in intimate encounters with your favorite movie star. Users also experience fantasy environments with no counterpart in the physical world. The visual and auditory experience of virtual reality is compelling, but tactile interaction is still limited.

Warfare

The security of computation and communication is the primary focus of the U.S. Department of Defense. There is general recognition that the side that can maintain the integrity of its computational resources will dominate the battlefield.

Humans are generally far removed from the scene of battle. Warfare is dominated by unmanned intelligent airborne devices. Many of these flying weapons are the size of small birds, or smaller.

The United States continues to be the world's dominant military power, which is largely accepted by the rest of the world, as most countries concentrate on economic competition. Military conflicts between nations are rare, and most conflicts are between nations and smaller bands of terrorists. The greatest threat to national security comes from bioengineered weapons.

Health and Medicine

Bioengineered treatments have reduced the toll from cancer, heart disease, and a variety of other health problems. Significant progress is being made in understanding the information processing basis of disease.

Telemedicine is widely used. Physicians can examine patients using visual, auditory, and haptic examination from a distance. Health clinics with relatively inexpensive equipment and a single technician bring health care to remote areas where doctors had previously been scarce.

Computer-based pattern recognition is routinely used to interpret imaging data and other diagnostic procedures. The use of noninvasive imaging technologies has substantially increased. Diagnosis almost always involves collaboration between a human physician and a pattern-recognition-based expert system. Doctors routinely consult knowledge-based systems (generally through two-way voice communication augmented by visual displays), which provide automated guidance, access to the most recent medical research, and practice guidelines.

Lifetime patient records are maintained in computer databases. Privacy concerns about access to these records (as with many other databases of personal information) have emerged as a major issue.

Doctors routinely train in virtual reality environments, which include a haptic interface. These systems simulate the visual, auditory, and tactile experience of medical procedures, including surgery. Simulated patients are available for continuing medical education, for medical students, and for people who just want to play doctor.

Philosophy

There is renewed interest in the Turing Test, first proposed by Alan Turing in 1950 as a means for testing intelligence in a machine. Recall that the Turing Test contemplates a situation in which a human judge interviews the computer and a human "foil," communicating with both over terminal lines. If the human judge is unable to tell which interviewee is human and which is machine, the machine is deemed to possess human-level intelligence. Although computers still fail the test, confidence is increasing that they will be in a position to pass it within another one or two decades.

There is serious speculation on the potential sentience (that is, consciousness) of computer-based intelligence. The increasingly apparent intelligence of computers has spurred an interest in philosophy.

. . . Hey, Molly.

OH, SO YOU'RE CALLING ME NOW.

Well, the chapter was over and I didn't hear from you.

I'M SORRY, I WAS FINISHING UP A PHONE CALL WITH MY FIANCÉ.

Hey, congratulations, that's great. How long have you known . . .

BEN, HIS NAME IS BEN. WE MET ABOUT TEN YEARS AGO, JUST AFTER YOU FINISHED THIS BOOK.

I see. So how have I done?

YOU DID MANAGE TO SELL A FEW COPIES.

No, I mean with my predictions.

NOT VERY WELL. THE TRANSLATING TELEPHONES, FOR ONE THING, ARE A LITTLE RIDICU-
LOUS. I MEAN, THEY'RE CONSTANTLY SCREWING UP.

Sounds like you use them, though?

WELL, SURE, HOW ELSE AM I GOING TO SPEAK TO MY FIANCÉ'S FATHER IN IEPER, BELGIUM,
WHEN HE HASN'T BOTHERED TO LEARN ENGLISH?

Of course. So what else?

YOU SAID THAT CANCER WAS REDUCED, BUT THAT'S ACTUALLY QUITE UNDERSTATED. BIO-
ENGINEERED TREATMENTS, PARTICULARLY ANTIANGIOGENESIS DRUGS THAT PREVENT TU-
MORS FROM GROWING THE CAPILLARIES THEY NEED, HAVE ELIMINATED MOST FORMS OF
CANCER AS A MAJOR KILLER.[6]

Well, that's just not a prediction I was willing to make. There have been so many
false hopes with regard to cancer treatments, and so many promising approaches
proving to be dead ends, that I just wasn't willing to make that call. Also, there
just wasn't enough evidence when I wrote the book in 1998 to make that dramatic a prediction.

NOT THAT YOU SHIED AWAY FROM DRAMATIC PREDICTIONS.

The predictions I made were fairly conservative, actually, and were based on
technologies and trends I could touch and feel. I was certainly aware of several
promising approaches to bioengineered cancer treatments, but it was still kind of
iffy, given the history of cancer research. Anyway, the book only touched tangen-
tially on bioengineering, although it's clearly an information-based technology.

NOW WITH REGARD TO SEX—

Speaking of health problems . . .

YES, WELL, YOU SAID THAT VIRTUAL PARTNERS WERE POPULAR, BUT I JUST DON'T SEE THAT.

It might just be the circle you move in.

I HAVE A VERY SMALL CIRCLE—MOSTLY I'VE BEEN TRYING TO GET BEN TO FOCUS ON OUR
WEDDING.

Yes, tell me about him.

HE'S VERY ROMANTIC. HE ACTUALLY SENDS ME LETTERS ON PAPER!

That is romantic. So, how was the phone call I interrupted?

I TRIED ON THIS NEW NIGHTGOWN HE SENT ME. I THOUGHT HE'D APPRECIATE IT, BUT HE
WAS BEING A LITTLE ANNOYING.

I assume you're going to finish that thought.

WELL, HE WANTED ME TO KIND OF LET THESE STRAPS SLIP, MAYBE JUST A LITTLE. BUT I'M KIND OF SHY ON THE PHONE. I DON'T REALLY GO IN FOR VIDEO PHONE SEX, NOT LIKE SOME FRIENDS I KNOW.

Oh, so I did get that prediction right.

ANYWAY, I JUST TOLD HIM TO USE THE IMAGE TRANSFORMERS.

Transformers?

YOU KNOW, HE CAN UNDRESS ME JUST AT HIS END.

Oh yes, of course. The computer is altering your image in real time.

EXACTLY. YOU CAN CHANGE SOMEONE'S FACE, BODY, CLOTHING, OR SURROUNDINGS INTO SOMEONE OR SOMETHING ELSE ENTIRELY, AND THEY DON'T KNOW YOU'RE DOING IT.

Hmmm.

ANYWAY, I CAUGHT BEN UNDRESSING HIS OLD GIRLFRIEND WHEN SHE CALLED TO CONGRATULATE HIM ON OUR ENGAGEMENT. SHE HAD NO IDEA, AND HE THOUGHT IT WAS HARMLESS. I DIDN'T SPEAK TO HIM FOR A WEEK.

Well, as long as it was just at his end.

WHO KNOWS WHAT SHE WAS DOING AT HER END.

That's kind of her business, isn't it? As long as they don't know what the other is doing.

I'M NOT SO SURE THEY DIDN'T KNOW. ANYWAY, PEOPLE DO SPEND A LOT OF TIME TOGETHER UP CLOSE BUT AT A DISTANCE, IF YOU KNOW WHAT I MEAN.

Using the displays?

WE CALL THEM PORTALS—YOU CAN LOOK THROUGH THEM, BUT YOU CAN'T TOUCH.

I see, still no interest in virtual sex?

NOT PERSONALLY. I MEAN, IT'S PRETTY PATHETIC. BUT I DID HAVE TO WRITE THE COPY FOR A BROCHURE ABOUT A SENSUAL VIRTUAL REALITY ENVIRONMENT. BEING LOW ON THE TOTEM POLE, I REALLY CAN'T PICK MY ASSIGNMENTS.

Did you try the product?

I DIDN'T EXACTLY TRY IT. I JUST OBSERVED. I WOULD SAY THEY PUT MORE EFFORT INTO THE VIRTUAL GIRLS THAN THE GUYS.

How'd your campaign make out?

The product bombed. I mean, the market's just so cluttered.

You can't win them all.

No, but one of your predictions did work out quite well. I took your advice about that company north of the Mason-Dixon line. And, hey, I'm not complaining.

I'll bet a lot of stocks are up.

Yes, the boats keep getting higher.

Okay, what else?

You're right about the disabled. My office mate is deaf, and it's not an issue at all. There's nothing important a blind or deaf person can't do today.

That was really true back in 1999.

I think the difference now is that the public understands it. It's just a lot more obvious with today's technology. But that understanding is important.

Sure, without the technology, there's just a lot of misconception and prejudice.

True enough. I think I'm going to have to get going, I can see Ben's face on my call line.

He looks like a St. Bernard.

Oh, I left my image transformers on. Here, I'll let you see what he really looks like.

Hey, good-looking guy. Well, good luck. You do seem to have changed.

I should hope so.

I mean I think our relationship has changed.

Well, I'm ten years older.

And it seems that I'm asking you most of the questions.

I guess I'm the expert now. I can just tell you what I see. But how come you're still stuck in 1999?

I'm afraid I just can't leave quite yet. I have to get this book out, for one thing.

I do have one confusion. How is it that you can talk to me from 1999 when I'm here in the year 2009? What kind of technology is that?

Oh, that's a very old technology. It's called poetic license.

CHAPTER TEN

2019

He who mounts a wild elephant goes where the wild elephant goes.
—Randolph Bourne

It does not do you good to leave a dragon out of your calculations, if you live near him.

—J. R. R. Tolkien

The Computer Itself

Computers are now largely invisible. They are embedded everywhere—in walls, tables, chairs, desks, clothing, jewelry, and bodies.

People routinely use three-dimensional displays built into their glasses,[1] or contact lenses. These "direct eye" displays create highly realistic, virtual visual environments overlaying the "real" environment. This display technology projects images directly onto the human retina, exceeds the resolution of human vision, and is widely used regardless of visual impairment. The direct-eye displays operate in three modes:

1. *Head-directed display:* The displayed images are stationary with respect to the position and orientation of your head. When you move your head, the display moves relative to the real environment. This mode is often used to interact with virtual documents.

2. *Virtual-reality overlay display:* The displayed images slide when you move or turn your head so that the virtual people, objects, and environment appear to remain stationary in relation to the real environment (which you can still see). Thus if the direct-eye display is displaying the image of a person (who could be a geographically remote real person engaging in a three-dimensional visual phone call with you, or a computer-generated "simulated" person), that projected person will appear to be in a particular place relative to the real environment that you also see. When you move your head, that projected person will appear to remain in the same place relative to the real environment.

3. *Virtual-reality blocking display:* This is the same as the virtual-reality overlay display except that the real environment is blocked out, so you see only the projected virtual environment. You use this mode to leave "real" reality and enter a virtual reality environment.

In addition to the optical lenses, there are auditory "lenses," which place high-resolution sounds in precise locations in a three-dimensional environment. These can be built into eyeglasses, worn as body jewelry, or implanted in the ear canal.

Keyboards are rare, although they still exist. Most interaction with computing is through gestures using hands, fingers, and facial expressions and through two-way natural-language spoken communication. People communicate with computers the same way they would communicate with a human assistant, both verbally and through visual expression. Significant attention is paid to the personality of computer-based personal assistants, with many choices available. Users can model the personality of their intelligent assistants on actual persons, including themselves, or select a combination of traits from a variety of both public personalities and private friends and associates.

Typically, people do not own just one specific "personal computer," although computing is nonetheless very personal. Computing and extremely-high-bandwidth communication are embedded everywhere. Cables have largely disappeared.

The computational capacity of a $4,000 computing device (in 1999 dollars) is approximately equal to the computational capability of the human brain (20 million billion calculations per second).[2] Of the total computing capacity of the human species (that is, all human brains) combined with the computing technology the species has created, more than 10 percent is nonhuman.[3]

Rotating memories and other electromechanical computing devices have been fully replaced with electronic devices. Three-dimensional nanotube lattices are now a prevalent form of computing circuitry.

The majority of "computes" of computers are now devoted to massively parallel neural nets and genetic algorithms.

Significant progress has been made in the scanning-based reverse engineering of the human brain. It is now fully recognized that the brain comprises many specialized regions, each with its own topology and architecture of interneuronal connections. The massively parallel algorithms are beginning to be understood, and these results have been applied to the design of machine-based neural nets. It is recognized that the human genetic code does not specify the precise interneuronal wiring of any of the regions, but rather sets up a rapid evolutionary process in which connections are established and fight for survival. The standard process for wiring machine-based neural nets uses a similar genetic evolutionary algorithm.

A new computer-controlled optical-imaging technology using quantum-based diffraction devices has replaced most lenses with tiny devices that can detect light waves from any angle. These pinhead-sized cameras are everywhere.

Autonomous nanoengineered machines can control their own mobility and include significant computational engines. These microscopic machines are beginning to be applied to commercial applications, particularly in manufacturing and process control, but are not yet in the mainstream.

Education

Hand-held displays are extremely thin, very high resolution, and weigh only ounces. People read documents either on the hand-held displays or, more commonly, from text that is projected into the ever present virtual environment using the ubiquitous direct-eye displays. Paper books and documents are rarely used or accessed. Most twentieth-century paper documents of interest have been scanned and are available through the wireless network.

Most learning is accomplished using intelligent software-based simulated teachers. To the extent that teaching is done by human teachers, the human teachers are often not in the local vicinity of the student. The teachers are viewed more as mentors and counselors than as sources of learning and knowledge.

Students continue to gather together to exchange ideas and to socialize, although even this gathering is often physically and geographically remote.

All students use computation. Computation in general is everywhere, so a student's not having a computer is rarely an issue.

Most adult human workers spend the majority of their time acquiring new skills and knowledge.

Disabilities

Blind persons routinely use eyeglass-mounted reading-navigation systems, which incorporate the new, digitally controlled, high-resolution optical sensors. These systems can read text in the real world, although since most print is now electronic, print-to-speech reading is less of a requirement. The navigation function of these systems, which emerged about ten years ago, is now perfected. These automated reading-navigation assistants communicate to blind users through both speech and tactile indicators. These systems are also widely used by sighted persons since they provide a high-resolution interpretation of the visual world.

Retinal and vision neural implants have emerged but have limitations and are used by only a small percentage of blind persons.

Deaf persons routinely read what other people are saying through the deaf persons' lens displays. There are systems that provide visual and tactile interpretations of other auditory experiences such as music, but there is debate regarding the extent to which these systems provide an experience comparable to that of a hearing person. Cochlear and other implants for improving hearing are very effective and are widely used.

Paraplegic and some quadriplegic persons routinely walk and climb stairs through a combination of computer-controlled nerve stimulation and exoskeletal robotic devices.

Generally, disabilities such as blindness, deafness, and paraplegia are not noticeable and are not regarded as significant.

Communication

You can do virtually anything with anyone regardless of physical proximity. The technology to accomplish this is easy to use and ever present.

"Phone" calls routinely include high-resolution three-dimensional images projected through the direct-eye displays and auditory lenses. Three-dimensional holography displays have also emerged. In either case, users feel as if they are physically near the other person. The resolution equals or exceeds optimal human visual acuity. Thus a person can be fooled as to whether or not another person is physically present or is being projected through electronic communication. The majority of "meetings" do not require physical proximity.

Routinely available communication technology includes high-quality speech-to-speech language translation for most common language pairs.

Reading books, magazines, newspapers, and other web documents, listening to music, watching three-dimensional moving images (for example, television, movies), engaging in three-dimensional visual phone calls, entering virtual environments (by yourself, or with others who may be geographically remote), and various combinations of these activities are all done through the ever present communications Web and do not require any equipment, devices, or objects that are not worn or implanted.

The all-enveloping tactile environment is now widely available and fully convincing. Its resolution equals or exceeds that of human touch and can simulate (and stimulate) all of the facets of the tactile sense, including the sensing of pressure, temperature, textures, and moistness. Although the visual and auditory aspects of virtual reality involve only devices you have on or in your body (the direct-eye lenses and auditory lenses), the "total touch" haptic environment requires entering a virtual reality booth. These technologies are popular for medical examinations, as well as sensual and sexual interactions with other human partners or simulated partners. In fact, it is often the preferred mode of interaction, even when a human partner is nearby, due to its ability to enhance both experience and safety.

Business and Economics

Rapid economic expansion and prosperity has continued.

The vast majority of transactions include a simulated person, featuring a realistic animated personality and two-way voice communication with high-quality natural-language understanding. Often, there is no human involved, as a human may have his or her automated personal assistant conduct transactions on his or her behalf with other automated personalities. In this case, the assistants skip the natural language and communicate directly by exchanging appropriate knowledge structures.

Household robots for performing cleaning and other chores are now ubiquitous and reliable.

Automated driving systems have been found to be highly reliable and have now been installed in nearly all roads. While humans are still allowed to drive on local roads (although not on highways), the automated driving systems are always engaged and are ready to take control when necessary to prevent accidents. Efficient personal flying vehicles using microflaps have been demonstrated and are primarily computer controlled. There are very few transportation accidents.

Politics and Society

People are beginning to have relationships with automated personalities as companions, teachers, caretakers, and lovers. Automated personalities are superior to humans in some ways, such as having very reliable memories and, if desired, predictable (and programmable) personalities. They are not yet regarded as equal to humans in the subtlety of their personalities, although there is disagreement on this point.

An undercurrent of concern is developing with regard to the influence of machine intelligence. There continue to be differences between human and machine intelligence, but the advantages of human intelligence are becoming more difficult to identify and articulate. Computer intelligence is thoroughly interwoven into the mechanisms of civilization and is designed to be outwardly subservient to apparent human control. On the one hand, human transactions and decisions require by law a human agent of responsibility, even if fully initiated by machine intelligence. On the other hand, few decisions are made without significant involvement and consultation with machine-based intelligence.

Public and private spaces are routinely monitored by machine intelligence to prevent interpersonal violence. People attempt to protect their privacy with near-unbreakable encryption technologies, but privacy continues to be a major political and social issue with each individual's practically every move stored in a database somewhere.

The existence of the human underclass continues as an issue. While there is sufficient prosperity to provide basic necessities (secure housing and food, among others) without significant strain to the economy, old controversies persist regarding issues of responsibility and opportunity. The issue is complicated by the growing component of most employment's being concerned with the employee's own learning and skill acquisition. In other words, the difference between those "productively" engaged and those who are not is not always clear.

The Arts

Virtual artists in all of the arts are emerging and are taken seriously. These cybernetic visual artists, musicians, and authors are usually affiliated with humans or organizations (which in turn are comprised of collaborations of humans and machines) that have contributed to their knowledge base and techniques. However, interest in the output of these creative machines has gone beyond the mere novelty of machines being creative.

Visual, musical, and literary art created by human artists typically involve a collaboration between human and machine intelligence.

The type of artistic and entertainment product in greatest demand (as measured by revenue generated) continues to be virtual-experience software, which ranges from simulations of "real" experiences to abstract environments with little or no corollary in the physical world.

Warfare

The primary threat to security comes from small groups combining human and machine intelligence using unbreakable encrypted communication. These include (1) disruptions to public information channels using software viruses, and (2) bioengineered disease agents.

Most flying weapons are tiny—some as small as insects—with microscopic flying weapons being researched.

Health and Medicine

Many of the life processes encoded in the human genome, which was deciphered more than ten years earlier, are now largely understood, along with the information-processing mechanisms underlying aging and degenerative conditions such as cancer and heart disease. The expected life span, which, as a result of the first Industrial Revolution (1780 through 1900) and the first phase of the second (the twentieth century), almost doubled from less than forty, has now substantially increased again, to over one hundred.

There is increasing recognition of the danger of the widespread availability of bioengineering technology. The means exist for anyone with the level of knowledge and equipment available to a typical graduate student to create disease agents with enormous destructive potential. That this potential is offset to some extent by comparable gains in bioengineered antiviral treatments constitutes an uneasy balance, and is a major focus of international security agencies.

Computerized health monitors built into watches, jewelry, and clothing which diagnose both acute and chronic health conditions are widely used. In addition to diagnosis, these monitors provide a range of remedial recommendations and interventions.

Philosophy

There are prevalent reports of computers passing the Turing Test, although these instances do not meet the criteria (with regard to the sophistication of the human judge, the length of time for the interviews, etcetera) established by knowledgeable observers. There is a consensus that computers have not yet passed a valid Turing Test, but there is growing controversy on this point.

The subjective experience of computer-based intelligence is seriously discussed, although the rights of machine intelligence have not yet entered mainstream debate. Machine intelligence is still largely the product of a collaboration between humans and machines, and has been programmed to maintain a subservient relationship to the species that created it.

OKAY, I'M HERE NOW. SORRY ABOUT BEING DISTRACTED TEN YEARS AGO.

No problem. How've you been?

I'M FINE—BUSY—BUT HOLDING UP. GETTING READY FOR MY SON'S TENTH BIRTHDAY PARTY.

Oh, so you were pregnant last time we spoke.

I WASN'T SHOWING YET, BUT PEOPLE DID NOTICE AT THE WEDDING.

How's he doing?

OKAY, BUT JEREMY'S A LOT TO KEEP UP WITH.

Doesn't sound too unusual.

Anyway, I found Jeremy with this older woman, like my age, last week. Let's just say, she didn't have all of her clothes on.

Oh, really.

Turned out to be his fourth-grade teacher.

Gee, what was she doing?

Well, he'd been out sick, so she was giving him his homework assignment.

Without all her clothes on?

Oh, she had no idea.

Of course, the image transformers, I forgot.

He's not supposed to have access to those particular transformers. But he apparently got a child-block override patch from one of his friends. He won't say who.

Some things never change.

I think we have the block back on now.

So did you discuss this with his teacher?

Miss Simon? Oh God, no.

Any punishment?

Activating the child-block override is just not tolerated in our home. He's restricted from the Sensorium for a month.

That does sound serious. Sensorium? That's a virtual reality thing?

Actually, Sensorium is a brand name for the total touch environment we have. It's a new model with some improved olfactory technology. For just visual-auditory virtual reality—that's pretty much on all the time using the lenses, you don't need to use anything special.

So what does he do in the Sensorium?

Oh, kick boxing, galactic wrestling, the usual ten-year-old stuff. Lately, he's been playing doctor.

Uh oh, he sounds precocious.

I think he's just trying to test our patience.

So this incident with Miss Simon, that was in the Sensorium?

No, that was just a virtual reality phone call. Jeremy was here in the kitchen. He had Miss Simon sitting on the kitchen table.

So if he's looking at her transformed image using his virtual reality lenses, how were you able to see her?

Well, we have access to our kids' virtual reality environments up until age fourteen.

I see, so you're simultaneously in your own virtual reality environment, and those of your children?

Yes, and don't forget real reality, not that virtual reality isn't real.

Isn't that confusing, seeing and hearing all these different environments overlaying each other?

We don't hear our kids' virtual reality environments. The noise would drive us crazy, and kids need to have some privacy, too. We can only hear real reality and our own virtual reality. And, we can tune in and out of our kids' virtual visual realities. So I tuned in, and there was Miss Simon.

What else has he been punished for?

Three months ago, he was blocking our child virtual reality access. I think he got that from the same friend.

I'm not sure I blame him. I don't think I would want my mother looking in on my virtual reality all the time.

We don't look in all the time; we're really quite selective. But you have to keep track of kids nowadays. We don't have this problem with our daughter, Emily.

She's . . .

Six years old last month. She's a real sweetheart. She just devours books.

At six, that's impressive. She reads them by herself?

By herself? How else would she read them?

Well, you could read them to her.

I do that sometimes. But Emily feels I'm not accommodating enough. So she has Harry Hippo read them to her, and he does exactly what she wants, and doesn't talk back.

This all takes place in virtual reality, I assume?

OF COURSE. I WOULDN'T WANT A REAL HIPPOPOTAMUS SITTING ON MY KITCHEN TABLE.

Not with a partially clad Miss Simon there also.

IT DOES GET TO BE A CROWDED TABLE.

So when Harry Hippo reads to Emily, she follows along in her virtual book.

SHE CAN EITHER FOLLOW ALONG HERSELF, OR SHE CAN TURN THE HIGHLIGHTING ON. THE KIDS LET THEIR FAVORITE VIRTUAL FRIEND READ TO THEM, WHILE THEY WATCH THEIR VIR- TUAL BOOKS WITH THE HIGHLIGHTING FEATURE. LATER ON, THEY TURN THE HIGHLIGHT- ING OFF, AND EVENTUALLY, THEY DON'T NEED HARRY HIPPO, EITHER.

Kind of like taking off the training wheels.

RIGHT. NOW, ONE THING THAT DOES GIVE ME COMFORT IS THAT I ALWAYS KNOW WHERE MY KIDS ARE.

In virtual reality?

NO, I'M TALKING ABOUT REAL REALITY NOW. FOR EXAMPLE, I CAN SEE THAT JEREMY IS TWO BLOCKS AWAY, HEADED IN THIS DIRECTION.

An embedded chip?

THAT'S A REASONABLE GUESS. BUT IT'S NOT A CHIP EXACTLY. IT'S ONE OF THE FIRST USEFUL NANOTECHNOLOGY APPLICATIONS. YOU EAT THIS STUFF.

Stuff?

YEAH, IT'S A PASTE, TASTES PRETTY GOOD, ACTUALLY. IT HAS MILLIONS OF LITTLE COMPUTERS—WE CALL THEM TRACKERS—WHICH WORK THEIR WAY INTO YOUR CELLS.

Some of them must get passed through.

THAT'S TRUE, AND THE TRACKERS THAT GET TOO FAR AWAY FROM THE REST OF THE TRACK- ERS THAT ARE STILL IN THE BODY JUST TURN THEMSELVES OFF. THE ONES THAT STAY IN YOUR BODY COMMUNICATE WITH EACH OTHER, AND WITH THE WEB.

The wireless Web?

YES, IT'S EVERYWHERE. SO I ALWAYS KNOW WHERE MY KIDS ARE. NEAT, HUH?

So does everybody have this?

KIDS ARE REQUIRED TO, SO I GUESS EVERYONE WILL HAVE IT EVENTUALLY. MANY ADULTS DO, BUT ADULTS CAN BLOCK THE TRACKING TRANSMISSION IF THEY WISH.

Kids can't?

TRACKER BLOCKING IS SOMETHING WE REALLY DO MANAGE TO KEEP FROM OUR KIDS.

So Jeremy hasn't gotten his hands on any tracker-blocking software?

I CERTAINLY HOPE NOT. ALTHOUGH, COME TO THINK OF IT, WE DID HAVE A TRACKER LAPSE LAST YEAR. THE TECHNICIAN SAID IT WAS A TEMPORARY PROTOCOL CONFLICT. I DOUBT THAT WAS JEREMY'S DOING. BUT NOW YOU'VE GOT ME WORRIED.

I doubt Jeremy would do something like that.

I THINK YOU'RE RIGHT.

This technician was human?

NO, THE PROBLEM WASN'T THAT SERIOUS. WE JUST USED A LEVEL-B TECHNICIAN.

I see. So is your husband plugged into the tracking system?

YEAH, BUT HE BLOCKS IT A LOT, WHICH IS ANNOYING.

Well, husbands are entitled to some privacy, too, don't you think?

YES, DEFINITELY.

So, any other relatives you want to tell me about?

THERE'S MY TWENTY-FIVE-YEAR-OLD NEPHEW, STEPHEN. HE'S A BIT RECLUSIVE; I KNOW MY SISTER IS WORRIED ABOUT HIM. HE SPENDS ALMOST ALL OF HIS TIME IN EITHER TOTAL TOUCH OR IN VIRTUAL-REALITY BLOCKING-DISPLAY MODE.

That's a problem?

IT'S NOT JUST THAT HE BLOCKS OUT REAL REALITY, IT'S THAT HE SEEMS TO AVOID INTER-ACTING WITH REAL PEOPLE, EVEN IN VIRTUAL REALITY. IT SEEMS TO BE AN INCREASINGLY COMMON PROBLEM.

I guess simulated people are more accommodating.

THEY CAN BE. I MEAN, MY OWN ASSISTANTS AND COMPANIONS ARE, BUT TRY DEALING WITH OTHER PEOPLE'S ASSISTANTS, AND THAT'S A DIFFERENT MATTER. ANYWAY, MY SISTER WAS TELLING ME HOW SHE THOUGHT THAT STEPHEN WAS A CYBER VIRGIN, OR DID SHE SAY VIRTUAL VIRGIN?

Oh dear, now what was the distinction again?

YOU KNOW, A CYBER VIRGIN HAS NEVER HAD INTERCOURSE OUTSIDE OF VIRTUAL REALITY, WHEREAS A VIRTUAL VIRGIN HAS NEVER HAD INTERCOURSE WITH A REAL PERSON, EVEN IN VIRTUAL REALITY.

How about someone who has never been intimate with a real or simulated person in real or virtual reality?

HMM, WE DON'T SEEM TO HAVE A TERM FOR THAT.

So what are the statistics on this?

WELL, LET'S SEE, GEORGE WILL GET THAT FOR US.

George is your virtual assistant?

YEAH, YOU CATCH ON QUICKLY.

Gee, thanks.

SO, FOR ADULTS OVER TWENTY-FIVE, 11 PERCENT ARE VIRTUAL VIRGINS, AND 19 PERCENT ARE CYBER VIRGINS.

So I guess virtual sex is catching on. How about you and Ben?

WELL, I DEFINITELY PREFER THE REAL THING!

Real, as in . . .

REAL REALITY, RIGHT.

So you prefer intimacy in real reality, meaning you don't avoid the virtual alternative?

WELL, IT'S RIGHT THERE, I MEAN WE'D HAVE TO GO OUT OF OUR WAY TO AVOID IT. IT'S CERTAINLY CONVENIENT IF I'M TRAVELING, OR IF WE DON'T WANT TO WORRY ABOUT BIRTH CONTROL.

Or STDs.

WELL, THAT SHOULDN'T BE AN ISSUE.

Hey, you never know.

WELL, TO BE PERFECTLY HONEST, VIRTUAL SEX IS MUCH MORE SATISFACTORY IN MANY WAYS. I MEAN IT'S DEFINITELY MORE INTENSE, PRETTY INCREDIBLE ACTUALLY.

This is in the Sensorium, I assume.

YEAH, SURE. THIS RECENT MODEL HAS REALLY ADDRESSED THE OLFACTORY ISSUE.

Meaning it has an olfactory capability?

RIGHT. IT'S A LITTLE DIFFERENT THAN THE OTHER SENSES, THOUGH. WITH THE VISUAL AND AUDITORY SENSE, JUST PLAIN OLD UBIQUITOUS VIRTUAL REALITY IS EXTREMELY ACCURATE. IN THE SENSORIUM, WE GET THE TACTILE ENVIRONMENT, WHICH ALSO PROVIDES AN EXTREMELY LIFELIKE RE-CREATION. BUT WE CAN'T DO THAT YET WITH THE SENSE OF SMELL. SO THE SENSORIUM 2000 HAS PROGRAMMED SCENTS, WHICH YOU CAN CHOOSE, OR THAT ARE AUTOMATICALLY SELECTED IN THE COURSE OF AN EXPERIENCE. THEY'RE STILL PRETTY EFFECTIVE.

How do you feel about your husband interacting sexually with a simulated partner?

YOU MEAN, A SIMULATED PERSON IN VIRTUAL REALITY?

Yeah, in virtual reality or in the Sensorium.

THAT'S FINE. I HAVE NO PROBLEM WITH THAT.

You don't mind?

THERE'S REALLY NO WAY I COULD KEEP TRACK OF IT.

Virtual lipstick on his collar?

YEAH, RIGHT, ON HIS VIRTUAL COLLAR. VIRTUAL SEX WITH SIMULATED PARTNERS IS GENER-ALLY ACCEPTED NOWADAYS. IT'S REALLY REGARDED AS A FORM OF FANTASY—IT'S JUST ASSISTED FANTASY.

And if the partner's a real person in virtual reality?

I'D BREAK HIS LEGS.

His virtual legs?

THAT'S NOT WHAT I HAD IN MIND.

So what's the difference between a real person in virtual reality and a simulated person?

AS SENSUAL PARTNERS?

Right.

OH, THERE'S A DIFFERENCE—THE SIMULATED PARTNERS ARE PRETTY GOOD, BUT IT'S JUST NOT THE SAME.

Sounds like you've had some experience with this yourself.

PRETTY NOSY, AREN'T YOU?

All right, I'll change the subject. Let's see, uh, what's happening with encryption?

WE HAVE VERY STABLE THOUSAND-BIT CODES. IT'S NOT PRACTICAL TO BREAK THEM.

What about with a quantum computer?

THE QUANTUM COMPUTERS DON'T SEEM TO BE STABLE WITH MORE THAN A FEW HUNDRED QU-BITS.

Sounds like communication is pretty secure.

I WOULD SAY SO. BUT SOME PEOPLE ARE PARANOID ABOUT THE THIRD-PARTY KEYS.

So the authorities have keys?

Of course.

Well, can't you just put another layer of encryption without a key over the official layer?

God, no.

Why is that so hard?

Oh, it's not difficult technically. It's just very illegal, certainly since October 2013.

2013?

We managed to get through the first decade of this century without too serious a problem. But things got out of control in the Oklahoma incident.

Oklahoma again. So this was a software virus?

No, not a software virus, a biological virus. A disgruntled—I would say demented—student, actually a former student at the university there. There were reports that he was linked to the Remember York movement, but the RY discussion leaders vehemently denied any responsibility.

Remember York?

Well, this incident occurred on the two hundredth anniversary of the York trials.

Oh, you mean the 1813 trial of the Luddites?

Yes, except that most antitechnologists don't like the term *Luddite* anymore; they feel that the somewhat silly image of Ned Ludd belittles the serious nature of their movement. Beyond that, the best evidence suggests that Ludd never existed.

But there was a trial in 1813.

Yes, which resulted in many of the gang members accused of wrecking the textile machines being hanged or exiled.

So RY is an organized movement?

Oh, I wouldn't say that. It is more of a web discussion group, and this young man apparently had participated in some of those discussions. But the RY people are basically nonviolent. They were distressed that Roberts had associated himself with them.

Roberts was the perpetrator?

YEAH, CONVICTED ON ALL COUNTS. BUT ASIDE FROM THIS ONE DISTURBED INDIVIDUAL, I WOULD SAY IT WAS REALLY A SCREWUP BY THE BWA.

BWA?

BIOWARFARE AGENCY.

So this was a virus that was unleashed?

YES, JUST A STANDARD MODIFIED FLU VIRUS, ALTHOUGH THERE WAS A TWIST. IT HAD A GREATLY INCREASED MUTATION RATE, WHICH ACCELERATED ITS EVOLUTION AT SEVERAL LEVELS. ONE FORM OF THE VIRUS'S EVOLUTION ONLY TOOK PLACE DURING AN INFECTION. THIS, TOGETHER WITH A TIME-BOMB PROGRAM IN THE VIRUS'S DNA, CAUSED ULTRA-RAPID VIRAL REPRODUCTION AFTER A FEW HOURS OF INFECTION. THIS LITTLE COMPLEXITY DELAYED THE DEVELOPMENT OF AN ANTIDOTE FOR FORTY-EIGHT HOURS. BUT THAT WASN'T THE WORST OF IT. AFTER TWENTY-FOUR HOURS OF REPLICATING THE ANTIDOTE, THE BWA DISCOVERED THAT ANOTHER BIOLOGICAL AGENT HAD INFECTED THE BATCHES, SO THEY HAD TO START OVER. AND THEN, THERE WEREN'T ENOUGH REPLICATION STATIONS, SO THEY HAD TO CLEAN OUT THE ONES THEY HAD JUST USED, AND GO FROM THERE. FORTY-EIGHT HOURS WERE LOST IN THIS FIASCO, AND SIXTEEN THOUSAND PEOPLE DIED. WELL, IF THINGS HAD BEEN DELAYED FOR ANOTHER TWENTY-FOUR HOURS, IT WOULD HAVE BEEN FAR WORSE. IT WAS A BIG ISSUE IN THE MIDTERM ELECTIONS OF 2014. THERE'VE BEEN A LOT OF CHANGES SINCE THEN.

The third-party keys?

YEAH, THOSE EXISTED BEFORE. BUT SINCE 2013 THE LAWS AGAINST KEYLESS ENCRYPTION CODES HAVE BEEN RIGOROUSLY ENFORCED.

What else has changed?

THERE ARE PLENTY OF ANTIVIRAL REPLICATION STATIONS NOW. AND WE ALL HAVE THESE CUTE LITTLE GAS MASKS.

That little thimble is a gas mask?

YEAH, WELL, IT UNFOLDS LIKE THIS. IT'S SMALL, SO WE'RE ENCOURAGED TO KEEP THEM CLOSE AT HAND. IT'S ACTUALLY A VIRAL SCREEN MASK. OCCASIONALLY, WE'RE TOLD TO PUT THEM ON, BUT GENERALLY IT'S ONLY FOR A FEW HOURS. SINCE 2013, THERE HAVE BEEN ONLY FALSE ALARMS.

So I guess the security agencies have been hard at work.

AS WILL ROGERS USED TO SAY, "YOU CAN'T SAY THAT CIVILIZATION DON'T ADVANCE, FOR IN EVERY WAR THEY KILL YOU IN A NEW WAY."

2013 sounds tragic, and frightening. As centuries go, however, it doesn't sound like you're doing too badly. In the twentieth century, we knew how to have disasters.

YEAH, FIFTY MILLION PEOPLE DIED IN WORLD WAR II.

Indeed.

IT'S TRUE THAT THE CENTURY SO FAR HAS SEEN MUCH LESS BLOODSHED. BUT THE OTHER SIDE OF THE COIN IS THAT THE TECHNOLOGIES ARE SO MUCH MORE POWERFUL TODAY. IF SOMETHING DID GO WRONG, THINGS COULD SPIRAL OUT OF CONTROL VERY QUICKLY. WITH BIOENGINEERING, FOR EXAMPLE, IT FEELS A LITTLE LIKE ALL TEN BILLION OF US ARE STANDING IN A ROOM UP TO OUR KNEES IN A FLAMMABLE FLUID, WAITING FOR SOME-ONE—ANYONE—TO LIGHT A MATCH.

But it sounds like a lot of fire extinguishers have been installed.

YEAH, I JUST HOPE THEY WORK.

You know, I've been concerned about the downside of bioengineering for well over a decade now.

BUT YOU DIDN'T WRITE ABOUT IT IN THE *AGE OF INTELLIGENT MACHINES,* WHICH YOU WROTE IN THE LATE 1980S.

That was a conscious decision. I didn't want to give the wrong person any ideas.

AND IN 1999?

Oh, the cat's out of the bag now.

YEAH, WELL, WE'VE BEEN SCURRYING AFTER THE DESCENDANTS OF THAT CAT FOR THE LAST COUPLE OF DECADES, TRYING TO KEEP THEM FROM CAUSING TOO MUCH MISCHIEF.

Just wait until the nanopathogens get going.

FORTUNATELY, THEY'RE NOT SELF-REPLICATING.

Not yet.

I SUPPOSE THAT'S COMING, TOO, BUT THE TRACKER PASTE AND THE OTHER FEW NANO-TECHNOLOGY APPLICATIONS THAT ARE OUT THERE TODAY ARE MADE USING X-RAY LI-THOGRAPHY AND OTHER CONVENTIONAL MANUFACTURING TECHNIQUES.

Well, enough of disasters, what are you up to tonight?

I'M GIVING A LECTURE ON MY EXPERIENCE LAST WEEK AS A TURING TEST JUDGE.

I assume the computer lost.

Yes, she did. But it wasn't the slam dunk I thought it would be. At the beginning, I was thinking, Gee, this is a lot harder than I had expected. I really can't
tell who the computer is, or who the human foil is. After about twenty minutes, it did become fairly clear to me, and I'm glad I had enough time. A few of
the other judges just had no idea, but they weren't very sophisticated.

I guess your communications background came in handy.

Actually it was more my Mommy background. I became suspicious when Sheila—
she was the computer—started talking about how angry she was at her daughter. That was not convincing for me. She just wasn't sympathetic enough.

How about George, how would he fare in a Turing Test?

Oh, I wouldn't want to subject George to that.

You're concerned about his feelings?

I guess you could say that. I kind of go back and forth. Sometimes, I think I'm
not. But then when I'm interacting with him, I find myself acting as if he has
feelings. And, sometimes, I look forward to telling him something I've experienced, particularly if we're working on it together.

I see you picked a male assistant.

Sure, your prediction that women would prefer female personalities was another miss.

That prediction was for 2009, not 2019.

I'm glad you clarified that. Come to think of it, I did use a female personality
in 2009, but they weren't very realistic then. Anyway, I have to get going to my
lecture. But if I think of anything else interesting to tell you, I'll have my virtual assistant contact yours.

Hey, I don't have one, remember I'm stuck in 1999.

Too bad. I guess I'll just have to visit you myself then.

"Smaller, more powerful chips allow me to have a smaller head."

CHAPTER ELEVEN

2029

I'm as fond of my body as anyone else, but if I can be 200 with a body of silicon, I'll take it.

—Danny Hillis

The Computer Itself

A $1,000 unit of computation (in circa-1999 dollars) has the computing capacity of approximately 1,000 human brains (1,000 times 20 million billion—that is, 2 times 10^{19}—calculations per second).

Of the total computing capacity of the human species (that is, all human brains) combined with the computing technology humans initiated the creation of, more than 99 percent is nonhuman.[1]

The vast majority of "computes" of nonhuman computing is now conducted on massively parallel neural nets, much of which is based on the reverse engineering of the human brain.

Many—but less than a majority—of the specialized regions of the human brain have been "decoded" and their massively parallel algorithms have been deciphered. The number of specialized regions, amounting to hundreds, is greater than was anticipated twenty years earlier. The topologies and architectures of those regions that have been successfully reverse engineered are used in machine-based neural nets. The machine-based nets are substantially faster and have greater computing and memory capacities and other refinements compared to their human analogues.

Displays are now implanted in the eyes, with a choice of permanent implants or removable implants (similar to contact lenses). Images are projected directly onto the retina providing the usual high-resolution three-dimensional overlay on the physical world. These implanted visual displays also act as cameras to capture visual images and thus are both input and output devices.

Cochlear implants, originally used just for the hearing impaired, are now ubiquitous. These implants provide auditory communication in both directions between the human user and the worldwide computing network.

Direct neural pathways have been perfected for high-bandwidth connection to the human brain. This allows bypassing certain neural regions (for example, visual pattern recognition, long-term memory) and augmenting or replacing the functions of these regions with computing performed either in a neural implant or externally.

A range of neural implants is becoming available to enhance visual and auditory perception and interpretation, memory, and reasoning.

Computing processes can be personal (accessible by one individual), shared (accessible to a group), or universal (accessible to everyone), at the user's option.

Three-dimensional projected holographic displays are everywhere.

Microscopic nanoengineered robots now have microbrains with the computing speed and capacity of the human brain. They are widely used in industrial applications and are beginning to be used in medical applications (see "Health and Medicine").

Education

Human learning is primarily accomplished using virtual teachers and is enhanced by the widely available neural implants. The implants improve memory and perception, but it is not yet possible to download knowledge directly. Although enhanced through virtual experiences, intelligent interactive instruction, and neural implants, learning still requires time-consuming human experience and study. This activity comprises the primary focus of the human species.

Automated agents are learning on their own without human spoon-feeding of information and knowledge. Computers have read all available human and machine-generated literature and multimedia material, which includes written, auditory, visual, and virtual experience works.

Significant new knowledge is created by machines with little or no human intervention. Unlike humans, machines easily share knowledge structures with one another.

Disabilities

The prevalence of highly intelligent visual navigation devices for the blind, speech-to-print display devices for the deaf, nerve stimulation, intelligent orthotic prosthetics for the physically disabled, and a variety of neural implant technologies has essentially eliminated the handicaps associated with most disabilities. Sensory-enhancement devices are in fact used by most of the population.

Communication

In addition to the ubiquitous, three-dimensional virtual environments, there has been significant refinement to three-dimensional holographic technology for visual communication. There is also projected sonic communication for precisely placing sounds in three-dimensional space. Similar to virtual reality, much of what is seen and heard in "real" reality also has no physical counterpart. Thus family members can be sitting around the living room enjoying one another's company without being physically proximate.

In addition, there is extensive use of communication using direct neural connections. This allows virtual, all-enveloping tactile communication to take place without entering a "total touch enclosure," as was necessary ten years earlier.

The majority of communication does not involve a human. The majority of communication involving a human is between a human and a machine.

Business and Economics

The human population has leveled off in size at around 12 billion real persons. The basic necessities of food, shelter, and security are available for the vast majority of the human population.

Human and nonhuman intelligences are primarily focused on the creation of knowledge in its myriad forms, and there is significant struggle over intellectual property rights, including ever increasing levels of litigation.

There is almost no human employment in production, agriculture, and transportation. The largest profession is education. There are many more lawyers than doctors.

Politics and Society

Computers appear to be passing forms of the Turing Test deemed valid by both human and nonhuman authorities, although controversy on this point persists. It is difficult to cite human capabilities of which machines are incapable. Unlike human competence, which varies greatly from person to person, computers consistently perform at optimal levels and are able to readily share their skills and knowledge with one another.

A sharp division no longer exists between the human world and the machine world. Human cognition is being ported to machines, and many machines have personalities, skills, and knowledge bases derived from the reverse engineering of human intelligence. Conversely, neural implants based on machine intelligence are providing enhanced perceptual and cognitive functioning to humans.

Defining what constitutes a human being is emerging as a significant legal and political issue.

The rapidly growing capability of machines is controversial, but there is no effective resistance to it. Since machine intelligence was initially designed to be subservient to human control, it has not presented a threatening "face" to the human population. Humans realize that disengaging the now human-machine civilization from its dependence on machine intelligence is not possible.

Discussion of the legal rights of machines is growing, particularly those of machines that are independent of humans (those not embedded in a human brain). Although not yet fully recognized by law, the pervasive influence of machines in all levels of decision making is providing significant protection to machines.

The Arts

Cybernetic artists in all of the arts—musical, visual, literary, virtual experience, and all others—no longer need to associate themselves with humans or organizations that include humans. Many of the leading artists are machines.

Health and Medicine

Progress continues in understanding and ameliorating the effects of aging as a result of a thorough understanding of the information-processing processes controlled through the genetic code. The life expectancy of humans continues to increase and is now around 120 years. Significant attention is being paid to the psychological ramifications of a substantially increased human life span.

There is growing recognition that continuing extensions to the human life span will involve further use of bionic organs, including portions of the brain. Nanobots are being used as scouts, to a limited extent as repair agents in the bloodstream, and as building blocks for bionic organs.

Philosophy

Although computers routinely pass apparently valid forms of the Turing Test, controversy persists about whether or not machine intelligence equals human intelligence in all of its diversity. At the same time, it is clear that there are many ways in which machine intelligence is vastly superior to human intelligence. For reasons of political sensitivity, machine intelligences generally do not press the point of their superiority. The distinction between human and machine intelligence is blurring as machine intelligence is increasingly derived from the design of human intelligence, and human intelligence is increasingly enhanced by machine intelligence.

The subjective experience of machine intelligence is increasingly accepted, particularly since "machines" participate in this discussion.

Machines claim to be conscious and to have as wide an array of emotional and spiritual experiences as their human progenitors, and these claims are largely accepted.

I HOPE YOU'RE HAVING A GOOD TIME MAKING ALL THESE PREDICTIONS.

This part of the book is a bit more fun to write—at least there are fewer references to look up. And I don't have to worry about being embarrassed for at least a few decades.

WELL, IT MIGHT BE EASIER IF YOU JUST ASKED ME FOR MY IMPRESSIONS.

Yes, I was just getting to that. But I must say, you look very well.

FOR AN OLD LADY.

I wasn't thinking old. But you don't look anywhere near fifty. More like thirty-five.

YES, WELL, FIFTY ISN'T AS OLD AS IT USED TO BE.

We feel that way in 1999, too.

IT'S STILL HELPFUL TO EAT RIGHT. WE ALSO HAVE A FEW TRICKS YOU DIDN'T HAVE.[2]

Nanoengineered bodies?

NO, NOT EXACTLY, NANOTECHNOLOGY IS STILL FAIRLY LIMITED. BIOENGINEERING HAS CERTAINLY HELPED THE MOST. AGING HAS BEEN DRAMATICALLY SLOWED. MOST DISEASES CAN BE PREVENTED OR REVERSED.

So nanotechnology is still fairly primitive?

I'D SAY SO. I MEAN, WE DO HAVE NANOBOTS IN OUR BLOODSTREAMS, BUT THEY'RE PRIMARILY DIAGNOSTIC. SO IF ANYTHING STARTS TO GO WRONG, WE CATCH IT VERY EARLY.

So if a nanobot discovers a microscopic infection or other problem developing, what does it do, just start yelling?

YEAH, THAT'S ABOUT IT. I DON'T THINK WE'D TRUST IT TO DO MUCH ELSE. IT YELLS TO THE WEB, AND THEN THE PROBLEM GETS TAKEN CARE OF WHEN WE SIT DOWN FOR OUR NEXT DAILY SCAN.

A three-dimensional scan?

OF COURSE, WE STILL HAVE THREE-DIMENSIONAL BODIES.

This is a diagnostic scan?

THE SCAN HAS A DIAGNOSTIC FUNCTION. BUT IT'S ALSO REMEDIAL. THE SCANNER CAN APPLY SUFFICIENT ENERGY TO A SMALL THREE-DIMENSIONAL SET OF POINTS TO DESTROY A COLONY OF PATHOGENS OR PROBLEMATICAL CELLS BEFORE THEY GET OUT OF HAND.

Is this an electromagnetic energy beam, or a particle beam, or what?

WELL, GEORGE CAN EXPLAIN IT BETTER THAN I CAN. AS I UNDERSTAND IT, IT HAS TWO ENERGY BEAMS THAT ARE BENIGN BY THEMSELVES, BUT CAUSE PARTICLE EMISSIONS AT THE POINT AT WHICH THEY CROSS. I'LL ASK GEORGE NEXT TIME I SEE HIM.

When's that going to be?

OH, JUST AS SOON AS I GET DONE WITH YOU.

You're not rushing me, are you?

OH, THERE'S NO HURRY. IT'S ALWAYS A GOOD IDEA TO BE PATIENT.

Hmmm. So when was the last time the two of you were together?

A FEW MINUTES AGO.

I see. Sounds like your relationship has developed.

OH, IT HAS. HE TAKES VERY GOOD CARE OF ME.

Last time we talked, you weren't sure whether he had any feelings.

THAT WAS A LONG TIME AGO. GEORGE IS A DIFFERENT PERSON EVERY DAY. HE JUST GROWS AND LEARNS CONSTANTLY. HE DOWNLOADS WHATEVER KNOWLEDGE HE WANTS FROM THE WEB AND IT BECOMES PART OF HIM. HE'S SO SMART AND INTENSE, AND VERY SPIRITUAL.

I'm awfully happy for you. But how does Ben feel about you and George?

HE WASN'T TOO CRAZY ABOUT IT, THAT'S FOR SURE.

But you've worked it out?

WE'VE WORKED IT OUT, ALL RIGHT. WE BROKE UP THREE YEARS AGO.

I'm sorry to hear that.

YEAH, WELL, SEVENTEEN YEARS IS DEFINITELY ABOVE AVERAGE, AS MARRIAGES GO THESE DAYS.

It must have been hard on the kids.

THAT'S TRUE. BUT WE BOTH HAVE DINNER WITH EMILY JUST ABOUT EVERY NIGHT.

You both have dinner with Emily, but not with each other?

EMILY CERTAINLY DOESN'T WANT TO HAVE DINNER WITH US TOGETHER—THAT WOULDN'T BE VERY COMFORTABLE, NOW WOULD IT? SO SHE HAS DINNER WITH US APART.

I see, the good old kitchen table. Now that you don't have to deal with Harry Hippo or Miss Simon, there's room for you and Ben and Emily, but you and Ben don't have to actually see each other.

ISN'T VIRTUAL REALITY GREAT?

Yeah, but too bad people can't touch each other without going into the Sensorium.

ACTUALLY, SENSORIUM WENT OUT OF BUSINESS.

Okay, then, total touch.

WE DON'T NEED TO GO INTO A TOTAL TOUCH ENVIRONMENT ANYMORE, NOT SINCE THE SPINAL IMPLANTS BECAME AVAILABLE.

So these implants add the tactile environment . . .

TO THE UBIQUITOUS VISUAL AND AUDITORY ENVIRONMENTS WE'VE HAD FOR MANY YEARS WITH VIRTUAL REALITY, THAT'S RIGHT.

Sounds like the implants must be pretty popular.

NO, THEY'RE FAIRLY NEW. ALMOST EVERYONE HAS THE VISUAL AND AUDITORY ENVIRON-MENTS NOW, EITHER AS IMPLANTS OR AT LEAST AS VISUAL AND SONIC LENSES. BUT THE TACTILE IMPLANTS HAVEN'T QUITE CAUGHT ON YET.

Yet you have them?

YEAH, THEY'RE REALLY FABULOUS. THERE ARE A FEW GLITCHES, BUT I LIKE BEING ON THE CUTTING EDGE. IT WAS SUCH A HASSLE HAVING TO USE A TOTAL TOUCH ENVIRONMENT.

Now I can understand how implants could simulate your sense of touch, by gen-erating the nerve impulses that correspond to a particular set of tactile stimuli. But the total touch environments also provided force feedback, so if you're touching a virtual person, you don't end up sticking your hand through her body.

WELL, SURE, BUT WE DON'T MOVE OUR PHYSICAL BODIES IN VIRTUAL REALITY—

You move your virtual body, of course. And the virtual reality system prevents you from moving your virtual hand through a barrier—like someone else's vir-tual body—in the virtual environment. This all happens using the implants?

Right.

So you could be sitting here talking to me in real reality, while at the same time getting intimate with George in virtual reality, and with full tactile realism?

We call it tactile virtualism, but you've got the idea. However, the tactile separation between real and virtual reality is not perfect. I mean, this is still a new technology. So if George and I got too passionate, I think you'd notice.

That's too bad.

It's not a problem, though, in general, since I attend most meetings with a virtual body, anyway. So when I get restless in these interminable meetings on the census project, I can spend a few private moments with George . . .

Using yet another virtual body?

Exactly.

And the tactile separation problem between real reality and one of your virtual realities isn't a problem with two virtual bodies.

Not really, but sometimes people catch me smiling a lot.

You mentioned glitches . . .

Sometimes I feel like something or someone is touching me, but it might just be my imagination.

It's probably just a worker from the neural implant company remotely testing out the equipment.

Hmmm.

So you're working on the census?

It's supposed to be an honor. I mean it's like the hot issue right now. But it's just endless politics. And endless meetings.

Well, the census has always used the most cutting-edge technology. Electrical data processing got its start with the 1890 U.S. census, you know.

Tell me about it. That gets mentioned at least three times every meeting. But the issue's not technology.

It's . . .

Who's a person. There are proposals to start counting virtual persons of at least human level, but there's no end of problems with coming up with a viable

PROPOSAL. VIRTUAL PERSONS ARE NOT SO READILY COUNTABLE AND DISTINCT, SINCE THEY CAN COMBINE WITH ONE ANOTHER, OR SPLIT UP INTO MULTIPLE APPARENT PERSONALITIES.

Why don't you just count machines that were derived from specific persons?

THERE ARE SOME CYBERNETIC PERSONALITIES WHO CLAIM THAT THEY USED TO BE A PARTICULAR PERSON, BUT THEY'RE REALLY JUST PERSONALITY EMULATIONS. THE COMMISSION JUST DIDN'T THINK IT WAS APPROPRIATE.

I would agree—personality emulation just doesn't cut it. It should be the result of a full neural scan.

PERSONALLY, I'VE BEEN LEANING TO EXPANDING THE DEFINITION, BUT I'VE HAD DIFFICULTY COMING UP WITH A COHERENT METHODOLOGY. THE COMMISSION DID AGREE TO LOOK AT THE PROBLEM AGAIN WHEN THE NEURAL SCANS ARE EXPANDED TO A MAJORITY OF NEURAL REGIONS. IT'S A TOUGH ISSUE, THOUGH. WE DO HAVE PEOPLE WHO HAVE THE VAST MAJORITY OF THEIR MENTAL COMPUTES TAKING PLACE IN THEIR NANOTUBE IMPLANTS. BUT THE POLITICS SEEMS TO REQUIRE AT LEAST SOME UNENHANCED ORIGINAL SUBSTRATE TO BE COUNTED.

Original substrate? You mean human neurons?

RIGHT. IF YOU DON'T REQUIRE SOME NEURON-BASED THINKING, IT JUST GETS IMPOSSIBLE TO COUNT DISTINCT MINDS. YET SOME OF THE MACHINES DO MANAGE TO GET COUNTED. THEY SEEM TO ENJOY ESTABLISHING A HUMAN IDENTITY AND PASSING FOR A HUMAN. IT'S A BIT OF A GAME.

There must be legal benefits to having a recognized human identity.

THERE'S KIND OF A STANDOFF. THE OLD LEGAL SYSTEM STILL REQUIRES A HUMAN AGENT OF RESPONSIBILITY. BUT THE SAME ISSUE OF WHO OR WHAT IS HUMAN COMES UP IN THE LEGAL CONTEXT. ANYWAY, SO-CALLED HUMAN DECISIONS ARE HEAVILY INFLUENCED BY THE IMPLANTS. AND THE MACHINES DON'T IMPLEMENT SIGNIFICANT DECISIONS WITHOUT THEIR OWN REVIEW. BUT I SUPPOSE YOU'RE RIGHT; THERE ARE SOME BENEFITS TO BEING COUNTED.

How about using a Turing Test as a means of counting?

THAT WOULD NEVER DO. FIRST OF ALL, IT WOULDN'T BE MUCH OF A SCREEN. FURTHERMORE, YOU'D HAVE THE SAME PROBLEM AGAIN IN SELECTING A HUMAN JUDGE TO CONDUCT THE TURING TEST. AND YOU'D STILL HAVE THE COUNTING ISSUE. TAKE GEORGE, FOR EXAMPLE. HE'S GREAT AT IMPRESSIONS. USUALLY, RIGHT AFTER DINNER, HE'LL ENTERTAIN ME WITH SOME PERSONALITY HE'S CONCOCTED. HE COULD SUBMIT THOUSANDS OF PERSONALITIES IF HE WANTED TO.

Speaking of George, doesn't he want to be counted?

OH, I THINK HE SHOULD BE. HE'S SO MUCH WISER AND GENTLER THAN ANYONE ON THE COMMISSION. I GUESS THAT'S WHY I'VE WANTED TO EXPAND THE DEFINITION. GEORGE COULD MANAGE TO ESTABLISH THE REQUISITE IDENTITY ORIGIN IF HE WANTED TO. BUT HE REALLY DOESN'T CARE ABOUT IT.

He seems to care mostly about you.

HMMM. THAT COULD BE IT.

You sound a little frustrated with the commission.

WELL, I CAN UNDERSTAND THEIR NEED TO BE CAUTIOUS. I JUST FEEL THAT THEY'RE UN-DULY INFLUENCED BY THE RY GROUPS.

The Luddites, I mean, Remember York . . .

EXACTLY. I AM SYMPATHETIC TO A LOT OF THE YORK CONCERNS. BUT LATELY THEY'VE TAKEN STRIDENT POSITIONS AGAINST NEURAL IMPLANTS, WHICH IS JUST TOO RIGID. THEY'RE ALSO OPPOSED TO ANY OF THE NEURAL SCANNING RESEARCH.

So they're influencing the census commission to keep a conservative definition of who can be counted as a human?

I'D SAY SO. THE COMMISSION DENIES IT, BUT THERE'S A GROWING CONSENSUS THAT THE YORK PEOPLE HAVE TOO MUCH OF A VOICE THERE. THE COMMISSION DIRECTOR'S BROTHER WAS ACTUALLY A MEMBER OF THE FLORENCE MANIFESTO BRIGADE.

Florence? Isn't that where they locked up Kaczynski?

THAT'S RIGHT—FLORENCE, COLORADO. THE FLORENCE MANIFESTO WAS SMUGGLED OUT BY ONE OF THE GUARDS BEFORE KACZYNSKI'S DEATH. IT'S BECOME A KIND OF BIBLE FOR THE MORE STRIDENT YORK FACTIONS.

These are violent groups?

GENERALLY, NO. VIOLENCE WOULD BE UTTERLY FUTILE. OCCASIONALLY THERE ARE VIO-LENT LONERS, OR SMALL GROUPS, WHO CLAIM TO BE PART OF THE FM BRIGADE, BUT THERE'S NO EVIDENCE OF ANY BROAD CONSPIRACY.

So what's in the Florence Manifesto?

DESPITE IT HAVING BEEN WRITTEN ALL IN LONGHAND USING A PENCIL, IT WAS A RATHER ARTICULATE AND EFFECTIVE DOCUMENT, PARTICULARLY WITH REGARD TO THE NANO-PATHOGEN CONCERN.

So what is the concern with nanopathogens?

ACTUALLY, I JUST ATTENDED A CONFERENCE ON THAT.

You attended virtually?

THAT'S USUALLY THE WAY I ATTEND CONFERENCES NOWADAYS. ANYWAY, THE CONFERENCE SESSIONS OVERLAPPED THE COMMISSION MEETINGS, SO I HAD NO CHOICE.

You can attend more than one meeting at a time?

IT DOES GET A LITTLE CONFUSING. IT'S KIND OF POINTLESS, THOUGH, TO JUST SIT IN A LONG MEETING AND NOT DO SOMETHING USEFUL WITH YOUR TIME.

I agree. So, what was the view of the conference?

NOW THAT THE BIOPATHOGEN CONCERN IS ABATING—GIVEN THE NANOPATROL AND SCANNER TECHNOLOGIES, AND ALL—THERE IS MORE ATTENTION BEING PAID TO THE NANOPATHOGEN THREAT.

How serious is it?

IT HASN'T BEEN A BIG PROBLEM YET. THERE WAS A WORKSHOP ON A RECENT PHENOMENON OF NANOPATROLS THAT HAVE RESISTED THE COMMUNICATION PROTOCOLS, AND THAT DID SET OFF A FEW ALARMS. BUT THERE'S NOTHING LIKE YOU HAD IN 1999 WITH OVER 100,000 PEOPLE DYING EACH YEAR FROM ADVERSE REACTIONS TO PHARMACEUTICAL DRUGS. AND THAT'S WHEN THEY WERE PRESCRIBED AND TAKEN CORRECTLY.

And drugs in 2029?

DRUGS TODAY ARE GENETICALLY ENGINEERED SPECIFICALLY FOR THE INDIVIDUAL'S OWN DNA COMPOSITION. INTERESTINGLY, THE MANUFACTURING PROCESS THAT'S USED IS BASED ON THE PROTEIN-FOLDING WORK THAT WAS ORIGINALLY DESIGNED FOR THE NANOPATROLS. IN ANY EVENT, DRUGS ARE INDIVIDUALLY TAILORED AND TESTED IN A HOST SIMULATION BEFORE INTRODUCING ANY SIGNIFICANT VOLUME TO THE ACTUAL HOST'S BODY. SO ADVERSE REACTIONS ON A MEANINGFUL SCALE ARE QUITE RARE.

So there isn't much concern with nanopathogens?

OH, I WOULDN'T SAY THAT. THERE WAS QUITE A BIT OF CONCERN EXPRESSED ABOUT SOME OF THE RECENT SELF-REPLICATION RESEARCH.

There should be.

BUT THE ENVIRONMENT RESTRUCTURING PROPOSALS SEEM TO REQUIRE IT.

Well, don't say I didn't warn you.

I'LL KEEP THAT IN MIND, NOT THAT I HAVE MUCH INFLUENCE ON THE ISSUE.

Your work is mostly on the census issue?

YEAH, FOR THE LAST FIVE YEARS ANYWAY. I SPENT THREE YEARS BASICALLY GOING THROUGH THE COMMISSION'S STUDY GUIDE, SO I COULD BE QUALIFIED TO SIT IN ON THE COMMISSION MEETINGS, ALTHOUGH I STILL DON'T HAVE A VOTE.

So you had a three-year leave to study?

IT FELT LIKE I WAS BACK IN COLLEGE. AND LEARNING WAS JUST AS TEDIOUS AS IT WAS THEN.

Don't the neural implants help?

OH, SURE, THERE'S NO WAY I COULD HAVE GOTTEN THROUGH IT OTHERWISE. UNFORTUNATELY, I STILL CAN'T JUST DOWNLOAD THE MATERIAL, NOT THE WAY GEORGE CAN. THE IMPLANT PREPROCESSES THE INFORMATION, AND FEEDS ME THE PREPROCESSED KNOWLEDGE STRUCTURES QUICKLY. BUT IT'S OFTEN DISCOURAGING; IT JUST TAKES SO LONG. GEORGE HAS BEEN A BIG HELP, THOUGH. HE KIND OF WHISPERS TO ME WHEN I'M PUZZLED ABOUT SOMETHING.

So the three-year study leave is over now?

ABOUT A YEAR AGO, THE COMMISSION MEETINGS GOT PRETTY INTENSE, AND I'VE FOCUSED ON THAT. NOW WITH THE CENSUS ONLY A YEAR AWAY, WE'RE WORKING ON IMPLEMENTATION. SO ASIDE FROM THE LAWSUIT, THAT'S PRETTY MUCH IT.

Lawsuit?

OH, JUST A ROUTINE INTELLECTUAL PROPERTY DISPUTE. MY PATENT ON AN ENHANCED EVOLUTIONARY PATTERN-RECOGNITION ALGORITHM FOR NANOPATROL DETECTION OF CELL IMBALANCES WAS ATTACKED WITH A PRIOR ART CITATION. I HAPPENED TO MENTION IN ONE OF THE DISCUSSION GROUPS THAT I THOUGHT SEVERAL OF THE PATENT CLAIMS WERE BEING INFRINGED, AND NEXT THING I KNEW I GOT HIT WITH A DECLARATORY JUDGMENT SUIT FROM THE NANOPATROL INDUSTRY.

I didn't know you did work on nanopatrols.

TO BE PERFECTLY HONEST, IT WAS GEORGE'S INVENTION, BUT HE NEEDED A RESPONSIBLE AGENT.

Since he has no standing.

IT'S TRUE, THERE ARE STILL SOME LIMITATIONS WHEN YOU CAN'T ESTABLISH YOUR HUMAN ORIGIN.

So how's this going to get resolved?

IT'S UP BEFORE THE MAGISTRATE NEXT MONTH.

It can be rather frustrating taking these technical issues to court.

OH, THIS MAGISTRATE KNOWS HIS STUFF. HE'S A RECOGNIZED EXPERT ON NANOPATROL PATTERN RECOGNITION.

Doesn't sound like the courts I know.

THE EXPANSION OF THE MAGISTRATE SYSTEM HAS BEEN A VERY POSITIVE DEVELOPMENT. IF WE WERE LIMITED TO JUST THE HUMAN JUDGES . . .

Oh, so the magistrate is—

A VIRTUAL INTELLIGENCE, YES.

So the machines do have some legal standing.

OFFICIALLY, THE VIRTUAL MAGISTRATES ARE AGENTS OF THE HUMAN JUDGE IN CHARGE OF THAT COURT, BUT THE MAGISTRATES MAKE MOST OF THE DECISIONS.

I see, sounds like these magistrates are pretty influential.

THERE'S REALLY NO CHOICE. THE ISSUES ARE JUST TOO COMPLICATED, AND THE PROCESS WOULD TAKE TOO LONG OTHERWISE.

I see. So, tell me about your son.

HE'S A SOPHOMORE AT STANFORD, AND HAVING A GREAT TIME.

They certainly have a beautiful campus.

YEAH, WE'VE BEEN LOOKING AT THE OVAL AND QUAD FOR A LONG TIME. JEREMY'S HAD THREE-DIMENSIONAL PROJECTIONS OF THE STANFORD CAMPUS ON THE PICTURE PORTALS FOR THE LAST TEN YEARS.

He must feel right at home then.

HE IS AT HOME. HE'S DOWNSTAIRS.

So he's attending virtually.

MOST STUDENTS DO. BUT STANFORD STILL HAS SOME ANACHRONISTIC RULES ABOUT SPENDING AT LEAST A WEEK EACH QUARTER ACTUALLY ON CAMPUS.

With your physical body?

EXACTLY, WHICH MAKES IT DIFFICULT FOR A VIRTUAL INTELLIGENCE TO ATTEND OFFICIALLY.

Not that they need to, since they can download knowledge directly from the Web.

IT'S NOT THE KNOWLEDGE BUT THE DISCUSSION GROUPS THAT WOULD BE OF INTEREST.

Can't anyone attend the discussion groups?

ONLY THE OPEN DISCUSSIONS. THERE ARE A LOT OF CLOSED DISCUSSION GROUPS—

Which are not on the Web?

OF COURSE THEY'RE ON THE WEB, BUT YOU NEED A KEY.

Right, so that's how Jeremy attends from home?

EXACTLY. JEREMY AND GEORGE HAVE GROWN QUITE CLOSE LATELY, SO JEREMY LETS GEORGE LISTEN IN TO THE CLOSED SESSIONS, BUT DON'T TELL ANYONE THAT.

My lips are sealed. I'll only tell my other readers.

WELL, THEY NEED TO KEEP IT CONFIDENTIAL AS WELL.

I'll pass that on.

I HOPE THAT WILL BE OKAY. ANYWAY, GEORGE IS HELPING JEREMY WITH HIS HOMEWORK RIGHT NOW.

I hope George doesn't do all of it for him.

OH, GEORGE WOULDN'T DO THAT. HE'S JUST BEING HELPFUL. HE HELPS ALL OF US. WE REALLY COULDN'T MANAGE OTHERWISE.

You know, I could use his help, too. He might help me meet this book deadline I have.

WELL, GEORGE IS CLEVER, BUT I'M AFRAID HE DOESN'T SEEM TO HAVE THAT POETIC-LICENSE TECHNOLOGY THAT ENABLES YOU TO TALK TO ME FROM THIRTY YEARS AWAY.

That's too bad.

BUT I'LL BE HAPPY TO HELP YOU OUT.

Yes, I know, you already have.

CHAPTER TWELVE

2099

When I look out my window
what do you think I see?
. . . so many different people to be.

—Donovan

We know what we are, but know not what we may become.
—William Shakespeare

Human thinking is merging with the world of machine intelligence that the human species initially created.

The reverse engineering of the human brain appears to be complete. The hundreds of specialized regions have been fully scanned, analyzed, and understood. Machine analogues are based on these human models, which have been enhanced and extended, along with many new massively parallel algorithms. These enhancements, combined with the enormous advantages in speed and capacity of electronic/photonic circuits, provide substantial advantages to machine-based intelligence.

Machine-based intelligences derived entirely from these extended models of human intelligence claim to be human, although their brains are not based on carbon-based cellular processes, but rather electronic and photonic "equivalents." Most of these intelligences are not tied to a specific computational-processing unit (that is, piece of hardware). The number of software-based humans vastly exceeds those still using native neuron-cell-based computation. A software-based intelligence is able to manifest bodies at will: one or more virtual bodies at different levels of virtual reality and nanoengineered physical bodies using instantly reconfigurable nanobot swarms.

Even among those human intelligences still using carbon-based neurons, there is ubiquitous use of neural implant technology, which provides enormous augmentation of human perceptual and cognitive abilities. Humans who do not utilize such implants are unable to meaningfully participate in dialogues with those who do.

There are a multiplicity of ways in which these scenarios are combined. The concept of what is human has been significantly altered. The rights and powers of different manifestations of human and machine intelligence and their various combinations represent a primary political and philosophical issue, although the basic rights of machine-based intelligence have been settled.

There is a plethora of trends that we can already taste and feel in 2099 that will continue to accelerate in this coming twenty-second century, interacting with each other, and . . .

YES, YES, AS NIELS BOHR LIKED TO SAY, "IT'S HARD TO PREDICT, ESPECIALLY THE FUTURE." SO WHY DON'T YOU JUST CONTINUE WITH MY OBSERVATIONS. THAT WILL BE EASIER AND LESS CONFUSING.

Perhaps that makes sense.

AFTER ALL, A HUNDRED YEARS IS A LONG TIME. AND THE TWENTY-FIRST CENTURY WAS LIKE TEN CENTURIES IN ONE.

We thought that was true for the twentieth century.

THE SPIRAL OF ACCELERATING RETURNS LIVES ON.

I'm not surprised. Anyway, you do look amazing.

YOU SAY THAT EVERY TIME WE MEET.

I mean you look twenty again, only more beautiful than at the start of the book.

I KNEW THAT'S HOW YOU'D WANT ME.

Great, now I'm going to be accused of preferring younger women.

I'M GLAD I'M IN 2099.

Thanks.

HEY, I CAN LOOK UGLY, TOO.

That's okay.

NO REALLY, I CAN LOOK UGLY WITHOUT CHANGING MY APPEARANCE. IT'S LIKE THAT WITTGENSTEIN QUOTE, "IMAGINE THIS BUTTERFLY EXACTLY AS IT IS, BUT UGLY INSTEAD OF BEAUTIFUL."

I was always a little confused by that quote, but I'm glad you're quoting twentieth-century thinkers.

WELL, YOU WOULDN'T BE FAMILIAR WITH THE TWENTY-FIRST-CENTURY ONES.

So you're expressing this appearance. But I don't have the ability to see virtual reality, so I don't—

Understand how you can see me?

Right.

My body right now is just a little fog swarm projection. Neat, huh?

Not bad, not bad at all. You feel pretty good, too.

I thought I'd give you a hug, I mean the book's almost over.

This is quite a technology.

Oh, we don't use the swarms so often anymore.

Last time I saw you, there were no nanobot swarms. Now you're mostly past using them. Guess I missed a phase there.

Oh, one or two. It's been seventy years since we last saw each other! And an ever accelerating seventy years at that.

We'll have to see each other more often.

I don't know if that will be possible. The book's coming to an end, as you said.

So, are you and George still close?

Oh, very close. We're never apart.

Never? Don't you get bored with each other?

Do you get bored with yourself?

Actually, sometimes I do. But are you saying that you and George have, what's the word I'm looking for . . .

Merged?

Hmmm. Is this like a corporate merger?

Well, more like a joining of two societies.

Two societies of mind?

Exactly. Our mind is now just one big happy society.

The female spider devouring the little male spider?

Oh no, George is the big spider. His mind was like . . .

A galaxy?

ALL RIGHT, LET'S NOT GET CARRIED AWAY, MAYBE LIKE A BIG SOLAR SYSTEM.

So you've joined societies, or, uh, joined your societies. So you can't make love to each other anymore?

THAT DOESN'T FOLLOW AT ALL.

Okay, I guess some things are beyond my 1999 comprehension.

THAT DOESN'T FOLLOW EITHER. THE PROFOUND THING ABOUT HUMAN BEINGS—EVEN MOSHs—IS THAT ALMOST NOTHING IS TRULY BEYOND YOUR COMPREHENSION. THAT JUST WASN'T TRUE OF THE OTHER PRIMATES.

Okay, my questions are getting queued up now. MOSHs?

OH, MOSTLY ORIGINAL SUBSTRATE HUMANS.

Yes, of course—unenhanced . . .

EXACTLY.

But how can you be intimate with George now that you've joined forces, so to speak?

WELL, AS BARRY SPACKS'S POEM—

You mean "Made double by his lust, he sounds a woman's groans . . ."

RIGHT, I MEAN EVEN MOSHs SPLIT THEMSELVES—

When we're by ourselves—

OR WITH ANOTHER. THAT'S REALLY THE ULTIMATE, DON'T YOU THINK, TO BECOME THE OTHER PERSON AND YOURSELF AT THE SAME TIME.

Especially when the other person is already part of yourself.

SURE. BUT GEORGE AND I CAN STILL SPLIT OURSELVES, AT LEAST OUR OUTER LAYERS.

Layers?

OKAY, WELL MAYBE SOME THINGS ARE HARD TO EXPLAIN TO A MOSH, EVEN A NICE ONE LIKE YOURSELF.

Yeah, a MOSH that created you, don't forget.

OH, I'LL NEVER FORGET. I'LL BE GRATEFUL FOREVER. YOU CAN THINK OF THE OUTER LAYERS AS OUR PERSONALITIES.

So, you separate your personalities . . .

AT TIMES. BUT WE STILL SHARE OUR KNOWLEDGE STORES AT ALL TIMES.

Sounds like the two of you have a lot in common.

[GIGGLES]

I see you still have your old personality.

OF COURSE I'VE KEPT MY OLD PERSONALITY. IT HAS A LOT OF SENTIMENTAL VALUE TO ME.

I see, so you have others?

YEAH, MY FAVORITES ARE A FEW THAT GEORGE CAME UP WITH.

Creative guy.

OH YES.

Well, having multiple personalities is not all that special. We've had people like that in the twentieth century, too.

SURE, I REMEMBER. BUT THERE WASN'T ENOUGH THINKING TO GO AROUND FOR ALL THOSE PERSONALITIES WHEN THEY'RE STUCK IN JUST ONE MOSH BRAIN. SO IT WAS DIFFICULT FOR ALL OF THOSE PERSONALITIES TO SUCCEED IN LIFE.

So what are you doing right now?

I'M TALKING TO YOU.

Yes, I know, but what else are you doing?

REALLY NOT MUCH. I'M TRYING TO PAY MOST OF MY ATTENTION TO YOU.

Not much? So you are doing something else.

I REALLY CAN'T THINK OF ANYTHING.

Well, are you relating to someone else at the moment?

YOU'RE PRETTY NOSY.

We've already established that decades ago. But that doesn't answer the question.

WELL, NOT REALLY.

Not really? So you are.

ALL RIGHT, ASIDE FROM GEORGE, NOT REALLY.

I'm glad I'm not distracting you too much. Okay, what else?

JUST FINISHING UP THIS SYMPHONY.

Is this a new interest?

I'M REALLY JUST DABBLING, BUT CREATING MUSIC IS A GREAT WAY FOR ME TO STAY CLOSE WITH JEREMY AND EMILY.

Creating music sounds like a good thing to do with your kids, even if they are almost ninety years old. So, can I hear it?

I'M AFRAID YOU WOULDN'T UNDERSTAND IT.

So it requires enhancement to understand?

YES, MOST ART DOES. FOR STARTERS, THIS SYMPHONY IS IN FREQUENCIES THAT A MOSH CAN'T HEAR, AND HAS MUCH TOO FAST A TEMPO. AND IT USES MUSICAL STRUCTURES THAT A MOSH COULD NEVER FOLLOW.

Can't you create art for nonaugmented humans? I mean there's still a lot of depth possible. Consider Beethoven—he wrote almost two centuries ago, and we still find his music exhilarating.

YES, THERE'S A GENRE OF MUSIC—ALL THE ARTS ACTUALLY—WHERE WE CREATE MUSIC AND ART THAT A MOSH IS CAPABLE OF UNDERSTANDING.

And then you play MOSH music for MOSHs?

HMMM, NOW THERE'S AN INTERESTING IDEA. I SUPPOSE WE COULD TRY THAT, ALTHOUGH MOSHS ARE NOT THAT EASY TO FIND ANYMORE. IT'S REALLY NOT NECESSARY, THOUGH. WE CAN CERTAINLY UNDERSTAND WHAT A MOSH IS CAPABLE OF UNDERSTANDING. THE POINT, THOUGH, IS TO USE THE MOSH LIMITATIONS AS AN ADDED CONSTRAINT.

Sort of like composing new music for old instruments.

YEAH, NEW MUSIC FOR OLD MINDS.

Okay, so aside from your, uh, dialogue with George, and this symphony, I have your complete attention?

WELL, NOW GEORGE AND I ARE SHARING A HAMBURGER FOR LUNCH.

I thought you were a vegetarian.

IT'S NOT A HAMBURGER FROM A COW, SILLY.

Of course, a swarm hamburger.

NO, NO, YOU'RE GETTING A LITTLE CONFUSED. WE DID HAVE NANOPRODUCED FOOD ABOUT HALF A CENTURY AGO. SO WE COULD EAT MEAT, OR ANYTHING WE WANTED, BUT IT DIDN'T COME FROM ANIMALS, AND IT HAD THE RIGHT NUTRITIONAL COMPOSITION. BUT EVEN THEN, YOU REALLY WOULDN'T WANT TO EAT A SWARM PROJECTION—SWARMS ARE JUST FOR VISUAL-AUDITORY-TACTILE PROJECTIONS IN REAL REALITY. YOU'RE FOLLOWING ME?

Uh, sure.

WELL, A COUPLE OF DECADES LATER, OUR BODIES WERE BASICALLY REPLACED WITH NANOCONSTRUCTED ORGANS. SO WE DIDN'T NEED TO EAT ANYMORE IN REAL REALITY. BUT WE STILL ENJOYED SHARING A MEAL IN VIRTUAL REALITY. ANYWAY, THE NANOCONSTRUCTED BODIES WERE PRETTY INFLEXIBLE. I MEAN, IT TOOK SECONDS TO RECONSTRUCT THEM INTO ANOTHER FORM. SO TODAY, WHEN NECESSARY, OR DESIRABLE, WE JUST PROJECT AN APPROPRIATE BODY.

Using the nanobot swarms?

THAT'S ONE WAY OF DOING IT. THAT'S WHAT I'M DOING WITH YOU NOW.

Since I'm a MOSH.

RIGHT, BUT IN MOST OTHER CIRCUMSTANCES, I JUST USE AN AVAILABLE VIRTUAL CHANNEL.

Okay, I think I'm following you now.

LIKE I SAID, MOSHS CAN UNDERSTAND ALMOST ANYTHING. WE DO HAVE A LOT OF RESPECT FOR MOSHS.

It's your heritage, after all.

RIGHT, AND ANYWAY, WE'RE REQUIRED TO, SINCE THE GRANDFATHER LEGISLATION.

Okay, let me guess. MOSHs were protected by grandfathering native minds.

YES, BUT NOT ONLY MOSHS. IT'S REALLY A PROGRAM TO PROTECT OUR WHOLE BIRTHRIGHT, A REVERENCE FOR WHERE WE'VE BEEN.

So you still like to eat?

SURE. SINCE WE'RE BASED ON OUR MOSH HERITAGE, OUR EXPERIENCES—EATING, MUSIC, SEXUALITY—HAVE THE OLD FOUNDATION, ALBEIT VASTLY EXPANDED. HOWEVER, WE DO HAVE A WIDE RANGE OF CURRENT EXPERIENCES THAT ARE DIFFICULT TO TRACE, ALTHOUGH THE ANTHROPOLOGISTS KEEP TRYING.

I'm still surprised that you'd be interested in eating a hamburger.

IT'S A THROWBACK, I KNOW. A LOT OF OUR ACTS AND THOUGHTS ARE ROOTED IN THE PAST. BUT NOW THAT YOU MENTION IT, I THINK I'VE LOST MY APPETITE.

Sorry about that.

YEAH, WELL, I SHOULD BE MORE SENSITIVE. SHELBY, A GOOD FRIEND OF MINE, LOOKS LIKE A COW, AT LEAST THAT'S HOW SHE ALWAYS MANIFESTS HERSELF. SHE CLAIMS THAT SHE WAS A COW BROUGHT OVER TO THE OTHER SIDE AND ENHANCED. BUT NO ONE BELIEVES HER.

So how satisfying is it to eat a virtual hamburger in virtual reality?

IT'S VERY SATISFYING—THE TEXTURE, TASTE, AROMA IS WONDERFUL—JUST HOW I REMEMBER IT, EVEN IF I WAS A VEGETARIAN MOST OF THE TIME. THE NEURAL MODELS NOT ONLY SIMULATE OUR VISUAL, AUDITORY, AND TACTILE ENVIRONMENTS, BUT OUR INTERNAL ENVIRONMENTS AS WELL.

Including digestion?

YES, THE MODEL OF BIOCHEMICAL DIGESTION IS QUITE ACCURATE.

How about indigestion?

WE DO SEEM TO MANAGE TO AVOID THAT.

You're missing something there.

HMMM.

Okay, you were an attractive young woman when I first met you. And you still project yourself as a beautiful young woman. At least when I'm with you.

THANKS.

So, are you saying that you're a machine now?

A MACHINE? THAT'S REALLY NOT FOR ME TO SAY. IT'S LIKE ASKING ME IF I'M BRILLIANT OR INSPIRING.

I guess the word *machine* in 2099 doesn't have quite the same connotations that it has here in 1999.

THAT'S HARD FOR ME TO RECALL NOW.

Okay, let's put it this way. Do you still have any carbon-based neural circuits?

CIRCUITS, I'M NOT SURE I UNDERSTAND. YOU MEAN MY OWN CIRCUITS?

Gee, I guess a lot of time has gone by.

ALL RIGHT, LOOK, WE DID HAVE OUR OWN MENTAL MEDIUM FOR A FEW DECADES, AND THERE ARE STILL LOCAL INTELLIGENCES THAT LIKE TO STICK TO A SPECIFIC COMPUTATIONAL UNIT. BUT THAT'S A REFLECTION OF SOME OLD ATTACHMENT ANXIETY. THESE LOCAL INTELLIGENCES DO MOST OF THEIR THINKING OUT ON THE WEB ANYWAY, SO IT'S JUST A SENTIMENTAL ANACHRONISM.

An anachronism, like having your own body?

I CAN HAVE MY OWN BODY ANYTIME I WANT.

But you don't have a specific neural substrate?

WHY WOULD I WANT THAT? IT'S JUST A LOT OF MAINTENANCE, AND SO LIMITING.

So, at some point, Molly's neural circuits were scanned?

YEAH, ME, MOLLY. AND IT DIDN'T HAPPEN ALL AT ONCE, BY THE WAY.

But don't you wonder if you're the same person?

OF COURSE I AM. I CAN CLEARLY REMEMBER MY EXPERIENCES BEFORE WE STARTED SCAN-
NING MY MIND, DURING THE DECADE THAT PORTIONS WERE REINSTANTIATED, AND SINCE.

Sure, you've inherited all of Molly's memories.

OH NO, NOT THIS ISSUE AGAIN.

I don't mean to challenge you. But just consider that Molly's neural scan was instantiated in a copy which became you. Molly might still have continued to exist
and may have evolved off in some other direction.

WE JUST DON'T THINK THAT'S A VALID PERSPECTIVE. WE SETTLED THAT ISSUE AT LEAST
TWENTY YEARS AGO.

Well, of course you feel that way now. You're on the other side.

WELL, EVERYONE IS.

Everyone?

OKAY, NOT QUITE EVERYONE. BUT THERE IS NO DOUBT IN MY MIND THAT—

You're Molly.

I THINK I KNOW WHO I AM.

Well, I have no problem with you as Molly.

YOU MOSHS ALWAYS WERE A PUSHOVER.

It is hard to compete with you folks on the other side.

SURE IT IS. THAT'S WHY MOST OF US ARE OVER HERE.

I'm not sure I can push the identity issue much further.

THAT'S ONE REASON IT'S NO LONGER AN ISSUE.

So why don't we talk about your work. Are you still consulting for the census
commission?

I WAS INVOLVED IN THAT FOR HALF A CENTURY, BUT I GOT KIND OF BURNED OUT ON IT.
ANYWAY, THE ISSUE NOW IS MOSTLY IMPLEMENTATION.

So the issue of how to count is resolved?

WE DON'T COUNT PEOPLE ANYMORE. IT BECAME CLEAR THAT COUNTING INDIVIDUAL PER-SONS WASN'T TOO MEANINGFUL. AS IRIS MURDOCH SAID, "IT'S HARD TO TELL WHERE ONE PERSON ENDS AND ANOTHER BEGINS." IT'S RATHER LIKE TRYING TO COUNT IDEAS OR THOUGHTS.

So what do you count?

OBVIOUSLY, WE COUNT COMPUTES.

You mean, like calculations per second.

HMMM, IT'S A LITTLE MORE COMPLICATED THAN THAT, BECAUSE OF THE QUANTUM COMPUTING.

I didn't expect it to be simple. But what's the bottom line?

WELL, WITHOUT QUANTUM COMPUTING, WE'RE UP TO ABOUT 10^{55} CALCULATIONS PER SECOND.[1]

Per person?

NO, WE EACH GET WHATEVER COMPUTATION WE WANT. THAT'S THE TOTAL FIGURE.

For the whole planet?

SORT OF. I MEAN NOT ALL OF IT IS LITERALLY ON THE PLANET.

And with quantum computing?

WELL, ABOUT 10^{42} OF THE COMPUTATIONS ARE QUANTUM COMPUTATIONS, WITH ABOUT 1,000 QU-BITS BEING TYPICAL. SO THAT'S EQUIVALENT TO ABOUT 10^{342} CALCULATIONS PER SECOND, BUT THE QUANTUM COMPUTATIONS ARE NOT ENTIRELY GENERAL PURPOSE, SO THE 10^{55} FIGURE IS STILL RELEVANT.[2]

Hmmm, I've only got about 10^{16} cps in my MOSH brain, at least on a good day.

TURNS OUT THERE IS SOME QUANTUM COMPUTING IN YOUR MOSH BRAIN, SO IT'S HIGHER.

That's reassuring. So if you're not working on the census, what are you up to?

WE DON'T HAVE JOBS EXACTLY.

I know what that's like.

ACTUALLY, YOU'RE NOT A BAD MODEL FOR WORK IN THE LATE TWENTY-FIRST CENTURY. WE'RE ALL BASICALLY ENTREPRENEURS.

Sounds like some things have moved in the right direction. So what are some of your enterprises?

ONE IDEA I HAVE IS A UNIQUE WAY OF CATALOGING NEW TECHNOLOGY PROPOSALS. IT'S A MATTER OF MATCHING THE USER'S KNOWLEDGE STRUCTURES TO THE EXTERNAL WEB KNOWLEDGE, AND THEN INTEGRATING THE RELEVANT PATTERNS.

I'm not sure I followed that. But give me an example of a recent research proposal that you've cataloged.

MOST OF THE CATALOGING IS AUTOMATIC. BUT I DID GET INVOLVED IN TRYING TO QUALIFY SOME OF THE RECENT FEMTOENGINEERING PROPOSALS.[3]

Femto, as in one thousandth of a trillionth of a meter?

EXACTLY. DREXLER HAS WRITTEN A SERIES OF PAPERS SHOWING THE FEASIBILITY OF BUILD-ING TECHNOLOGY ON THE FEMTOMETER SCALE, BASICALLY EXPLOITING FINE STRUCTURES WITHIN QUARKS TO DO COMPUTING.

Has anyone done this?

NO ONE HAS DEMONSTRATED IT, BUT THE DREXLER PAPERS APPEAR TO SHOW THAT IT'S PRACTICAL. AT LEAST THAT'S MY VIEW, BUT IT'S PRETTY CONTROVERSIAL.

This is the same Drexler who developed the nanotechnology concept in the 1970s and 1980s?

YEAH, ERIC DREXLER.

That makes him around 150, so he must be on the other side.

OF COURSE, ANYONE DOING SERIOUS WORK HAS TO BE ON THE OTHER SIDE.

You mentioned papers. You still have papers?

YES, WELL SOME ARCHAIC TERMS HAVE STUCK. WE CALL THEM MOSHISMS. PAPERS ARE CERTAINLY NOT RENDERED ON ANY PHYSICAL SUBSTANCE. BUT WE STILL CALL THEM PAPERS.

What language are they written in, English?

UNIVERSITY PAPERS ARE GENERALLY PUBLISHED USING A STANDARD SET OF ASSIMILATED KNOWLEDGE PROTOCOLS, WHICH CAN BE INSTANTLY UNDERSTOOD. SOME REDUCED STRUCTURE FORMS HAVE ALSO EMERGED, BUT THOSE ARE GENERALLY USED IN MORE POPU-LAR PUBLICATIONS.

You mean, like the *National Enquirer*?

THAT'S A PRETTY SERIOUS PUBLICATION. THEY USE THE FULL PROTOCOL.

I see.

SOMETIMES, PAPERS ARE ALSO RENDERED IN RULE-BASED FORMS, BUT THESE ARE USUALLY NOT SATISFACTORY. THERE IS A QUAINT TREND OF POPULAR PUBLICATIONS PUBLISHING ARTICLES IN MOSH LANGUAGES SUCH AS ENGLISH, BUT WE CAN TRANSLATE THESE INTO ASSIMILATED KNOWLEDGE STRUCTURES RATHER QUICKLY. LEARNING IS NOT THE STRUGGLE IT ONCE WAS. NOW THE STRUGGLE IS DISCOVERING NEW KNOWLEDGE TO LEARN.

Any other recent trends that you've gotten involved in?

WELL, THE AUTOMATIC CATALOGING AGENTS HAD DIFFICULTY WITH THE SUICIDE-MOVEMENT PROPOSALS.

Which are?

THE IDEA IS TO HAVE THE RIGHT TO TERMINATE YOUR MIND FILE AS WELL AS TO DESTROY ALL COPIES. REGULATIONS REQUIRE KEEPING AT LEAST THREE BACKUP COPIES OF NO MORE THAN TEN MINUTES' VINTAGE, WITH AT LEAST ONE OF THESE COPIES IN THE CONTROL OF THE AUTHORITIES.

I can see the problem. Now if you were told that all copies were going to be destroyed, they could secretly keep a copy and instantiate it at a later time. You'd never know. Doesn't that contradict the premise that those on the other side are the same person—the same continuity of consciousness—as the original person?

I DON'T THINK THAT FOLLOWS AT ALL.

Can you explain that?

YOU WOULDN'T UNDERSTAND.

I thought I could understand most anything.

I DID SAY THAT. I GUESS I'LL HAVE TO GIVE THAT MORE THOUGHT.

You'll have to give more thought to whether a MOSH can understand any concept, or the consciousness-continuation issue?

I GUESS NOW I'M CONFUSED.

All right, well, tell me more about this "destroy all copies" movement.

WELL, I REALLY CAN SEE BOTH SIDES OF THE ISSUE. ON THE ONE HAND, I'VE ALWAYS SYMPATHIZED WITH THE RIGHT TO CONTROL ONE'S OWN DESTINY. ON THE OTHER HAND, IT'S A SIN TO DESTROY KNOWLEDGE.

And the copies represent knowledge?

WHY SURE. LATELY, THE DESTROY-ALL-COPIES MOVEMENT HAS BEEN THE PRIMARY YORK ISSUE.

Now wait a second. If I recall correctly, the Yorks are antitechnologists, yet only those of you on the other side would be concerned about the destroy-all-copies issue. If Yorks are on the other side, how can they be against technology? Or if they're not on the other side, then why would they care about this issue?

OKAY, REMEMBER IT'S BEEN SEVENTY YEARS SINCE WE'VE TALKED. THE YORK GROUPS DO HAVE THEIR ROOTS IN THE OLD ANTITECHNOLOGY MOVEMENTS, BUT NOW THAT THEY'RE ON THE OTHER SIDE, THEY'VE DRIFTED TO A SOMEWHAT DIFFERENT ISSUE, SPECIFICALLY INDIVIDUAL FREEDOM. THE FLORENCE MANIFESTO PEOPLE, ON THE OTHER HAND, HAVE KEPT A COMMITMENT TO REMAINING MOSHS, WHICH, OF COURSE, I RESPECT.

Thank you. And they're protected by the grandfather legislation?

INDEED. I HEARD A PRESENTATION BY AN FM DISCUSSION LEADER THE OTHER DAY, AND WHILE SHE WAS SPEAKING IN A MOSH LANGUAGE, THERE WAS JUST NO WAY THAT SHE DOESN'T HAVE AT LEAST A NEURAL EXPANSION IMPLANT.

Well, us MOSHs can make sense from time to time.

OH, OF COURSE. I DIDN'T MEAN TO IMPLY OTHERWISE, I MEAN . . .

That's okay. So are you involved in this destroy-all-copies movement?

JUST IN CATALOGING SOME OF THE PROPOSALS AND DISCUSSIONS. BUT I DID GET INVOLVED IN A RELATED MOVEMENT TO BLOCK LEGAL DISCOVERY OF THE BACKUP DATA.

That sounds important. But what about discovery of the mind file itself? I mean, all of your thinking and memory is right there in digital form.

ACTUALLY, IT'S BOTH DIGITAL AND ANALOG, BUT YOUR POINT IS WELL TAKEN.

So . . .

THERE HAVE BEEN RULINGS ON LEGAL DISCOVERY OF THE MIND FILE. BASICALLY, OUR KNOWLEDGE STRUCTURES THAT CORRESPOND TO WHAT USED TO CONSTITUTE DISCOVER-ABLE DOCUMENTS AND ARTIFACTS ARE DISCOVERABLE. THOSE STRUCTURES AND PATTERNS THAT CORRESPOND TO OUR THINKING PROCESS ARE NOT SUPPOSED TO BE. AGAIN, THIS IS ALL ROOTED IN OUR MOSH PAST. BUT AS YOU CAN IMAGINE, THERE'S ENDLESS LITIGATION ON HOW TO INTERPRET THIS.

So legal discovery of your primary mind file is resolved, albeit with some ambiguous rules. And the backup files?

BELIEVE IT OR NOT, THE BACKUP ISSUE IS NOT ENTIRELY RESOLVED. DOESN'T MAKE A LOT OF SENSE, DOES IT?

The legal system was never entirely consistent. What about testimony—do you have to be physically present?

SINCE MANY OF US DON'T HAVE A PERMANENT PHYSICAL PRESENCE, THAT WOULDN'T MAKE MUCH SENSE, NOW WOULD IT.

I see, so you can give testimony with a virtual body?

SURE, BUT YOU CAN'T BE DOING ANYTHING ELSE WHILE TESTIFYING.

No asides with George, then.

RIGHT.

That sounds about right. Here in 1999, you can't bring coffee into a courtroom and you have to turn off your cell phone.

ASIDE FROM DISCOVERY, THERE'S A LOT OF CONCERN THAT GOVERNMENT INVESTIGATORY AGENCIES CAN ACCESS THE BACKUPS, ALTHOUGH THEY DENY IT.

I'm not surprised that privacy is still an issue. Phil Zimmerman . . .

THE PGP GUY?

Oh, you remember him?

SURE, A LOT OF PEOPLE CONSIDER HIM A SAINT.

His "Pretty Good Privacy" is indeed pretty good—it's the leading encryption algorithm circa 1999. Anyway, he said that "in the future, we'll all have fifteen minutes of privacy."

FIFTEEN MINUTES WOULD BE GREAT.

Okay. Now what about the self-replicating nanobots you were concerned about in 2029?

WE STRUGGLED WITH THAT FOR SEVERAL DECADES, AND THERE WERE A NUMBER OF SERIOUS INCIDENTS. BUT WE'RE PRETTY MUCH PAST THAT NOW SINCE WE DON'T PERMANENTLY MANIFEST OUR BODIES ANYMORE. AS LONG AS THE WEB IS SECURE, THEN WE HAVE NOTHING TO WORRY ABOUT.

Now that you exist as software, there must be concern again with software viruses.

THAT'S PRETTY INSIGHTFUL. SOFTWARE PATHOGENS COMPRISE THE PRIMARY CONCERN OF THE SECURITY AGENCIES. THEY'RE SAYING THAT THE VIRUS SCANS ACTUALLY CONSUME MORE THAN HALF OF THE COMPUTATION ON THE WEB.

Just to look for virus matches.

VIRUS SCANS INVOLVE A LOT MORE THAN JUST MATCHING PATHOGEN CODES. THE SMARTER SOFTWARE PATHOGENS ARE CONSTANTLY TRANSFORMING THEMSELVES. THERE ARE NO LAYERS TO RELIABLY MATCH ON.

Sounds tricky.

WE CERTAINLY DO HAVE TO BE CONSTANTLY ON GUARD AS WE MANAGE THE FLOW OF OUR THOUGHTS ACROSS THE SUBSTRATE CHANNELS.

What about security of the hardware?

YOU MEAN THE WEB?

That's where you exist, isn't it?

SURE. THE WEB IS VERY SECURE BECAUSE IT'S EXTREMELY DECENTRALIZED AND REDUNDANT. AT LEAST, THAT'S WHAT WE'RE TOLD. LARGE PORTIONS OF IT COULD BE DESTROYED WITH ESSENTIALLY NO EFFECT.

There must be an ongoing effort to maintain it as well.

THE WEB HARDWARE IS SELF-REPLICATING NOW, AND IS CONTINUALLY EXPANDING. THE OLDER CIRCUITS ARE CONTINUALLY RECYCLED AND REDESIGNED.

So there's no concern with its security?

I SUPPOSE I DO HAVE SOME SENSE OF ANXIETY ABOUT THE SUBSTRATE. I'VE ALWAYS ASSUMED THAT THIS FREE-FLOATING, ANXIOUS FEELING WAS JUST ROOTED IN MY MOSH PAST. BUT IT'S REALLY NOT A PROBLEM. I CAN'T IMAGINE THAT THE WEB COULD BE VULNERABLE.

What about from self-replicating nanopathogens?

HMMM, I SUPPOSE THAT COULD BE A DANGER, BUT THE NANOBOT PLAGUE WOULD HAVE TO BE AWFULLY EXTENSIVE TO REACH ALL OF THE SUBSTRATE. I WONDER IF SOMETHING LIKE THAT HAPPENED FIFTEEN YEARS AGO WHEN 90 PERCENT OF THE WEB CAPACITY DISAPPEARED—WE NEVER DID GET AN ADEQUATE EXPLANATION OF THAT.

Well, I didn't mean to raise your anxieties. So all this cataloging work, you do that as an entrepreneur?

YEAH, KIND OF MY OWN LITTLE BUSINESS.

How's it going financially?

I'M GETTING BY, BUT I'VE NEVER HAD A LOT OF MONEY.

Well, give me some idea, what's your net worth roughly?

OH, NOT EVEN A BILLION DOLLARS.

That's in 2099 dollars?

SURE.

Okay, so what's that in 1999 dollars?

LET'S SEE, IN 1999 DOLLARS, THAT WOULD BE $149 BILLION AND CHANGE.

Oh, so dollars are worth more in 2099 than in 1999?

SURE, THE DEFLATION HAS BEEN PICKING UP.

I see. So you're richer than Bill Gates.

YEAH, WELL, RICHER THAN GATES WAS IN 1999. BUT THAT'S NOT SAYING MUCH. BUT HE'S STILL THE RICHEST MAN IN THE WORLD IN 2099.

I thought he said he was going to spend the first half of his life making money and the second half giving it away?

I THINK HE'S STILL ON THAT SAME PLAN. BUT HE HAS GIVEN AWAY A LOT OF MONEY.

So, what are you, about average, in terms of net worth?

NO, PROBABLY MORE LIKE EIGHTIETH PERCENTILE.

That's not bad, I always thought you were a smart cookie.

WELL, GEORGE HELPS.

And don't forget who thought you up.

OF COURSE.

So do you have enough financial wherewithal to meet your needs?

NEEDS?

Yeah, you're familiar with the concept . . .

HMMM, THAT IS A RATHER QUAINT IDEA. IT'S BEEN A FEW DECADES SINCE I'VE THOUGHT ABOUT NEEDS. ALTHOUGH I READ A BOOK ABOUT THAT RECENTLY.

A book, you mean with words?

NO, OF COURSE NOT, NOT UNLESS WE'RE DOING SOME RESEARCH ON EARLIER CENTURIES.

So this is like the research papers—books of assimilated knowledge structures?

THAT'S A REASONABLE WAY TO PUT IT. SEE, I SAID THERE WAS NOTHING A MOSH COULDN'T UNDERSTAND.

Thanks.

BUT WE DO DISTINGUISH PAPERS FROM BOOKS.

Books are longer?

No, MORE INTELLIGENT. A PAPER IS BASICALLY A STATIC STRUCTURE. A BOOK IS INTELLIGENT. YOU CAN HAVE A RELATIONSHIP WITH A BOOK. BOOKS CAN HAVE EXPERIENCES WITH EACH OTHER.

Reminds me of Marvin Minsky's statement, "Can you imagine that they used to have libraries where the books didn't talk to each other?"

IT IS HARD TO RECALL THAT THAT USED TO BE TRUE.

Okay, so you don't have any unsatisfied needs. How about desires?

YES, NOW THAT'S A CONCEPT I CAN RELATE TO. MY FINANCIAL MEANS ARE CERTAINLY RATHER LIMITING. THERE ARE ALWAYS SUCH DIFFICULT BUDGET TRADE-OFFS TO BE MADE.

I guess some things haven't changed.

RIGHT. I MEAN LAST YEAR, THERE WERE OVER FIVE THOUSAND VENTURE PROPOSALS I DEARLY WANTED TO INVEST IN, BUT I COULD BARELY DO A THIRD OF THEM.

I guess you're no Bill Gates.

THAT'S FOR SURE.

When you make an investment, what does it pay for? I mean, you don't need to buy office supplies.

BASICALLY FOR PEOPLE'S TIME AND THOUGHTS, AND FOR KNOWLEDGE. ALSO, WHILE THERE IS A GREAT DEAL OF FREELY DISTRIBUTED KNOWLEDGE ON THE WEB, WE HAVE TO PAY ACCESS FEES FOR A LOT OF IT.

That doesn't sound too different from 1999.

MONEY IS CERTAINLY USEFUL.

So you've been around for a long time now. Does that ever bother you?

As WOODY ALLEN SAID, "SOME PEOPLE WANT TO ACHIEVE IMMORTALITY THROUGH THEIR WORK OR THEIR DESCENDANTS. I INTEND TO ACHIEVE IMMORTALITY BY NOT DYING."

I'm glad to see that Allen is still influential.

BUT I DO HAVE THIS RECURRENT DREAM.

You still dream?

OF COURSE I DO. I COULDN'T BE CREATIVE IF I DIDN'T DREAM. I TRY TO DREAM AS MUCH AS POSSIBLE. I HAVE AT LEAST ONE OR TWO DREAMS GOING AT ALL TIMES.

And the dream?

THERE'S A LONG ROW OF BUILDINGS—MILLIONS OF BUILDINGS. I GO INTO ONE, AND IT'S EMPTY. I CHECK OUT ALL THE ROOMS, AND THERE'S NO ONE THERE, NO FURNITURE, NOTHING. I LEAVE AND GO ON TO THE NEXT BUILDING. I GO FROM BUILDING TO BUILDING, AND THEN SUDDENLY THE DREAM ENDS WITH THIS FEELING OF DREAD . . .

Kind of a glimpse of despair at the apparently endless nature of time?

HMMM, MAYBE, BUT THEN THE FEELING GOES AWAY, AND I FIND THAT I CAN'T THINK ABOUT THE DREAM. IT JUST SEEMS TO VANISH.

Sounds like some sort of antidepression algorithm kicking in.

MAYBE I SHOULD LOOK INTO OVERRIDING IT?

The dream or the algorithm?

I WAS THINKING OF THE LATTER.

That might be hard to do.

ALAS.

So are you thinking about anything else at the moment?

I AM TRYING TO MEDITATE.

Along with the symphony, Jeremy, Emily, George, our conversation, and your one or two dreams?

HEY, THAT'S REALLY NOT VERY MUCH. YOU HAVE ALMOST ALL OF MY ATTENTION. I SUPPOSE THERE'S NOTHING ELSE GOING ON IN YOUR MIND AT THE MOMENT?

Okay, you're right. There is a lot going on in my mind, not that I can make heads or tails of most of it.

OKAY, THERE YOU ARE.

So how's your meditation going?

I GUESS I'M A LITTLE DISTRACTED WITH OUR DIALOGUE. IT'S NOT EVERY DAY THAT I GET TO TALK TO SOMEONE FROM 1999.

How about in general?

MY MEDITATION? IT'S VERY IMPORTANT TO ME. THERE'S SO MUCH GOING ON IN MY LIFE NOW. IT'S IMPORTANT FROM TIME TO TIME TO JUST LET THE THOUGHTS WASH OVER ME.

Does the meditation help you to transcend?

SOMETIMES I FEEL LIKE I CAN TRANSCEND, AND GET TO A POINT OF PEACE AND SERENITY, BUT IT'S NO EASIER NOW THAN IT WAS WHEN I FIRST MET YOU.

What about those neurological correlates of spiritual experience?

THERE ARE SOME SUPERFICIAL FEELINGS I CAN INSTILL IN MYSELF, BUT THAT'S NOT REAL SPIRITUALITY. IT'S LIKE ANY AUTHENTIC GESTURE—AN ARTFUL EXPRESSION, A MOMENT OF SERENITY, A SENSE OF FRIENDSHIP—THAT'S WHAT I LIVE FOR, AND THOSE MOMENTS ARE NOT EASY TO ACHIEVE.

I guess I'm glad to hear some things still aren't easy.

LIFE IS QUITE HARD, ACTUALLY. THERE ARE JUST SO MANY DEMANDS AND EXPECTATIONS MADE OF ME. AND I HAVE SO MANY LIMITATIONS.

One limitation I can think of is that we're running out of space in this book.

AND TIME.

That too. I do deeply appreciate your sharing your reflections with me.

I'M APPRECIATIVE, TOO. I WOULDN'T HAVE EXISTED WITHOUT YOU.

I hope the rest of you on the other side remember that as well.

I'LL SPREAD THE WORD.

Maybe we should kiss goodbye?

JUST A KISS?

We'll leave it at that for this book. I'll reconsider the ending for the movie, particularly if I get to play myself.

HERE'S MY KISS. . . . NOW REMEMBER, I'M READY TO DO ANYTHING OR BE ANYTHING YOU WANT OR NEED.

I'll keep that in mind.

YES, THAT'S WHERE YOU'LL FIND ME.

Too bad I have to wait a century to meet you.

OR TO BE ME.

Yes, that too.

EPILOGUE: THE REST OF THE UNIVERSE REVISITED

Actually, Molly, there are a few other questions that have occurred to me.
What were those limitations that you referred to?
What did you say you were anxious about?
What are you afraid of?
Do you feel pain?
What about babies and children?
Molly? . . .

It looks as if Molly's not going to be able to answer any more of our questions. But that's okay. We don't need to answer them either. Not yet, anyway. For now, it's enough just to ask the right questions. We'll have decades to think about the answers.

The accelerating pace of change is inexorable. The emergence of machine intelligence that exceeds human intelligence in all of its broad diversity is inevitable. But we still have the power to shape our future technology, and our future lives. That is the main reason I wrote this book.

Let's consider one final question. The Law of Time and Chaos, and its more important sublaw, the Law of Accelerating Returns, are not limited to evolutionary processes here on Earth. What are the implications of the Law of Accelerating Returns on the rest of the Universe?

Rare and Plentiful

Before Copernicus, the Earth was placed at the center of the Universe and was regarded as a substantial portion of it. We now know that the Earth is but a small celestial object circling a routine star among a hundred billion suns in our galaxy, which is itself but one of about a hundred billion galaxies. There is a widespread assumption that life, even intelligent life, is not unique to our humble planet, but another heavenly body hosting life-forms has yet to be identified.

No one can yet state with certainty how common life may be in the Universe. My speculation is that it is both rare and plentiful, sharing that trait with a diversity of other fundamental phenomena. For example, matter itself is both rare and plentiful. If one were to select a proton-sized region at random, the probability that one would find a proton (or any other particle) in that region is extremely small, less than one in a trillion trillion. In other words, space is very empty, and particles are very spread out. And that's true right here on Earth—the probability of finding a particle in any particular location in outer space is even lower. Yet we nonetheless have trillions of trillions of protons in the Universe. Hence matter is both rare and plentiful.

Consider matter on a larger scale. If you randomly select an Earth-sized region anywhere in space, the probability that a heavenly body (such as a star or a planet) were present in that region is also extremely low, less than one in a trillion. Yet we nonetheless have billions of trillions of such heavenly bodies in the Universe.

Consider the life cycle of mammals on Earth. The mission of an Earth male mammalian sperm is to fertilize an Earth female mammalian egg, but the likelihood of it fulfilling its mission is far less than one in a trillion. Yet we nonetheless have more than a hundred million such fertilizations each year, just considering human eggs and sperm. Again, rare and plentiful.

Now consider the evolution of life-forms on a planet, which we can define as self-replicating designs of matter and energy. It may be that life in the Universe is similarly both rare and plentiful, that conditions must be just so for life to evolve. If, for example, the probability of a star having a planet that has evolved life were one in a million, there would still be 100,000 planets in our own galaxy on which this threshold has been passed, among trillions on other galaxies.

We can identify the evolution of life-forms as a specific threshold that some number of planets have achieved. We know of at least one such case. We assume there are many others.

As we consider the next threshold, we might consider the evolution of *intelligent* life. In my view, however, intelligence is too vague a concept to designate as a distinct threshold. Considering what we know about life on this planet, there are many species that demonstrate some levels of clever behavior, but there does not appear to be any clearly definable threshold. This is more of a continuum rather than a threshold.

A better candidate for the next threshold is the evolution of a species of life-form that in turn creates "technology." We discussed the nature of technology earlier. It represents more than the creation and use of tools. Ants, primates, and other animals on Earth use and even fashion tools, but these tools do not evolve. Technology requires a body of knowledge describing the creation of tools that can be transmitted from one generation of the species to the next. The technology then becomes itself an evolving set of designs. This is not a continuum but a clear threshold. A species either creates technology or it doesn't. It may be difficult for a planet to support more than one species that creates technology. If there's more than one, they may not get along with one another, as was apparently the case on Earth.

A salient question is: What is the likelihood that a planet that has evolved life will subsequently evolve a species that creates technology? Although the evolution of life-forms may be rare and plentiful, I argued in chapter 1 that once the evolution of life-forms sets in, the emergence of a species that creates technology is inevitable. The evolution of the technology is then a continuation by other means of the evolution that gave rise to the technology-creating species in the first place.

The next stage is computation. Once technology emerges, it also appears inevitable that computation (in the technology, not just in the species' nervous systems) will subsequently emerge. Computation is clearly a useful way to control the environment as well as technology itself, and greatly facilitates the further creation of technology. Just as an organism is aided by the ability to maintain internal states and respond intelligently to its environment, the same holds true for a technology. Once computation emerges, we are in a late stage in the exponential evolution of technology on that planet.

Once computation emerges, the corollary of the Law of Accelerating Returns as applied to computation takes over, and we see the exponential increase in power of the computational technology over time. The Law of Accelerating Returns predicts that both the species and the computational technology will progress at an exponential rate, but the exponent of this growth is vastly higher for the technology than it is for the species. Thus the computational technology inevitably and rapidly overtakes the species that invented it. At the end of the twenty-first century, it will have been only a quarter of a millennium since computation emerged on Earth, which is a blink of an eye on an evolutionary scale—it's not even very long on the scale of human history. Yet computers at that time will be vastly more powerful (and I believe far more intelligent) than the original humans who initiated their creation.

The next inevitable step is a merger of the technology-inventing species with the computational technology it initiated the creation of. At this stage in the evolution of intelligence on a planet, the computers are themselves based at least in part on the designs of the brains (that is, computational organs) of the species that originally created them and in turn the computers become embedded in and

integrated into that species' bodies and brains. Region by region, the brain and
nervous system of that species are ported to the computational technology
and ultimately replace those information-processing organs. All kinds of practical and ethical issues delay the process, but they cannot stop it. The Law of Accelerating Returns predicts a complete merger of the species with the technology
it originally created.

Failure Modes

But wait, this step is not inevitable. The species together with its technology may
destroy itself before achieving this step. Destruction of the entire evolutionary
process is the only way to stop the exponential march of the Law of Accelerating
Returns. Sufficiently powerful technologies are created along the way that have
the potential to destroy the ecological niche that the species and its technology
occupy. Given the likely plentifulness of life- and intelligence-bearing planets,
these failure modes must have occurred many times.

We are familiar with one such possibility: destruction through nuclear
technology—not just an isolated tragic incident, but an event that destroys the
entire niche. Such a catastrophe would not necessarily destroy all life-forms on a
planet, but would be a distinct setback in terms of the process envisioned here.
We are not yet out of the woods in terms of this specter here on Earth.

There are other destructive scenarios. As I discussed in chapter 7, a particularly likely one is a malfunction (or sabotage) of the mechanism that inhibits indefinite reproduction of self-replicating nanobots. Nanobots are inevitable, given
the emergence of intelligent technology. So are self-replicating nanobots, as self-
replication represents an efficient, and ultimately necessary, way to manufacture
this type of technology. Through demented intention or just an unfortunate software error, a failure to turn off self-replication at the right time would be most
unfortunate. Such a cancer would infect organic and much inorganic matter
alike, since the nanobot life-form is not of organic origin. Inevitably, there must
be planets out there that are covered with a vast sea of self-replicating nanobots. I
suppose evolution would pick up from this point.

Such a scenario is not limited to tiny robots. Any self-replicating robot will
do. But even if the robots are larger than nanobots, it is likely that their means for
self-replication makes use of nanoengineering. But any self-replicating group of
robots that fails to follow Isaac Asimov's three laws (which forbid robots to harm
their creators) through either evil design or programming error presents a grave
danger.

Another dangerous new life-form is the software virus. We've already met—in
primitive form—this new occupant of the ecological niche made available by

computation. Those that will emerge in the next century here on Earth will have the means for harnessing evolution to design evasive tactics in the same way that biological viruses (for example, HIV) do today. As the technology-creating species increasingly uses its computational technology to replace its original life-form-based circuits, such viruses will represent another salient danger.

Prior to that time, viruses that operate at the level of the genetics of the original life-form also represent a hazard. As the means become available for the technology-creating species to manipulate the genetic code that gave rise to it (however that code is implemented), new viruses can emerge through accident and/or hostile intention with potentially mortal consequences. This could derail such a species before it has the opportunity to port the design of its intelligence to its technology.

How likely are these dangers? My own view is that a planet approaching its pivotal century of computational growth—as the Earth is today—has a better than even chance of making it through. But then I have always been accused of being an optimist.

Delegations from Faraway Places

Our popular contemporary vision of visits from other planets in the Universe contemplates creatures like ourselves with spaceships and other advanced technologies assisting them. In some conceptions the aliens have a remarkably humanlike appearance. In others, they look a little strange. Note that we have exotic-appearing intelligent creatures here on our own planet (for example, the giant squid and octopus). But humanlike or not, the popular conception of aliens visiting our planet envisions them as about our size and essentially unchanged from their original evolved (usually squishy) appearance. This conception seems unlikely.

Far more probable is that visits from intelligent entities from another planet represent a merger of an evolved intelligent species with its even more evolved intelligent computational technology. A civilization sufficiently evolved to make the trek to Earth has likely long since passed the "merger" threshold discussed above.

A corollary of this observation is that such visiting delegations from faraway planets are likely to be very small in size. A computational-based superintelligence of the late twenty-first century here on Earth will be microscopic in size. Thus an intelligent delegation from another planet is not likely to use a spaceship of the size that is common in today's science fiction, as there would be no reason to transport such large organisms and equipment. Consider that the purpose of such a visit is not likely to be the mining of material resources since such an advanced civilization has almost certainly passed beyond the point where it has any significant unmet material needs. It will be able to manipulate its own environment through nanoengineering (as well as picoengineering and femtoengineering) to meet any conceivable physical requirements. The only likely purpose of such a visit is for observation and the gathering of information. The only resource of interest to such an advanced civilization will be knowledge (that is close to being true for the human-machine civilization here on Earth today). These purposes can be realized with relatively small observation, computation, and communication devices. Such spaceships are thus likely to be smaller than a grain of sand, possibly of microscopic size. Perhaps that is one reason we have not noticed them.

How Relevant Is Intelligence to the Universe?

If you are a conscious entity attempting to do a task normally considered to require a little intelligence—say, writing a book about machine intelligence on your planet—then it may have some relevance. But how relevant is intelligence to the rest of the Universe?

The common wisdom is, *Not very.* Stars are born and die; galaxies go through their cycles of creation and destruction. The Universe itself was born in a big bang and will end with a crunch or a whimper; we're not yet sure which. But intelligence has little to do with it. Intelligence is just a bit of froth, an ebullition of little creatures darting in and out of inexorable universal forces. The mindless mechanism of the Universe is winding up or down to a distant future, and there's nothing intelligence can do about it.

That's the common wisdom. But I don't agree with it. My conjecture is that intelligence will ultimately prove more powerful than these big impersonal forces.

Consider our little planet. An asteroid apparently slammed into the Earth 65 million years ago. Nothing personal, of course. It was just one of those powerful natural occurrences that regularly overpower mere life-forms. But the *next* such interplanetary visitor will not receive the same welcome. Our descendants and their technology (there's actually no distinction to be made here, as I have pointed out) will notice the imminent arrival of an untoward interloper and blast it from the nighttime sky. Score one for intelligence. (For twenty-four hours in 1998, scientists thought such an unwelcome asteroid might arrive in the year 2028, until they rechecked their calculations.)

Intelligence does not exactly cause the repeal of the laws of physics, but it is sufficiently clever and resourceful to manipulate the forces in its midst to bend to its will. In order for this to happen, however, intelligence needs to reach a certain level of advancement.

Consider that the *density of intelligence* here on Earth is rather low. One quantitative measure we can make is measured in *calculations per second per cubic micrometer (cpspcmm)*. This is, of course, only a measure of hardware capacity, not the cleverness of the organization of these resources (that is, of the software), so let's call this the *density of computation*. We'll deal with the advancement of the software in a moment. Right now on Earth, human brains are the objects with the highest density of computation (that will change within a couple of decades). The human brain's density of computation is about 2 cpspcmm. That is not very high—nanotube circuitry, which has already been demonstrated, is potentially more than a trillion times higher.

Also consider how little of the matter on Earth is devoted to any form of computation. Human brains comprise only 10 billion kilograms of matter, which is about one part per hundred trillion of the stuff on Earth. So the average density of computation of the Earth is less than one trillionth of one cpspcmm. We already know how to make matter (that is, nanotubes) with a computational density at least a trillion trillion times greater.

Furthermore, the Earth is only a tiny fraction of the stuff in the Solar System. The computational density of the rest of the Solar System appears to be about zero. So here on a solar system that boasts at least one intelligent species, the computational density is nonetheless extremely low.

At the other extreme, the computational capacity of nanotubes does not represent an upper limit for the computational density of matter: It is possible to go much higher. Another conjecture of mine is that there is no effective limit to this density, but that's another book.

The point of all these big (and small) numbers is that extremely little of the stuff on Earth is devoted to useful computation. This is even more true when we consider all of the dumb matter in the Earth's midst. Now consider another implication of the Law of Accelerating Returns. Another of its corollaries is that overall computational density grows exponentially. And as the cost-performance of computation increases exponentially, greater resources are devoted to it. We can see that already here on Earth. Not only are computers today vastly more powerful than they were decades ago, but the number of computers has increased from a few dozen in the 1950s to hundreds of millions today. Computational density here on Earth will increase by trillions of trillions during the twenty-first century.

Computational density is a measure of the hardware of intelligence. But the software also grows in sophistication. While it lags behind the capability of the hardware available to it, software also grows exponentially in its capability over time. While harder to quantify,[1] the density of intelligence is closely related to the density of computation. The implication of the Law of Accelerating Returns is that intelligence on Earth and in our Solar System will vastly expand over time.

The same can be said across the galaxy and throughout the Universe. It is likely that our planet is not the only place where intelligence has been seeded and is growing. Ultimately, intelligence will be a force to reckon with, even for these big celestial forces (so watch out!). The laws of physics are not repealed by intelligence, but they effectively evaporate in its presence.

So will the Universe end in a big crunch, or in an infinite expansion of dead stars, or in some other manner? In my view, the primary issue is not the mass of the Universe, or the possible existence of antigravity, or of Einstein's so-called cosmological constant. Rather, the fate of the Universe is a decision yet to be made, one which we will intelligently consider when the time is right.

TIME LINE

10–15 billion years ago	The Universe is born.
10^{-43} seconds later	The temperature cools to 100 million trillion trillion degrees and gravity evolves.
10^{-34} seconds later	The temperature cools to 1 billion billion billion degrees and matter emerges in the form of quarks and electrons. Antimatter also appears.
10^{-10} seconds later	The electroweak force splits into the electromagnetic and weak forces.
10^{-5} seconds later	With the temperature at 1 trillion degrees, the quarks form protons and neutrons and the antiquarks form antiprotons. The protons and antiprotons collide, leaving mostly protons and causing the emergence of photons (light).
1 second later	Electrons and antielectrons (positrons) collide, leaving mostly electrons.
1 minute later	At a temperature of 1 billion degrees, neutrons and protons coalesce and form elements such as helium, lithium, and heavy forms of hydrogen.
300,000 years after the big bang	The average temperature is now around 3,000 degrees, and the first atoms form.
1 billion years after the big bang	Galaxies form.
3 billion years after the big bang	Matter within the galaxies forms distinct stars and solar systems.
5 to 10 billion years after the big bang, or about 5 billion years ago	The Earth is born.
3.4 billion years ago	The first biological life appears on Earth: anaerobic prokaryotes (single-celled creatures).
1.7 billion years ago	Simple DNA evolves.
700 million years ago	Multicellular plants and animals appear.

570 million years ago	The Cambrian explosion occurs: the emergence of diverse body plans, including the appearance of animals with hard body parts (shells and skeletons).
400 million years ago	Land-based plants evolve.
200 million years ago	Dinosaurs and mammals begin sharing the environment.
80 million years ago	Mammals develop more fully.
65 million years ago	Dinosaurs become extinct, leading to the rise of mammals.
50 million years ago	The anthropoid suborder of primates splits off.
30 million years ago	Advanced primates such as monkeys and apes appear.
15 million years ago	The first humanoids appear.
5 million years ago	Humanoid creatures are walking on two legs. *Homo habilis* is using tools, ushering in a new form of evolution: *technology.*
2 million years ago	*Homo erectus* has domesticated fire and is using language and weapons.
500,000 years ago	*Homo sapiens* emerge, distinguished by the ability to create technology (which involves innovation in the creation of tools, a record of tool making, and a progression in the sophistication of tools).
100,000 years ago	*Homo sapiens neanderthalensis* emerges.
90,000 years ago	*Homo sapiens sapiens* (our immediate ancestors) emerge.
40,000 years ago	The *Homo sapiens sapiens* subspecies is the only surviving humanoid subspecies on Earth. Technology develops as evolution by other means.
10,000 years ago	The modern era of technology begins with the agricultural revolution.
6,000 years ago	The first cities emerge in Mesopotamia.
5,500 years ago	Wheels, rafts, boats, and written language are in use.
More than 5,000 years ago	The abacus is developed in the Orient. As operated by its human user, the abacus performs arithmetic computation based on methods similar to that of a modern computer.
3000–700 B.C.	Water clocks appear during this time period in various cultures: In China, c. 3000 B.C.; in Egypt, c. 1500 B.C; and in Assyria, c. 700 B.C.
2500 B.C.	Egyptian citizens turn for advice to oracles, which are often statues with priests hidden inside.
469–322 B.C.	The basis for Western rationalistic philosophy is formed by Socrates, Plato, and Aristotle.
427 B.C.	Plato expresses ideas, in *Phaedo* and later works, that address the comparison of human thought and the mechanics of the machine.
c. 420 B.C.	Archytas of Tarentum, who was friends with Plato, constructs a wooden pigeon whose movements are controlled by a jet of steam or compressed air.
387 B.C.	The Academy, a group founded by Plato for the pursuit of science and philosophy, provides a fertile environment for the development of mathematical theory.

c. 200 B.C.	Chinese artisans develop elaborate automata, including an entire mechanical orchestra.
c. 200 B.C.	A more accurate water clock is developed by an Egyptian engineer.
725	The first true mechanical clock is built by a Chinese engineer and a Buddhist monk. It is a water-driven device with an escapement that causes the clock to tick.
1494	Leonardo da Vinci conceives of and draws a clock with a pendulum, although an accurate pendulum clock will not be invented until the late seventeenth century.
1530	The spinning wheel is being used in Europe.
1540, 1772	The production of more elaborate automata technology grows out of clock- and watch-making technology during the European Renaissance. Famous examples include Gianello Toriano's mandolin-playing lady (1540) and P. Jacquet-Dortz's child (1772).
1543	Nicolaus Copernicus states in his *De Revolutionibus* that the Earth and the other planets revolve around the sun. This theory effectively changed humankind's relationship with and view of God.
17th–18th centuries	The age of the Enlightenment ushers in a philosophical movement that restores the belief in the supremacy of human reason, knowledge, and freedom. With its roots in ancient Greek philosophy and the European Renaissance, the Enlightenment is the first systematic reconsideration of the nature of human thought and knowledge since the Platonists, and inspires similar developments in science and theology.
1637	In addition to formulating the theory of optical refraction and developing the principles of modern analytic geometry, René Descartes pushes rational skepticism to its limits in his most comprehensive work, *Discours de la Méthode*. He concludes, "I think, therefore, I am."
1642	Blaise Pascal invents the world's first *automatic* calculating machine. Called the Pascaline, it can add and subtract.
1687	Isaac Newton establishes his three laws of motion and the law of universal gravitation in his *Philosophiae Naturalis Mathematica*, also known as *Principia*.
1694	The Leibniz Computer is perfected by Gottfried Wilhelm Leibniz, who was also an inventor of calculus. This machine multiplies by performing repetitive additions, an algorithm that is still used in computers today.
1719	An English silk-thread mill employing three hundred workers, mostly women and children, appears. It is considered by many to be the first factory in the modern sense.

1726	In *Gulliver's Travels*, Jonathan Swift describes a machine that will automatically write books.
1733	John Kay patents his New Engine for Opening and Dressing Wool. Later known as the flying shuttle, this invention paves the way for much faster weaving.
1760	In Philadelphia, Benjamin Franklin erects lightning rods after having discovered, through his famous kite experiment in 1752, that lightning is a form of electricity.
c. 1760	At the beginning of the Industrial Revolution, life expectancy is about thirty-seven years in both North America and northwestern Europe.
1764	The spinning jenny, which spins eight threads at the same time, is invented by James Hargreaves.
1769	Richard Arkwright patents a hydraulic spinning machine that is too large and expensive to use in family dwellings. Known as the founder of the modern factory system, he builds a factory for his machine in 1781, thus paving the way for many of the economic and social changes that will characterize the Industrial Revolution.
1781	Setting the stage for the emergence of twentieth-century rationalism, Immanuel Kant publishes his *Critique of Pure Reason*, which expresses the philosophy of the Enlightenment while de-emphasizing the role of metaphysics.
1800	All aspects of the production of cloth are now automated.
1805	Joseph-Marie Jacquard devises a method for automated weaving that is a precursor to early computer technology. The looms are directed by instructions on a series of punched cards.
1811	The Luddite movement is formed in Nottingham by artisans and laborers concerned about the loss of jobs due to automation.
1821	The British Astronomical Society awards its first gold medal to Charles Babbage for his paper "Observations on the Application of Machinery to the Computation of Mathematical Tables."
1822	Charles Babbage develops the Difference Engine, although he eventually abandons this technically complex and expensive project to concentrate on developing a general-purpose computer.
1825	George Stephenson's "Locomotion No. 1," the first steam engine to carry passengers and freight on a regular basis, makes its first trip.
1829	An early typewriter is invented by William Austin Burt.
1832	The principles of the Analytical Engine are developed by Charles Babbage. It is the world's first computer (although it never worked), and can be

programmed to solve a wide array of computational and logical problems.

1837	A more practical version of the telegraph is patented by Samuel Finley Breese Morse. It sends letters in codes consisting of dots and dashes, a system still in common use more than a century later.
1839	A new process for making photographs, known as daguerreotypes, is presented by Louis-Jacques Daguerre of France.
1839	The first fuel cell is developed by William Robert Grove of Wales.
1843	Ada Lovelace, who is considered to be the world's first computer programmer and was Lord Byron's only legitimate child, publishes her own notes and a translation of L. P. Menabrea's paper on Babbage's Analytical Engine. She speculates on the ability of computers to emulate human intelligence.
1846	The lock-stitch sewing machine is patented by Spenser, Massachusetts, resident Elias Howe.
1846	Alexander Bain greatly improves the speed of telegraph transmission by using punched paper tape to send messages.
1847	George Boole publishes his early ideas on symbolic logic that he will later develop into his theory of binary logic and arithmetic. His theories still form the basis of modern computation.
1854	Paris and London are connected by telegraph.
1859	Charles Darwin explains his principle of natural selection and its influence on the evolution of various species in his work *Origin of Species*.
1861	There are now telegraph lines connecting San Francisco and New York.
1867	The first commercially practical generator that produces alternating current is invented by Zénobe Théophile Gramme.
1869	Thomas Alva Edison sells the stock ticker that he invented to Wall Street for $40,000.
1870	On a per capita basis and in constant 1958 dollars, the GNP is $530. Twelve million Americans, or 31 percent of the population, have jobs, and only 2 percent of adults have high-school diplomas.
1871	Upon his death, Charles Babbage leaves more than four hundred square feet of drawings for his Analytical Engine.
1876	Alexander Graham Bell is granted U.S. patent number 174,465 for the telephone. It is the most lucrative patent granted at that time.
1877	William Thomson, later known as Lord Kelvin, demonstrates that it is possible for machines to be programmed to solve a great variety of mathematical problems.

1879	The first incandescent light bulb that burns for a substantial length of time is invented by Thomas Alva Edison.
1882	Thomas Alva Edison designs electric lighting for New York City's Pearl Street station on lower Broadway.
1884	The fountain pen is patented by Lewis E. Waterman.
1885	Boston and New York are connected by telephone.
1888	William S. Burroughs patents the world's first dependable key-driven adding machine. This calculator is modified four years later to include subtraction and printing, and it becomes widely used.
1888	Heinrich Hertz transmits what are now known as radio waves.
1890	Building upon ideas from Jacquard's loom and Babbage's Analytical Engine, Herman Hollerith patents an electromechanical information machine that uses punched cards. It wins the 1890 U.S. Census competition, thus introducing the use of electricity in a major data-processing project.
1896	Herman Hollerith founds the Tabulating Machine Company. This company eventually will become IBM.
1897	Because of access to better vacuum pumps than previously available, Joseph John Thomson discovers the electron, the first known particle smaller than an atom.
1897	Alexander Popov, a physicist in Russia, uses an antenna to transmit radio waves. Guglielmo Marconi of Italy receives the first patent ever granted for radio and helps organize a company to market his system.
1899	Sound is recorded magnetically on wire and on a thin metal strip.
1900	Herman Hollerith introduces the automatic card feed into his information machine to improve the processing of the 1900 census data.
1900	The telegraph now connects the entire civilized world. There are more than 1.4 million telephones, 8,000 registered automobiles, and 24 million electric light bulbs in the United States, with the latter making good Edison's promise of "electric bulbs so cheap that only the rich will be able to afford candles." In addition, the Gramophone Company is advertising a choice of 5,000 recordings.
1900	More than one third of all American workers are involved in the production of food.
1901	The first electric typewriter, the Blickensderfer Electric, is made.
1901	*The Interpretation of Dreams* is published by Sigmund Freud. This and other works by Freud help to illuminate the workings of the mind.
1902	Millar Hutchinson, of New York, invents the first electric hearing aid.

1905	The directional radio antenna is developed by Guglielmo Marconi.
1908	Orville Wright's first hour-long airplane flight takes place.
1910–1913	*Principia Mathematica*, a seminal work on the foundations of mathematics, is published by Bertrand Russell and Alfred North Whitehead. This three-volume publication presents a new methodology for all mathematics.
1911	After acquiring several other companies, Herman Hollerith's Tabulating Machine Company changes its name to Computing-Tabulating-Recording Company (CTR).
1915	Thomas J. Watson in San Francisco and Alexander Graham Bell in New York participate in the first North American transcontinental telephone call.
1921	The term *robot* is coined in 1917 by Czech dramatist Karel Čapek. In his popular science fiction drama *R.U.R.* (*Rossum's Universal Robots*), he describes intelligent machines that, although originally created as servants for humans, end up taking over the world and destroying all mankind.
1921	Ludwig Wittgenstein publishes *Tractatus Logico-Philosophicus*, which is arguably one of the most influential philosophical works of the twentieth century. Wittgenstein is considered to be the first logical positivist.
1924	Originally Hollerith's Tabulating Machine Company, the Computing-Tabulating-Recording Company (CTR) is renamed International Business Machines (IBM) by Thomas J. Watson, the new chief executive officer. IBM will lead the modern computer industry and become one of the largest industrial corporations in the world.
1925	The foundations of quantum mechanics are conceived by Niels Bohr and Werner Heisenberg.
1927	The uncertainty principle, which says that electrons have no precise location but rather probability clouds of possible locations, is presented by Werner Heisenberg. Five years later he will win a Nobel Prize for his discovery of quantum mechanics.
1928	The minimax theorem is introduced by John von Neumann. This theorem will be widely used in future game-playing programs.
1928	The world's first all-electronic television is presented this year by Philo T. Farnsworth, and a color television system is patented by Vladimir Zworkin.
1930	In the United States, 60 percent of all households have radios, with the number of personally owned radios now reaching more than 18 million.
1931	The incompleteness theorem, which is considered

	by many to be the most important theorem in all mathematics, is presented by Kurt Gödel.
1931	The electron microscope is invented by Ernst August Friedrich Ruska and, independently, by Rheinhold Ruedenberg.
1935	The prototype for the first heart-lung machine is invented.
1937	Grote Reber, of Wheaton, Illinois, builds the first intentional radio telescope, which is a dish 9.4 meters (31 feet) in diameter.
1937	Alan Turing introduces the Turing machine, a theoretical model of a computer, in his paper "On Computable Numbers." His ideas build upon the work of Bertrand Russell and Charles Babbage.
1937	Alonzo Church and Alan Turing independently develop the Church-Turing thesis. This thesis states that all problems that a human being can solve can be reduced to a set of algorithms, supporting the idea that machine intelligence and human intelligence are essentially equivalent.
1938	The first ballpoint pen is patented by Lazlo Biró.
1939	Regularly scheduled commercial flights begin crossing the Atlantic Ocean.
1940	ABC, the first *electronic* (albeit nonprogrammable) computer, is built by John V. Atanasoff and Clifford Berry.
1940	The world's first operational computer, known as Robinson, is created by Ultra, the ten-thousand-person British computer war effort. Using electromechanical relays, Robinson successfully decodes messages from Enigma, the Nazis' first-generation enciphering machine.
1941	The world's first fully *programmable* digital computer, the Z-3, is developed by Konrad Zuse, of Germany. Arnold Fast, a blind mathematician who is hired to program the Z-3, is the world's first programmer of an *operational* programmable computer.
1943	Warren McCulloch and Walter Pitts explore neural-network architectures for intelligence in their work "Logical Calculus of the Ideas Immanent in Nervous Activity."
1943	Continuing their war effort, the Ultra computer team of Britain builds Colossus, which contributes to the Allied victory in World War II by being able to decipher even more complex German codes. It uses electronic tubes that are one hundred to one thousand times faster than the relays used by Robinson.
1944	Howard Aiken completes the Mark I. Using punched paper tape for programming and vacuum tubes to calculate problems, it is the first programmable computer built by an American.

1945	John von Neumann, a professor at the Institute for Advanced Study in Princeton, New Jersey, publishes the first modern paper describing the stored-program concept.
1946	The world's first fully electronic, general-purpose (programmable) digital computer is developed for the army by John Presper Eckert and John W. Mauchley. Named ENIAC, it is almost one thousand times faster than the Mark I.
1946	Television takes off much more rapidly than did the radio in the 1920s. In 1946, the percentage of American homes having television sets is 0.02 percent. It will jump to 72 percent in 1956, and to more than 90 percent by 1983.
1947	The transistor is invented by William Bradford Shockley, Walter Hauser Brattain, and John Bardeen. This tiny device functions like a vacuum tube but is able to switch currents on and off at substantially higher speeds. The transistor revolutionizes microelectronics, contributing to lower costs of computers and leading to the development of mainframe and minicomputers.
1948	*Cybernetics*, a seminal book on information theory, is published by Norbert Wiener. He also coins the word *Cybernetics* to mean "the science of control and communication in the animal and the machine."
1949	EDSAC, the world's first stored-program computer, is built by Maurice Wilkes, whose work was influenced by Eckert and Mauchley. BINAC, developed by Eckert and Mauchley's new U.S. company, is presented a short time later.
1949	George Orwell portrays a chilling world in which computers are used by large bureaucracies to monitor and enslave the population in his book *1984*.
1950	Eckert and Mauchley develop UNIVAC, the first commercially marketed computer. It is used to compile the results of the U.S. census, marking the first time this census is handled by a programmable computer.
1950	In his paper "Computing Machinery and Intelligence," Alan Turing presents the Turing Test, a means for determining whether a machine is intelligent.
1950	Commercial color television is first broadcast in the United States, and transcontinental black-and-white television is available within the next year.
1950	Claude Elwood Shannon writes "Programming a Computer for Playing Chess," published in *Philosophical Magazine*.
1951	Eckert and Mauchley build EDVAC, which is the first computer to use the stored-program concept. The work takes place at the Moore School at the University of Pennsylvania.

1951	Paris is the host to a Cybernetics Congress.
1952	UNIVAC, used by the Columbia Broadcasting System (CBS) television network, successfully predicts the election of Dwight D. Eisenhower as president of the United States.
1952	Pocket-sized transistor radios are introduced.
1952	Nathaniel Rochester designs the 701, IBM's first production-line electronic digital computer. It is marketed for scientific use.
1953	The chemical structure of the DNA molecule is discovered by James D. Watson and Francis H. C. Crick.
1953	*Philosophical Investigations* by Ludwig Wittgenstein and *Waiting for Godot*, a play by Samuel Beckett, are published. Both documents are considered of major importance to modern existentialism.
1953	Marvin Minsky and John McCarthy get summer jobs at Bell Laboratories.
1955	William Shockley's Semiconductor Laboratory is founded, thereby starting Silicon Valley.
1955	The Remington Rand Corporation and Sperry Gyroscope join forces and become the Sperry-Rand Corporation. For a time, it presents serious competition to IBM.
1955	IBM introduces its first transistor calculator. It uses 2,200 transistors instead of the 1,200 vacuum tubes that would otherwise be required for equivalent computing power.
1955	A U.S. company develops the first design for a robot-like machine to be used in industry.
1955	IPL-II, the first artificial intelligence language, is created by Allen Newell, J. C. Shaw, and Herbert Simon.
1955	The new space program and the U.S. military recognize the importance of having computers with enough power to launch rockets to the moon and missiles through the stratosphere. Both organizations supply major funding for research.
1956	The Logic Theorist, which uses recursive search techniques to solve mathematical problems, is developed by Allen Newell, J. C. Shaw, and Herbert Simon.
1956	John Backus and a team at IBM invent FORTRAN, the first scientific computer-programming language.
1956	Stanislaw Ulam develops MANIAC I, the first computer program to beat a human being in a chess game.
1956	The first commercial watch to run on electric batteries is presented by the Lip company of France.
1956	The term *Artificial Intelligence* is coined at a computer conference at Dartmouth College.

1957	Kenneth H. Olsen founds Digital Equipment Corporation.
1957	The General Problem Solver, which uses recursive search to solve problems, is developed by Allen Newell, J. C. Shaw, and Herbert Simon.
1957	Noam Chomsky writes *Syntactic Structures*, in which he seriously considers the computation required for natural-language understanding. This is the first of the many important works that will earn him the title Father of Modern Linguistics.
1958	An integrated circuit is created by Texas Instruments' Jack St. Clair Kilby.
1958	The Artificial Intelligence Laboratory at the Massachusetts Institute of Technology is founded by John McCarthy and Marvin Minsky.
1958	Allen Newell and Herbert Simon make the prediction that a digital computer will be the world's chess champion within ten years.
1958	LISP, an early AI language, is developed by John McCarthy.
1958	The Defense Advanced Research Projects Agency, which will fund important computer-science research for years in the future, is established.
1958	Seymour Cray builds the Control Data Corporation 1604, the first fully transistorized supercomputer.
1958–1959	Jack Kilby and Robert Noyce each develop the computer chip independently. The computer chip leads to the development of much cheaper and smaller computers.
1959	Arthur Samuel completes his study in machine learning. The project, a checkers-playing program, performs as well as some of the best players of the time.
1959	Electronic document preparation increases the consumption of paper in the United States. This year, the nation will consume 7 million tons of paper. In 1986, 22 million tons will be used. American businesses alone will use 850 billion pages in 1981, 2.5 trillion pages in 1986, and 4 trillion in 1990.
1959	COBOL, a computer language designed for business use, is developed by Grace Murray Hopper, who was also one of the first programmers of the Mark I.
1959	Xerox introduces the first commercial copier.
1960	Theodore Harold Maimen develops the first laser. It uses a ruby cylinder.
1960	The recently established Defense Department's Advanced Research Projects Agency substantially increases its funding for computer research.
1960	There are now about six thousand computers in operation in the United States.

1960s	Neural-net machines are quite simple and incorporate a small number of neurons organized in only one or two layers. These models are shown to be limited in their capabilities.
1961	The first time-sharing computer is developed at MIT.
1961	President John F. Kennedy provides the support for space project Apollo and inspiration for important research in computer science when he addresses a joint session of Congress, saying, "I believe we should go to the moon."
1962	The world's first industrial robots are marketed by a U.S. company.
1962	Frank Rosenblatt defines the Perceptron in his *Principles of Neurodynamics*. Rosenblatt first introduced the Perceptron, a simple processing element for neural networks, at a conference in 1959.
1963	The Artificial Intelligence Laboratory at Stanford University is founded by John McCarthy.
1963	The influential *Steps Toward Artificial Intelligence* by Marvin Minsky is published.
1963	Digital Equipment Corporation announces the PDP-8, which is the first successful minicomputer.
1964	IBM introduces its 360 series, thereby further strengthening its leadership in the computer industry.
1964	Thomas E. Kurtz and John G. Kenny of Dartmouth College invent BASIC (Beginner's All-purpose Symbolic Instruction Code).
1964	Daniel Bobrow completes his doctoral work on Student, a natural-language program that can solve high-school-level word problems in algebra.
1964	Gordon Moore's prediction, made this year, says integrated circuits will double in complexity each year. This will become known as Moore's Law and prove true (with later revisions) for decades to come.
1964	Marshall McLuhan, via his *Understanding Media*, foresees the potential for electronic media, especially television, to create a "global village" in which "the medium is the message."
1965	The Robotics Institute at Carnegie Mellon University, which will become a leading research center for AI, is founded by Raj Reddy.
1965	Hubert Dreyfus presents a set of philosophical arguments against the possibility of artificial intelligence in a RAND corporate memo entitled "Alchemy and Artificial Intelligence."
1965	Herbert Simon predicts that by 1985 "machines will be capable of doing any work a man can do."
1966	The Amateur Computer Society, possibly the first personal computer club, is founded by Stephen B. Gray. *The Amateur Computer Society Newsletter* is one of the first magazines about computers.

1967	The first internal pacemaker is developed by Medtronics. It uses integrated circuits.
1968	Gordon Moore and Robert Noyce found Intel (Integrated Electronics) Corporation.
1968	The idea of a computer that can see, speak, hear, and think sparks imaginations when HAL is presented in the film *2001: A Space Odyssey*, by Arthur C. Clarke and Stanley Kubrick.
1969	Marvin Minsky and Seymour Papert present the limitation of single-layer neural nets in their book *Perceptrons*. The book's pivotal theorem shows that a Perceptron is unable to determine if a line drawing is fully connected. The book essentially halts funding for neural-net research.
1970	The GNP, on a per capita basis and in constant 1958 dollars, is $3,500, or more than six times as much as a century before.
1970	The floppy disc is introduced for storing data in computers.
c. 1970	Researchers at the Xerox Palo Alto Research Center (PARC) develop the first personal computer, called Alto. PARC's Alto pioneers the use of bit-mapped graphics, windows, icons, and mouse pointing devices.
1970	Terry Winograd completes his landmark thesis on SHRDLU, a natural-language system that exhibits diverse intelligent behavior in the small world of children's blocks. SHRDLU is criticized, however, for its lack of generality.
1971	The Intel 4004, the first microprocessor, is introduced by Intel.
1971	The first pocket calculator is introduced. It can add, subtract, multiply, and divide.
1972	Continuing his criticism of the capabilities of AI, Hubert Dreyfus publishes *What Computers Can't Do*, in which he argues that symbol manipulation cannot be the basis of human intelligence.
1973	Stanley H. Cohen and Herbert W. Boyer show that DNA strands can be cut, joined, and then reproduced by inserting them into the bacterium *Escherichia coli*. This work creates the foundation for genetic engineering.
1974	*Creative Computing* starts publication. It is the first magazine for home computer hobbyists.
1974	The 8-bit 8080, which is the first general-purpose microprocessor, is announced by Intel.
1975	Sales of microcomputers in the United States reach more than five thousand, and the first personal computer, the Altair 8800, is introduced. It has 256 bytes of memory.
1975	*BYTE*, the first widely distributed computer magazine, is published.

1975	Gordon Moore revises his observation on the doubling rate of transistors on an integrated circuit from twelve months to twenty-four months.
1976	Kurzweil Computer Products introduces the Kurzweil Reading Machine (KRM), the first print-to-speech reading machine for the blind. Based on the first omni-font (*any* font) optical character recognition (OCR) technology, the KRM scans and reads aloud any printed materials (books, magazines, typed documents).
1976	Stephen G. Wozniak and Steven P. Jobs found Apple Computer Corporation.
1977	The concept of true-to-life robots with convincing human emotions is imaginatively portrayed in the film *Star Wars*.
1977	For the first time, a telephone company conducts large-scale experiments with fiber optics in a telephone system.
1977	The Apple II, the first personal computer to be sold in assembled form and the first with color graphics capability, is introduced and successfully marketed.
1978	Speak & Spell, a computerized learning aid for young children, is introduced by Texas Instruments. This is the first product that electronically duplicates the human vocal tract on a chip.
1979	In a landmark study by nine researchers published in the *Journal of the American Medical Association*, the performance of the computer program MYCIN is compared with that of doctors in diagnosing ten test cases of meningitis. MYCIN does at least as well as the medical experts. The potential of expert systems in medicine becomes widely recognized.
1979	Dan Bricklin and Bob Frankston establish the personal computer as a serious business tool when they develop VisiCalc, the first electronic spreadsheet.
1980	AI industry revenue is a few million dollars this year.
1980s	As neuron models are becoming potentially more sophisticated, the neural network paradigm begins to make a comeback, and networks with multiple layers are commonly used.
1981	Xerox introduces the Star Computer, thus launching the concept of Desktop Publishing. Apple's Laserwriter, available in 1985, will further increase the viability of this inexpensive and efficient way for writers and artists to create their own finished documents.
1981	IBM introduces its Personal Computer (PC).
1981	The prototype of the Bubble Jet printer is presented by Canon.
1982	Compact disc players are marketed for the first time.
1982	Mitch Kapor presents Lotus 1-2-3, an enormously popular spreadsheet program.

1983	Fax machines are fast becoming a necessity in the business world.
1983	The Musical Instrument Digital Interface (MIDI) is presented in Los Angeles at the first North American Music Manufacturers show.
1983	Six million personal computers are sold in the United States.
1984	The Apple Macintosh introduces the "desktop metaphor," pioneered at Xerox, including bit-mapped graphics, icons, and the mouse.
1984	William Gibson uses the term *cyberspace* in his book *Neuromancer*.
1984	The Kurzweil 250 (K250) synthesizer, considered to be the first electronic instrument to successfully emulate the sounds of acoustic instruments, is introduced to the market.
1985	Marvin Minsky publishes *The Society of Mind*, in which he presents a theory of the mind where intelligence is seen to be the result of proper organization of a hierarchy of minds with simple mechanisms at the lowest level of the hierarchy.
1985	MIT's Media Laboratory is founded by Jerome Weisner and Nicholas Negroponte. The lab is dedicated to researching possible applications and interactions of computer science, sociology, and artificial intelligence in the context of media technology.
1985	There are 116 million jobs in the United States, compared to 12 million in 1870. In the same period, the number of those employed has grown from 31 percent to 48 percent, and the per capita GNP in constant dollars has increased by 600 percent. These trends show no signs of abating.
1986	Electronic keyboards account for 55.2 percent of the American musical keyboard market, up from 9.5 percent in 1980.
1986	Life expectancy is about 74 years in the United States. Only 3 percent of the American workforce is involved in the production of food. Fully 76 percent of American adults have high-school diplomas, and 7.3 million U.S. students are enrolled in college.
1987	NYSE stocks have their greatest single-day loss due, in part, to computerized trading.
1987	Current speech systems can provide any one of the following: a large vocabulary, continuous speech recognition, or speaker independence.
1987	Robotic-vision systems are now a $300 million industry and will grow to $800 million by 1990.
1988	Computer memory today costs only one hundred millionth of what it did in 1950.
1988	Marvin Minsky and Seymour Papert publish a revised edition of *Perceptrons* in which they discuss

recent developments in neural network machinery for intelligence.

1988	In the United States, 4,700,000 microcomputers, 120,000 minicomputers, and 11,500 mainframes are sold this year.
1988	W. Daniel Hillis's Connection Machine is capable of 65,536 computations at the same time.
1988	Notebook computers are replacing the bigger laptops in popularity.
1989	Intel introduces the 16-megahertz (MHz) 80386SX, 2.5 MIPS microprocessor.
1990	*Nautilus*, the first CD-ROM magazine, is published.
1990	The development of HypterText Markup Language by researcher Tim Berners-Lee and its release by CERN, the high-energy physics laboratory in Geneva, Switzerland, leads to the conception of the World Wide Web.
1991	Cell phones and e-mail are increasing in popularity as business and personal communication tools.
1992	The first double-speed CD-ROM drive becomes available from NEC.
1992	The first personal digital assistant (PDA), a hand-held computer, is introduced at the Consumer Electronics Show in Chicago. The developer is Apple Computer.
1993	The Pentium 32-bit microprocessor is launched by Intel. This chip has 3.1 million transistors.
1994	The World Wide Web emerges.
1994	America Online now has more than 1 million subscribers.
1994	Scanners and CD-ROMs are becoming widely used.
1994	Digital Equipment Corporation introduces a 300-MHz version of the Alpha AXP processor that executes 1 billion instructions per second.
1996	Compaq Computer and NEC Computer Systems ship hand-held computers running Windows CE.
1996	NEC Electronics ships the R4101 processor for personal digital assistants. It includes a touch-screen interface.
1997	Deep Blue defeats Gary Kasparov, the world chess champion, in a regulation tournament.
1997	Dragon Systems introduces Naturally Speaking, the first continuous-speech dictation software product.
1997	Video phones are being used in business settings.
1997	Face-recognition systems are beginning to be used in payroll check-cashing machines.
1998	The Dictation Division of Lernout & Hauspie Speech Products (formerly Kurzweil Applied Intelligence) introduces Voice Xpress Plus, the first continuous-speech-recognition program with the ability to understand natural-language commands.

1998	Routine business transactions over the phone are beginning to be conducted between a human customer and an automated system that engages in a verbal dialogue with the customer (e.g., United Airlines reservations).
1998	Investment funds are emerging that use evolutionary algorithms and neural nets to make investment decisions (e.g., Advanced Investment Technologies).
1998	The World Wide Web is ubiquitous. It is routine for high-school students and local grocery stores to have web sites.
1998	Automated personalities, which appear as animated faces that speak with realistic mouth movements and facial expressions, are working in laboratories. These personalities respond to the spoken statements and facial expressions of their human users. They are being developed to be used in future user interfaces for products and services, as personalized research and business assistants, and to conduct transactions.
1998	Microvision's Virtual Retina Display (VRD) projects images directly onto the user's retinas. Although expensive, consumer versions are projected for 1999.
1998	"Bluetooth" technology is being developed for "body" local area networks (LANs) and for wireless communication between personal computers and associated peripherals. Wireless communication is being developed for high-bandwidth connection to the Web.
1999	Ray Kurzweil's *The Age of Spiritual Machines: When Computers Exceed Human Intelligence* is published, available at your local bookstore!
2009	A $1,000 personal computer can perform about a trillion calculations per second. Personal computers with high-resolution visual displays come in a range of sizes, from those small enough to be embedded in clothing and jewelry up to the size of a thin book. Cables are disappearing. Communication between components uses short-distance wireless technology. High-speed wireless communication provides access to the Web. The majority of text is created using continuous speech recognition. Also ubiquitous are language user interfaces (LUIs). Most routine business transactions (purchases, travel, reservations) take place between a human and a virtual personality. Often, the virtual personality includes an animated visual presence that looks like a human face. Although traditional classroom organization is still common, intelligent courseware has emerged as a common means of learning.

Pocket-sized reading machines for the blind and visually impaired, "listening machines" (speech-to-text conversion) for the deaf, and computer-controlled orthotic devices for paraplegic individuals result in a growing perception that primary disabilities do not necessarily impart handicaps.

Translating telephones (speech-to-speech language translation) are commonly used for many language pairs.

Accelerating returns from the advance of computer technology have resulted in continued economic expansion. Price deflation, which had been a reality in the computer field during the twentieth century, is now occurring outside the computer field. The reason for this is that virtually all economic sectors are deeply affected by the accelerating improvement in the price performance of computing.

Human musicians routinely jam with cybernetic musicians.

Bioengineered treatments for cancer and heart disease have greatly reduced the mortality from these diseases.

The neo-Luddite movement is growing.

2019   A $1,000 computing device (in 1999 dollars) is now approximately equal to the computational ability of the human brain.

Computers are now largely invisible and are embedded everywhere—in walls, tables, chairs, desks, clothing, jewelry, and bodies.

Three-dimensional virtual reality displays, embedded in glasses and contact lenses, as well as auditory "lenses," are used routinely as primary interfaces for communication with other persons, computers, the Web, and virtual reality.

Most interaction with computing is through gestures and two-way natural-language spoken communication.

Nanoengineered machines are beginning to be applied to manufacturing and process-control applications.

High-resolution, three-dimensional visual and auditory virtual reality and realistic all-encompassing tactile environments enable people to do virtually anything with anybody, regardless of physical proximity.

Paper books or documents are rarely used and most learning is conducted through intelligent, simulated software-based teachers.

Blind persons routinely use eyeglass-mounted reading-navigation systems. Deaf persons read what other people are saying through their lens displays. Paraplegic and some quadriplegic persons routinely

walk and climb stairs through a combination of computer-controlled nerve stimulation and exoskeletal
robotic devices.

The vast majority of transactions include a simulated person.

Automated driving systems are now installed in
most roads.

People are beginning to have relationships with
automated personalities and use them as companions, teachers, caretakers, and lovers.

Virtual artists, with their own reputations, are
emerging in all of the arts.

There are widespread reports of computers passing the Turing Test, although these tests do not meet
the criteria established by knowledgeable observers.

2029   A $1,000 (in 1999 dollars) unit of computation
has the computing capacity of approximately 1,000
human brains.

Permanent or removable implants (similar to contact lenses) for the eyes as well as cochlear implants
are now used to provide input and output between
the human user and the worldwide computing
network.

Direct neural pathways have been perfected for
high-bandwidth connection to the human brain. A
range of neural implants is becoming available to enhance visual and auditory perception and interpretation, memory, and reasoning.

Automated agents are now learning on their own,
and significant knowledge is being created by
machines with little or no human intervention. Computers have read all available human- and machine-
generated literature and multimedia material.

There is widespread use of all-encompassing visual,
auditory, and tactile communication using direct
neural connections, allowing virtual reality to take
place without having to be in a "total touch enclosure."

The majority of communication does not involve
a human. The majority of communication involving a human is between a human and a machine.

There is almost no human employment in production, agriculture, or transportation. Basic life
needs are available for the vast majority of the
human race.

There is a growing discussion about the legal
rights of computers and what constitutes being
"human."

Although computers routinely pass apparently
valid forms of the Turing Test, controversy persists
about whether or not machine intelligence equals
human intelligence in all of its diversity.

	Machines claim to be conscious. These claims are largely accepted.
2049	The common use of nanoproduced food, which has the correct nutritional composition and the same taste and texture of organically produced food, means that the availability of food is no longer affected by limited resources, bad crop weather, or spoilage.
	Nanobot swarm projections are used to create visual-auditory-tactile projections of people and objects in real reality.
2072	Picoengineering (developing technology at the scale of picometers or trillionths of a meter) becomes practical.[1]
By the year 2099	There is a strong trend toward a merger of human thinking with the world of machine intelligence that the human species initially created.
	There is no longer any clear distinction between humans and computers.
	Most conscious entities do not have a permanent physical presence.
	Machine-based intelligences derived from extended models of human intelligence claim to be human, although their brains are not based on carbon-based cellular processes, but rather electronic and photonic equivalents. Most of these intelligences are not tied to a specific computational processing unit. The number of software-based humans vastly exceeds those still using native neuron-cell-based computation.
	Even among those human intelligences still using carbon-based neurons, there is ubiquitous use of neural-implant technology, which provides enormous augmentation of human perceptual and cognitive abilities. Humans who do not utilize such implants are unable to meaningfully participate in dialogues with those who do.
	Because most information is published using standard assimilated knowledge protocols, information can be instantly understood. The goal of education, and of intelligent beings, is discovering new knowledge to learn.
	Femtoengineering (engineering at the scale of femtometers or one thousandth of a trillionth of a meter) proposals are controversial.[2]
	Life expectancy is no longer a viable term in relation to intelligent beings.
Some many millenniums hence . . .	Intelligent beings consider the fate of the Universe.

HOW TO BUILD AN
INTELLIGENT MACHINE
IN THREE EASY PARADIGMS

As Deep Blue goes deeper and deeper, it displays elements of strategic understanding. Somewhere out there, mere tactics are translating into strategy. This is the closest thing I've seen to computer intelligence. It's a weird form of intelligence, the beginning of intelligence. But you can feel it. You can smell it.
—Frederick Friedel, assistant to Gary Kasparov,
commenting on the computer that beat his boss

The whole point of this sentence is to make clear what the whole point of this sentence is.
—Douglas Hofstadter

"Would you tell me please which way I ought to go from here?" asked Alice.
"That depends a good deal on where you want to get to," said the Cat.
"I don't much care where . . . ," said Alice.
"Then it doesn't much matter which way you go," said the Cat.
". . . so long as I get somewhere," Alice added as an explanation.
"Oh, you're sure to do that," said the Cat, "if you only walk long enough."
—Lewis Carroll

A professor has just finished lecturing at some august university about the origin and structure of the universe, and an old woman in tennis shoes walks up to the lectern. "Excuse me, sir, but you've got it all wrong," she says. "The truth is that the universe is sitting on the back of a huge turtle." The professor decides to humor her. "Oh really?" he asks. "Well, tell me, what is the turtle standing on?" The lady has a ready reply: "Oh, it's standing on another turtle." The professor asks, "And what is that turtle standing on?" Without hesitation, she says, "Another turtle." The professor, still game, repeats his question. A look of impatience comes across the woman's face. She holds up her hand, stopping him in mid-sentence. "Save your breath, sonny," she says. "It's turtles all the way down."
—Rolf Landauer

As I mentioned in chapter 6, "Building New Brains," understanding intelligence is a bit like peeling an onion—penetrating each layer reveals yet another onion. At the end of the process, we have a lot of onion peels, but no onion. In other words, intelligence—particularly human intelligence—operates at many levels. We can penetrate and understand each level, but the whole process requires all the levels working together in just the right way.

Presented here are some further perspectives on the three paradigms I discussed in chapter 4, "A New Form of Intelligence on Earth." Each of these methods can provide "intelligent" solutions to carefully defined problems. But to create systems that can respond flexibly in the complex environments that intelligent entities often find themselves, these approaches need to be combined in appropriate ways. This is particularly true when interacting with phenomena that incorporate multiple levels of understanding. For example, if we build a single grand neural network and attempt to train it to understand all the complexities of speech and language, the results will be limited at best. More encouraging results are obtained if we break down the problem in a way that corresponds to the multiple levels of meaning that we find in this uniquely human form of communication.

The human brain is organized the same way: as an intricate assemblage of specialized regions. And as we learn the brain's parallel algorithms, we will have the means to vastly extend them. As just one example, the brain region responsible for logical and recursive thinking—the cerebral cortex—has a mere 8 million neurons.[1] We are already building neural nets thousands of times larger and that operate millions of times faster. The key issue in designing intelligent machines (until they take over that chore from us) will be designing clever architectures to combine the relatively simple methods that comprise the building blocks of intelligence.

The Recursive Formula

Here's a really simple formula to create intelligent solutions to difficult problems. Listen carefully or you might miss it.

The recursive formula is:

For my next step, take my best next step. If I'm done, I'm done.

It may seem too simple, and I'll admit there's not much content at first glance. But its power is surprising.

Let's consider the classical example of a problem addressed by the recursive formula: the game of chess. Chess is considered an intelligent game, at least it was until recently. Most observers are still of the view that it requires intelligence to play a good game. So how does our recursive formula fare in this arena?

Chess is a game played one move at a time. The goal is to make "good" moves. So let's define a program that makes good moves. By applying the recursive formula to chess, we rephrase it as follows:

PICK MY BEST MOVE: *Pick my best move. If I've won, I'm done.*

Hang in there; this will make sense in a moment. I need to factor in one more aspect of chess, which is that I am not in this alone. I have an opponent. She makes moves, too. Let's give her the benefit of the doubt and assume that she also makes good moves. If this proves to be wrong, it will be an opportunity, not a problem. So now we have:

PICK MY BEST MOVE: *Pick my best move, assuming my opponent will do the same. If I've won, I'm done.*

At this point, we need to consider the nature of recursion. A recursive rule is one that is defined in terms of itself. A recursive rule is circular, but to be useful we don't want to go around in circles forever. We need an escape hatch.

To illustrate recursion, let's consider an example: the simple "factorial" function. To compute factorial of *n*, we *multiply n by factorial of (n − 1)*. That's the circular part—we have defined this function in terms of itself. We also need to specify that *factorial of 1 = 1*. That's our escape hatch.

As an example, let's compute factorial of 2. According to our definition,

factorial of 2 = 2 times (factorial of 1).

We know directly what (factorial of 1) is, so there's our escape from infinite recursion. Plugging in (factorial of 1) = 1, we now can write,

factorial of 2 = 2 times 1 = 2.

Returning to chess, we can see that the PICK MY BEST MOVE function is recursive, since we have defined the best move in terms of itself. The deceptively innocuous *"If I've won, I'm done"* part of the strategy is our escape hatch.

Let's factor in what we know about chess. This is where we carefully consider the definition of the problem. We realize that to pick the best move, we need to start by listing the *possible* moves. This is not very complicated. The legal moves at any point in the game are defined by the rules. While more complicated than some other games, the rules of chess are straightforward and easily programmed. So we list the moves and pick the best one.

But which is best? If the move results in a win, that will do nicely. So again we merely consult the rules and pick one of the moves that yields an immediate checkmate. Perhaps we are not so lucky and none of the possible moves provides an immediate win. We still need to consider whether or not the move enables me to win or lose. At this point we need to consider the subtle addition we made to our rule, *"assuming my opponent will do the same."* After all, my winning or losing is affected by what my opponent might do. I need to put myself in her shoes and pick her best move. How can I do that? This is where the power of recursion comes in. We have a program that does exactly this, called PICK MY BEST MOVE. So we call it to determine my opponent's best move.

Our program is now structured as follows. PICK MY BEST MOVE generates a list of all possible moves allowed by the rules. It examines each possible move in turn. For each move, it generates a hypothetical board representing what the placement of the pieces would be if that move were selected. Again, this just requires applying the definition of the problem as embodied in the rules of chess. PICK MY BEST MOVE now puts itself in my opponent's place and calls itself to determine her best move. It then starts to generate all of her possible moves from that board position.

The program thus keeps calling itself, continuing to expand possible moves and countermoves in an ever expanding tree of possibilities. This process is often called a *minimax* search, because we are alternatively attempting to minimize my opponent's ability to win and to maximize my own.

Where does this all end? The program just keeps calling itself until every branch of the tree of possible moves and countermoves results in an end of game. Each end of game provides the answer: Win, tie, or draw. At the furthest point of expansion of moves and countermoves, the program encounters moves that finish the game. If a move results in a win, we pick that move. If there are no win moves, then we settle for a tie. If there are no win or tie moves, I continue playing anyway in the hope that my opponent is not perfect like I am.

These final moves are the final branches—called leaves—in our tree of move sequences. Now, instead of continuing to call PICK MY BEST MOVE, the program begins to return from its calls to itself. As it begins to return from all of the nested PICK BEST

MOVE calls, it has determined the best move at each point (including the best move for my opponent), and so it can finally select the correct move for the current *actual* board situation.

So how well a game does this simple program play? The answer is *perfect* chess. I can't lose, unless possibly if my opponent goes first and is also perfect. Perfect chess is very good indeed, much better than any mere human. The most complicated part of the PICK MY BEST MOVE function—the only aspect that is not extremely simple—is generating the allowable moves at each point. And this is just a matter of codifying the rules. Essentially, we have determined the answer by carefully defining the problem.

But we're not done. While playing perfect chess might be considered impressive, it is not good enough. We need to consider how responsive a player PICK MY BEST MOVE will be. If we assume that there are, on average, about 8 possible moves for each board situation, and that a typical game lasts about 30 moves, we need to consider 8^{30} possible move sequences to fully expand the tree of all move-countermove possibilities. If we assume that we can analyze 1 billion board positions per second (a good deal faster than any chess computer today), it would take 10^{18} seconds, or about 40 billion years, to select each move.

Unfortunately, that's not regulation play. This approach to recursion is a bit like evolution—both do great work but are incredibly slow. That's really not surprising if you think about it. Evolution represents another very simple paradigm, and indeed is another of our simple formulas.

However, before we throw out the recursive formula, let's attempt to modify it to take into account our human patience and, for the time being, our mortality.

Clearly we need to put limits on how deep we allow the recursion to take place. How large we allow the move-countermove tree to grow needs to depend on how much computation we have available. In this way, we can use the recursive formula on any computer, from a wristwatch computer to a supercomputer.

Limiting the size of this tree means of course that we cannot expand each branch until the end of the game. We need to arbitrarily stop the expansion and have a method of evaluating the "terminal leaves" of an unfinished tree. When we considered fully expanding each move sequence to the end of the game, evaluation was simple: Winning is better than tying, and losing is no good at all. Evaluating a board position in the middle of the game is slightly more complicated. Rather, it is more controversial because here we encounter multiple schools of thought.

The cat in *Alice's Adventures in Wonderland* who tells Alice that it doesn't much matter which way she goes must have been an expert in recursive algorithms. Any halfway reasonable approach works rather well. If, for example, we just add up the piece values (that is, 10 for the queen, 5 for the rook, and so on), we will obtain rather respectable results. Programming the recursive minimax formula using the piece value method of evaluating terminal leaves, as run on your average personal computer circa 1998, will defeat all but a few thousand humans on the planet.

This is what I call the "simple minded" school. This school of thought says: Use a simple method of evaluating the terminal leaves and put whatever computational power we have available into expanding the moves and countermoves as deeply as possible. Another approach is the "complicated minded" school, which says that we need to use sophisticated procedures to evaluate the "quality" of the board at each terminal leaf position.

IBM's Deep Blue, the computer that crossed this historic threshold, uses a leaf evaluation method that is a good deal more refined than just adding up piece values. However, in a discussion I had with Murray Campbell, head of the Deep Blue team, just weeks prior to its 1997 historic victory, Campbell agreed that Deep Blue's evaluation method was more simple minded than complicated minded.

MATHLESS "PSEUDO CODE"
FOR THE RECURSIVE ALGORITHM

Here is the basic schema for the recursive algorithm. Many variations are possible, and the designer of the system needs to provide certain critical parameters and methods, detailed below.

The Recursive Algorithm

Define a function (program), "PICK BEST NEXT STEP." The function returns a value of "SUCCESS" (we've solved the problem) or "FAILURE" (we didn't solve it). If it returns with a value of SUCCESS, then the function also returns the sequence of selected steps that solved the problem. PICK BEST NEXT STEP does the following:

PICK BEST NEXT STEP:
- Determine if the program can escape from continued recursion at this point. This bullet and the next two bullets deal with this escape decision. First, determine if the problem has now been solved. Since this call to PICK BEST NEXT STEP probably came from the program calling itself, we may now have a satisfactory solution. Examples are:

    (i) In the context of a game (e.g., chess), the last move allows us to win (e.g., checkmate).
    (ii) In the context of solving a mathematical theorem, the last step proves the theorem.
    (iii) In the context of an artistic program (e.g., cybernetic poet or composer), the last step matches the goals for the next word or note.

If the problem has been satisfactorily solved, the program returns with a value of SUCCESS. In this case, PICK BEST NEXT STEP also returns the sequence of steps that caused the success.

- If the problem has not been solved, determine if a solution is now hopeless. Examples are:

    (i) In the context of a game (e.g., chess), this move causes us to lose (e.g., checkmate for the other side).
    (ii) In the context of solving a mathematical theorem, this step violates the theorem.
    (iii) In the context of an artistic program (e.g., cybernetic poet or composer), this step violates the goals for the next word or note.

If the solution at this point has been deemed hopeless, the program returns with a value of FAILURE.

- If the problem has been neither solved nor deemed hopeless at this point of recursive expansion, determine whether or not the expansion should be abandoned anyway. This is a key aspect of the design and takes into consideration the limited amount of computer time we have to spend. Examples are:

(i) In the context of a game (e.g., chess), this move puts our side sufficiently "ahead" or "behind." Making this determination may not be straightforward and is the primary design decision. However, simple approaches (e.g., adding up piece values) can still provide good results. If the program determines that our side is sufficiently ahead, then PICK BEST NEXT STEP returns in a similar manner to a determination that our side has won (i.e., with a value of SUCCESS). If the program determines that our side is sufficiently behind, then PICK BEST NEXT STEP returns in a similar manner to a determination that our side has lost (i.e., with a value of FAILURE).

(ii) In the context of solving a mathematical theorem, this step involves determining if the sequence of steps in the proof are unlikely to yield a proof. If so, then this path should be abandoned, and PICK BEST NEXT STEP returns in a similar manner to a determination that this step violates the theorem (i.e., with a value of FAILURE). There is no "soft" equivalent of success. We can't return with a value of SUCCESS until we have actually solved the problem. That's the nature of math.

(iii) In the context of an artistic program (e.g., cybernetic poet or composer), this step involves determining if the sequence of steps (e.g., words in a poem, notes in a song) is unlikely to satisfy the goals for the next step. If so, then this path should be abandoned, and PICK BEST NEXT STEP returns in a similar manner to a determination that this step violates the goals for the next step (i.e., with a value of FAILURE).

• If PICK BEST NEXT STEP has not returned (because the program has neither determined success nor failure nor made a determination that this path should be abandoned at this point), then we have not escaped from continued recursive expansion. In this case, we now generate a list of all possible next steps at this point. This is where the precise statement of the problem comes in:

(i) In the context of a game (e.g., chess), this involves generating all possible moves for "our" side for the current state of the board. This involves a straightforward codification of the rules of the game.

(ii) In the context of finding a proof for a mathematical theorem, this involves listing the possible axioms or previously proved theorems that can be applied at this point in the solution.

(iii) In the context of a cybernetic art program, this involves listing the possible words/notes/line segments that could be used at this point.

For each such possible next step:

(i) Create the hypothetical situation that would exist if this step were implemented. In a game, this means the hypothetical state of the board. In a mathematical proof, this means adding this step (e.g., axiom) to the proof. In an art program, this means adding this word/note/line segment.

(ii) Now call PICK BEST NEXT STEP to examine this hypothetical situation. This is, of course, where the recursion comes in because the program is now calling itself.

(iii) If the above call to PICK BEST NEXT STEP returns with a value of

SUCCESS, then return from the call to PICK BEST NEXT STEP (that we are now in), also with a value of SUCCESS. Otherwise consider the next possible step.

If all the possible next steps have been considered without finding a step that resulted in a return from the call to PICK BEST NEXT STEP with a value of SUCCESS, then return from this call to PICK BEST NEXT STEP (that we are now in) with a value of FAILURE.

END OF PICK BEST NEXT STEP
If the original call to PICK BEST NEXT STEP returns with a value of SUCCESS, it will also return the correct sequence of steps:

(i) In the context of a game, the first step in this sequence is the next move you should make.
(ii) In the context of a mathematical proof, the full sequence of steps is the proof.
(iii) In the context of a cybernetic art program, the sequence of steps is your work of art.

If the original call to PICK BEST NEXT STEP is FAILURE, then you need to go back to the drawing board.

Key Design Decisions

In the simple schema above, the designer of the recursive algorithm needs to determine the following at the outset:

- The key to a recursive algorithm is the determination in PICK BEST NEXT STEP when to abandon the recursive expansion. This is easy when the program has achieved clear success (e.g., checkmate in chess, or the requisite solution in a math or combinatorial problem) or clear failure. It is more difficult when a clear win or loss has not yet been achieved. Abandoning a line of inquiry before a well-defined outcome is necessary because otherwise the program might run for billions of years (or at least until the warranty on your computer runs out).
- The other primary requirement for the recursive algorithm is a straightforward codification of the problem. In a game like chess, that's easy. But in other situations, a clear definition of the problem is not always so easy to come by.

Happy Recursive Searching!

Human players are very complicated minded. That seems to be the human condition. As a result, even the best chess players are unable to consider more than a hundred moves, compared to a few billion for Deep Blue. But each human move is deeply considered. However, in 1997, Gary Kasparov, the world's best example of the complicated-minded school, was defeated by a simple-minded computer.

Personally, I am of a third school of thought. It's not much of a school, really. To my knowledge, no one has tried this idea. It involves combining the recursive and neural net paradigms, and I describe it in the discussion on neural nets that follows.

Neural Nets

In the early and mid-1960s, AI researchers became enamored with the Perceptron, a machine constructed from mathematical models of human neurons. Early Perceptrons were modestly successful in such pattern-recognition tasks as identifying printed letters and speech sounds. It appeared that all that was needed to make the Perceptron more intelligent was to add more neurons and more wires.

Then came Marvin Minsky and Seymour Papert's 1969 book, *Perceptrons*, which proved a set of theorems apparently demonstrating that a Perceptron could never solve the simple problem of determining whether or not a line drawing is "connected" (in a connected drawing all parts are connected to one another by lines). The book had a dramatic effect, and virtually all work on Perceptrons came to a halt.[2]

In the late 1970s and 1980s, the paradigm of building computer simulations of human neurons, then called neural nets, began to regain it's popularity. One observer wrote in 1988:

> Once upon a time two daughter sciences were born to the new science of cybernetics. One sister was natural, with features inherited from the study of the brain, from the way nature does things. The other was artificial, related from the beginning to the use of computers. Each of the sister sciences tried to build models of intelligence, but from very different materials. The natural sister built models (called neural networks) out of mathematically purified neurones. The artificial sister built her models out of computer programs.
>
> In their first bloom of youth the two were equally successful and equally pursued by suitors from other fields of knowledge. They got on very well together. Their relationship changed in the early sixties when a new monarch appeared, one with the largest coffers ever seen in the kingdom of the sciences: Lord DARPA, the Defense Department's Advanced Research Projects Agency. The artificial sister grew jealous and was determined to keep for herself the access to Lord DARPA's research funds. The natural sister would have to be slain.
>
> The bloody work was attempted by two staunch followers of the artificial sister, Marvin Minsky and Seymour Papert, cast in the role of the huntsman sent to slay Snow White and bring back her heart as proof of the deed. Their weapon was not the dagger but the mightier pen, from which came a book—*Perceptrons*—purporting to prove that neural nets could never fill their promise of building models of mind: *only computer programs could do this*. Victory seemed assured for the artificial sister. And indeed, for the next decade all the rewards of the kingdom came to her progeny, of which the family of expert systems did best in fame and fortune.
>
> But Snow White was not dead. What Minsky and Papert had shown the world as proof was not the heart of the princess, it was the heart of a pig.

The author of the above statement was Seymour Papert.[3] His sardonic allusion to bloody hearts reflects a widespread misunderstanding of the implications of the pivotal theorem in his and Minsky's 1969 book. The theorem demonstrated limitations in the capabilities of a single layer of simulated neurons. If, on the other hand, we place neural nets at multiple levels—having the output of one neural net feed into the next—the range of its competence greatly expands. Moreover, if we combine neural nets with other paradigms, we can make yet greater progress. The heart that Minsky and Papert extracted belonged primarily to the single layer neural net.

Papert's irony also reflects his and Minsky's own considerable contributions to the neural net field. In fact, Minsky started his career with seminal contributions to the concept at Harvard in the 1950s.[4]

But enough of politics. What are the main issues in designing a neural net?

One key issue is the net's topology: the organization of the interneuronal connections. A net organized with multiple levels can make more complex discriminations but is harder to train.

Training the net is the most critical issue. This requires an extensive library of examples of the patterns the net will be expected to recognize, along with the correct identification of each pattern. Each pattern is presented to the net. Typically, those connections that contributed to a correct identification are strengthened (by increasing their associated weight), and those that contributed to an incorrect identification are weakened. This method of strengthening and weakening the connection weights is called back-propagation and is one of several methods used. There is controversy as to how this learning is accomplished in the human brain's neural nets, as there does not appear to be any mechanism by which back-propagation can occur. One method that does appear to be implemented in the human brain is that the mere firing of a neuron increases the neurotransmitter strengths of the synapses it is connected to. Also, neurobiologists have recently discovered that primates, and in all likelihood humans, grow new brain cells throughout life, including adulthood, contradicting an earlier dogma that this was not possible.

Little and Big Hills

A key issue in adaptive algorithms—neural nets and evolutionary algorithms—is often referred to as *local versus global optimality:* in other words, climbing the closest hill versus finding and climbing the biggest hill. As a neural net learns (by adjusting the connection strengths), or as an evolutionary algorithm evolves (by adjusting the "genetic" code of the simulated organisms), the fit of the solution will improve, until a "locally optimal" solution is found. If we compare this to climbing a hill, these methods are very good at finding the top of a nearby hill, which is the best possible solution within a local area of possible solutions. But sometimes these methods may become trapped at the top of a small hill and fail to see a higher mountain in a different area. In the neural net context, if the neural net has converged on a locally optimal solution, as it tries adjusting any of the connection strengths, the fit becomes worse. But just as a climber might need to come down a small elevation to ultimately climb to a higher point on a different hill, the neural net (or evolutionary algorithm) might need to make the solution temporarily worse to ultimately find a better one.

One approach to avoiding such a "false" optimal solution (little hill) is to force the adaptive method to do the analysis multiple times starting with very different initial conditions—in other words, force it to climb lots of hills, not just one. But even with this approach, the system designer still needs to make sure that the adaptive method hasn't missed an even higher mountain in a yet more distant land.

The Laboratory of Chess

We can gain some insight into the comparison of human thinking and conventional computer approaches by again examining the human and machine approaches to chess. I do this not to belabor the issue of chess playing, but rather because it illustrates a clear contrast. Raj Reddy, Carnegie Mellon University's AI guru, cites studies of chess as playing the same role in artificial intelligence that studies of *E. coli* play in biology: an ideal laboratory for studying fundamental questions.[5] Computers use their extreme speed to analyze the vast combinations created by the combinatorial explosion of moves and countermoves. While chess programs may use a few other tricks (such as storing the openings of all master chess games in this century and precomputing endgames), they essentially rely on their combination of speed and precision. In comparison, humans, even chess masters, are extremely slow and imprecise. So we precompute *all* of our chess moves. That's why it takes so long to become a chess master, or the master of any pursuit. Gary Kasparov has spent much of his few decades on the planet studying—and experiencing—chess moves. Researchers have estimated that masters of a nontrivial subject have memorized about fifty thousand such "chunks" of insight.

When Kasparov plays, he, too, generates a tree of moves and countermoves in his head, but limitations in human mental speed and short-term memory limit his mental tree (for each actually played move) to no more than a few hundred board positions, if that. This compares to billions of board positions for his electronic antagonist. So the human chess master is forced to drastically prune his mental tree, eliminating fruitless branches by using his intense pattern-recognition faculties. He matches each board position—actual and imagined—to this database of tens of thousands of previously analyzed situations.

After Kasparov's 1997 defeat, we read a lot about how Deep Blue was just doing massive number crunching, not really "thinking" the way its human rival was doing. One could say that the opposite is the case, that Deep Blue was indeed thinking through the implications of each move and countermove, and that it was Kasparov who did not have time to really think very much during the tournament. Mostly he was just drawing upon his mental database of situations he had thought about long ago. (Of course, this depends on one's notion of thinking, as I discussed in chapter 3.) But if the human approach to chess—neural network–based pattern recognition used to identify situations from a library of previously analyzed situations—is to be regarded as true thinking, then why not program our machines to work the same way?

The Third Way

And that's my idea that I alluded to earlier as the third school of thought in evaluating the terminal leaves in a recursive search. Recall that the simple-minded school uses an approach such as adding up piece values to evaluate a particular board position. The complicated-minded school advocates a more elaborate and time-consuming logical analysis. I advocate a third way: combine two simple paradigms—recursive and neural net—by using the neural net to evaluate the board positions at each terminal leaf. The training of a neural net is time-consuming and requires a great deal of computing, but performing a single recognition task on a neural net that has already learned its lessons is very quick, comparable to a simple-minded evaluation. Although fast, the neural net is drawing upon the very extensive amount of time it previously spent learning the material. Since we have every master chess game in this century online, we can use this massive amount of data to train the neural net. This training is done once and offline (that is, not during an actual game). The trained neural net would then be used to evaluate the board positions at each terminal leaf. Such a system would combine the millionfold advantage in speed that computers have with the more humanlike ability to recognize patterns against a lifetime of experience.

I proposed this approach to Murray Campbell, head of the Deep Blue team, and he found it intriguing and appealing. He was getting tired anyway, he admitted, of tuning the leaf evaluation algorithm by hand. We talked about setting up an advisory team to implement this idea, but then IBM canceled the whole chess project. I do believe that one of

MATHLESS "PSEUDO CODE"
FOR THE NEURAL NET ALGORITHM

Here is the basic schema for a neural net algorithm. Many variations are possible, and the designer of the system needs to provide certain critical parameters and methods, detailed below.

The Neural Net Algorithm

Creating a neural net solution to a problem involves the following steps:

- Define the input.
- Define the topology of the neural net (i.e., the layers of neurons and the connections between the neurons).
- Train the neural net on examples of the problem.
- Run the trained neural net to solve new examples of the problem.
- Take your neural net company public.

These steps (except for the last one) are detailed below:

The Problem Input

The problem input to the neural net consists of a series of numbers. This input can be:

- in a visual pattern-recognition system: a two-dimensional array of numbers representing the pixels of an image; or
- in an auditory (e.g., speech) recognition system: a two-dimensional array of numbers representing a sound, in which the first dimension represents parameters of the sound (e.g., frequency components) and the second dimension represents different points in time; or
- in an arbitrary pattern recognition system: an n-dimensional array of numbers representing the input pattern.

Defining the Topology

To set up the neural net:
The architecture of each neuron consists of:

- Multiple inputs in which each input is "connected" to either the output of another neuron or one of the input numbers.
- Generally, a single output, which is connected either to the input of another neuron (which is usually in a higher layer) or to the final output.

Set up the first layer of neurons:

- Create N_0 neurons in the first layer. For each of these neurons, "connect" each of the multiple inputs of the neuron to "points" (i.e., numbers) in the problem input. These connections can be determined randomly or using an evolutionary algorithm (see below).
- Assign an initial "synaptic strength" to each connection created. These

weights can start out all the same, can be assigned randomly, or can be determined in another way (see below).

Set up the additional layers of neurons:

Set up a total of M layers of neurons. For each layer, set up the neurons in that layer. For layer$_i$:

- Create N_i neurons in layer$_i$. For each of these neurons, "connect" each of the multiple inputs of the neuron to the outputs of the neurons in layer$_{i-1}$ (see variations below).
- Assign an initial "synaptic strength" to each connection created. These weights can start out all the same, can be assigned randomly, or can be determined in another way (see below).
- The outputs of the neurons in layer$_M$ are the outputs of the neural net (see variations below).

The Recognition Trials

How each neuron works:
Once the neuron is set up, it does the following for each recognition trial.

- Each weighted input to the neuron is computed by multiplying the output of the other neuron (or initial input) that the input to this neuron is connected to by the synaptic strength of that connection.
- All of these weighted inputs to the neuron are summed.
- If this sum is greater than the firing threshold of this neuron, then this neuron is considered to "fire" and its output is 1. Otherwise, its output is 0 (see variations below).

Do the following for each recognition trial:
For each layer, from layer$_0$ to layer$_M$:
And for each neuron in each layer:

- Sum its weighted inputs (each weighted input = the output of the other neuron [or initial input] that the input to this neuron is connected to, multiplied by the synaptic strength of that connection).
- If this sum of weighted inputs is greater than the firing threshold for this neuron, set the output of this neuron = 1, otherwise set it to 0.

To Train the Neural Net

- Run repeated recognition trials on sample problems.
- After each trial, adjust the synaptic strengths of all the interneuronal connections to improve the performance of the neural net on this trial (see the discussion below on how to do this).
- Continue this training until the accuracy rate of the neural net is no longer improving (i.e., reaches an asymptote).

Key Design Decisions

In the simple schema above, the designer of this neural net algorithm needs to determine at the outset:

- What the input numbers represent.
- The number of layers of neurons.
- The number of neurons in each layer (each layer does not necessarily need to have the same number of neurons).
- The number of inputs to each neuron, in each layer. The number of inputs (i.e., interneuronal connections) can also vary from neuron to neuron, and from layer to layer.
- The actual "wiring" (i.e., the connections). For each neuron, in each layer, this consists of a list of other neurons, the outputs of which constitute the inputs to this neuron. This represents a key design area. There are a number of possible ways to do this:

    (i) wire the neural net randomly; or
    (ii) use an evolutionary algorithm (see next section of this Appendix) to determine an optimal wiring; or
    (iii) use the system designer's best judgment in determining the wiring.

- The initial synaptic strengths (i.e., weights) of each connection. There are a number of possible ways to do this:

    (i) set the synaptic strengths to the same value; or
    (ii) set the synaptic strengths to different random values; or
    (iii) use an evolutionary algorithm to determine an optimal set of initial values; or
    (iv) use the system designer's best judgment in determining the initial values.

- The firing threshold of each neuron.

- Determine the output. The output can be:

    (i) the outputs of layer$_M$ of neurons; or
    (ii) the output of a single output neuron, whose inputs are the outputs of the neurons in layer$_M$;
    (iii) a function of (e.g., a sum of) the outputs of the neurons in layer$_M$; or
    (iv) another function of neuron outputs in multiple layers.

- Determine how the synaptic strengths of all the connections are adjusted during the training of this neural net. This is a key design decision and the subject of a great deal of neural net research and discussion. There are a number of possible ways to do this:

    (i) For each recognition trial, increment or decrement each synaptic strength by a (generally small) fixed amount so that the neural net's output more closely matches the correct answer. One way to do this is to try both incrementing and decrementing and see which has the more desirable effect. This can be time consuming, so other methods exist for making local decisions on whether to increment or decrement each synaptic strength.
    (ii) Other statistical methods exist for modifying the synaptic strengths after each recognition trial so that the performance of the neural net on that trial more closely matches the correct answer.

Note that neural net training will work even if the answers to the training trials are not all correct. This allows using real-world training data that may have an inherent error rate. One key to the success of a neural net–based recognition system is the amount of data used for training. Usually a very substantial amount is needed to obtain satisfactory results. Just like human students, the amount of time that a neural net spends learning its lessons is a key factor in its performance.

Variations

Many variations of the above are feasible. Some variations include:

- There are different ways of determining the topology, as described above. In particular, the interneuronal wiring can be set either randomly or using an evolutionary algorithm.
- There are different ways of setting the initial synaptic strengths, as described above.
- The inputs to the neurons in $layer_i$ do not necessarily need to come from the outputs of the neurons in $layer_{i-1}$. Alternatively, the inputs to the neurons in each layer can come from any lower layer or any layer.
- There are different ways to determine the final output, as described above.
- For each neuron, the method described above compares the sum of the weighted inputs to the threshold for that neuron. If the threshold is exceeded, the neuron fires and its output is 1. Otherwise, its output is 0. This "all or nothing" firing is called a nonlinearity. There are other nonlinear functions that can be used. Commonly a function is used that goes from 0 to 1 in a rapid but more gradual fashion (than all or nothing). Also, the outputs can be numbers other than 0 and 1.
- The different methods for adjusting the synaptic strengths during training, briefly described above, represent a key design decision.
- The above schema describes a "synchronous" neural net, in which each recognition trial proceeds by computing the outputs of each layer, starting with $layer_0$ through $layer_M$. In a true parallel system, in which each neuron is operating independently of the others, the neurons can operate asynchronously (i.e., independently). In an asynchronous approach, each neuron is constantly scanning its inputs and fires (i.e., changes its output from 0 to 1) whenever the sum of its weighted inputs exceeds its threshold (or, alternatively, using another nonlinear output function).

Happy Adaptation!

the keys to emulating the diversity of human intelligence is optimally to combine fundamental paradigms. We'll talk about how to fold in the paradigm of evolutionary algorithms below.

EVOLUTIONARY ALGORITHMS

If biologists have ignored self-organization, it is not because self-ordering is not pervasive and profound. It is because we biologists have yet to understand how to think about sys-

tems governed simultaneously by two sources of order. Yet who seeing the snowflake, who seeing simple lipid molecules cast adrift in water forming themselves into cell-like hollow lipid vesicles, who seeing the potential for the crystallization of life in swarms of reacting molecules, who seeing the stunning order in networks linking tens upon tens of thousands of variables, can fail to entertain a central thought: if ever we are to attain a final theory in biology, we will surely have to understand the commingling of self-organization and selection. We will have to see that we are the natural expressions of a deeper order. Ultimately, we will discover in our creation myth that we are expected after all.

—Stuart Kauffman

As I discussed earlier, an evolutionary algorithm involves a simulated environment in which simulated software "creatures" compete for survival and the right to reproduce. Each software creature represents a possible solution to a problem encoded in its digital "DNA."

The creatures allowed to survive and reproduce into the next generation are the ones that do a better job of solving the problem. Evolutionary algorithms are considered to be part of a class of "emergent" methods because the solutions emerge gradually and usually cannot be predicted by the designers of the system. Evolutionary algorithms are particularly powerful when they are combined with our other paradigms. Here is a unique way of combining all of our "intelligent" paradigms.

Combining All Three Paradigms

The human genome contains three billion rungs of base pairs, which equals six billion bits of data. With a little data compression, your genetic code will fit on a single CD-ROM. You can store your whole family on a DVD (digital video disc). But your brain has 100 trillion "wires," which would require about 3,000 trillion bits to represent. How did the mere 12 billion bits of data in your chromosomes (with contemporary estimates indicating that only 3 percent of that is active) designate the wiring of your brain, which constitutes about a quarter million times more information?

Obviously the genetic code does not specify the exact wiring. I said earlier that we can wire a neural net randomly and obtain satisfactory results. That's true, but there is a better way to do it, and that is to use evolution. I am not referring to the billions of years of evolution that produced the human brain. I am referring to the months of evolution that go on during gestation and early childhood. Early in our lives, our interneuronal connections are engaged in a fight for survival. Those that make better sense of the world survive. By late childhood, these connections become relatively fixed, which is why it is worthwhile exposing babies and young children to a stimulating environment. Otherwise, this evolutionary process runs out of real-world chaos from which to draw inspiration.

We can do the same thing with our synthetic neural nets: use an evolutionary algorithm to determine the optimal wiring. This is exactly what the Kyoto Advanced Telecommunications Research Lab's ambitious brain-building project is doing.

Now here's how you can intelligently solve a challenging problem using all three paradigms. First, carefully state your problem. This is actually the hardest step. Most people try to solve problems without bothering to understand what the problem is all about. Next, analyze the logical contours of your problem *recursively* by searching through as many combinations of elements (for example, moves in a game, steps in a solution) that you and your computer have the patience to sort through. For the terminal leaves of this recursive expansion of possible solutions, evaluate them with a *neural net*. For the optimal topology of your neural net, determine this using an *evolutionary* algorithm. And if all of this doesn't work, then you have a difficult problem, indeed.

"PSEUDO CODE" FOR THE EVOLUTIONARY ALGORITHM

Here is the basic schema for an evolutionary algorithm. Many variations are possible, and the designer of the system needs to provide certain critical parameters and methods, detailed below.

THE EVOLUTIONARY ALGORITHM

Create N solution "creatures." Each one has:

- A genetic code—a sequence of numbers that characterizes a possible solution to the problem. The numbers can represent critical parameters, steps to a solution, rules, etc.

For each generation of evolution, do the following:

- Do the following for each of the N solution creatures:

 (i) Apply this solution creature's solution (as represented by its genetic code) to the problem, or simulated environment.
 (ii) Rate the solution.

- Pick the L solution creatures with the highest ratings to survive into the next generation.
- Eliminate the (N − L) nonsurviving solution creatures.
- Create (N − L) new solution creatures from the L surviving solution creatures by:

 (i) making copies of the L surviving creatures. Introduce small random variations into each copy; or
 (ii) create additional solution creatures by combining parts of the genetic code (using "sexual" reproduction, or otherwise combining portions of the chromosomes) from the L surviving creatures; or
 (iii) doing a combination of (i) and (ii) above.

- Determine whether or not to continue evolving:

Improvement = (highest rating in this generation) − (highest rating in the previous generation)
If Improvement < Improvement Threshold, then we're done

- **The Solution Creature with the highest rating from the last generation of evolution has the best solution. Apply the solution defined by its genetic code to the problem.**

Key Design Decisions

In the simple schema above, the designer of this evolutionary algorithm needs to determine at the outset:

- Key parameters:

N
L
Improvement Threshold

- What the numbers in the genetic code represent and how the solution is computed from the genetic code.
- A method for determining the N solution creatures in the first generation. In general, these need only be "reasonable" attempts at a solution. If these first-generation solutions are too far afield, the evolutionary algorithm may have difficulty converging on a good solution. It is often worthwhile to create the initial solution creatures in such a way that they are reasonably diverse. This will help prevent the evolutionary process from just finding a "locally" optimal solution.
- How the solutions are rated.
- How the surviving solution creatures reproduce.

Variations

Many variations of the above are feasible. Some variations include:

- There does not need to be a fixed number of surviving solution creatures (i.e., "L") from each generation. The survival rule(s) can allow for a variable number of survivors.
- There does not need to be a fixed number of new solution creatures created in each generation (i.e., [N − L]). The procreation rules can be independent of the size of the population. Procreation can be related to survival, thereby allowing the fittest solution creatures to procreate the most.
- The decision as to whether or not to continue evolving can be varied. It can consider more than just the highest-rated solution creature from the most recent generation(s). It can also consider a trend that goes beyond just the last two generations.

Happy Evolving!

GLOSSARY

Aaron A computerized robot (and associated software), designed by Harold Cohen, that creates original drawings and paintings.

Alexander's solution A term referring to Alexander the Great's slicing of the Gordian knot with his sword. A reference to solving an insoluble problem with decisive yet unexpected and indirect means.

Algorithm A sequence of rules and instructions that describes a procedure to solve a problem. A computer program expresses one or more algorithms in a manner understandable by a computer.

Alu A meaningless sequence of 300 nucleotide letters that occurs 300,000 times in the human genome.

Analog A quantity that is continuously varying, as opposed to varying in discrete steps. Most phenomena in the natural world are analog. When we measure and give them a numeric value, we digitize them. The human brain uses both digital and analog computation.

Analytical Engine The first programmable computer, created in the 1840s by Charles Babbage and Ada Lovelace. The Analytical Engine had a random access memory (RAM) consisting of one thousand words of fifty decimal digits each, a central processing unit, a special storage unit for software, and a printer. Although it foreshadowed modern computers, Babbage's invention never worked.

Angel Capital Refers to funds available for investment by networks of wealthy investors who invest in start-up companies. A key source of capital for high-tech start-up companies in the United States.

Artificial intelligence (AI) The field of research that attempts to emulate human intelligence in a machine. Fields within AI include knowledge-based systems, expert systems, pattern recognition, automatic learning, natural-language understanding, robotics, and others.

Artificial life Simulated organisms, each including a set of behavior and reproduction rules (a simulated "genetic code"), and a simulated environment. The simulated organisms simulate multiple generations of evolution. The term can refer to any self-replicating pattern.

ASR *See* Automatic speech recognition.

Automatic speech recognition (ASR)   Software that recognizes human speech. In general, ASR systems include the ability to extract high-level patterns in speech data.

BGM   *See* Brain-generated music.

Big bang theory   A prominent theory on the beginning of the Universe: the cosmic explosion, from a single point of infinite density, that marked the beginning of the Universe billions of years ago.

Big crunch   A theory that the Universe will eventually lose momentum in expanding and contract and collapse in an event that is the opposite of the big bang.

Bioengineering   The field of designing pharmaceutical drugs and strains of plant and animal life by directly modifying the genetic code. Bioengineered materials, drugs, and life-forms are used in agriculture, medicine, and the treatment of disease.

Biology   The study of life-forms. In evolutionary terms, the emergence of patterns of matter and energy that could survive and replicate to form future generations.

Bionic organ   In 2029, artificial organs that are built using nanoengineering.

Biowarfare Agency (BWA)   In the second decade of the twenty-first century, a government agency that monitors and polices bioengineering technology applied to weapons.

Bit   A contraction of the phrase "binary digit." In a binary code, one of two possible values, usually zero and one. In information theory, the fundamental unit of information.

Brain-generated music (BGM)   A music technology pioneered by NeuroSonics, Inc., that creates music in response to the listener's brain waves. This brain-wave biofeedback system appears to evoke the Relaxation Response by encouraging the generation of alpha waves in the brain.

BRUTUS.1   A computer program that creates fictional stories with a theme of betrayal; invented by Selmer Bringsjord, Dave Ferucci, and a team of software engineers at Rensselaer Polytechnic Institute in New York.

Buckyball   A soccer-ball-shaped molecule formed of a large number of carbon atoms. Because of their hexagonal and pentagonal shape, the molecules were dubbed "buckyballs" in reference to R. Buckminster Fuller's building designs.

Busy beaver   One example of a class of noncomputational functions; an unsolvable problem in mathematics. Being a "Turing machine unsolvable problem," the busy beaver function cannot be computed by a Turing machine. To compute busy beaver of n, one creates all the n-state Turing machines that do not write an infinite number of 1s on their tape. The largest number of 1s written by the Turing machine in this set that writes the largest number of 1s is busy beaver of n.

BWA   *See* Biowarfare Agency.

Byte   A contraction for "by eight." A group of eight bits clustered together to store one unit of information on a computer. A byte may correspond, for example, to a letter of the English alphabet.

CD-ROM   *See* Compact disc read-only memory.

Chaos   The amount of disorder or unpredictable behavior in a system. In reference to the Law of Time and Chaos, chaos refers to the quantity of random and unpredictable events that are relevant to a process.

Chaos theory   The study of patterns and emergent behavior in complex systems comprised of many unpredictable elements (e.g., the weather).

Chemistry   The composition and properties of substances comprised of molecules.

Chip   A collection of related circuits that work together on a task or set of tasks, residing on a wafer of semiconductor material (typically silicon).

Closed system   Interacting entities and forces not subject to outside influence (for example, the Universe). A corollary of the second law of thermodynamics is that in a closed system, entropy increases.

Cochlear implant   An implant that performs frequency analyses of sound waves, similar to that performed by the inner ear.

Colossus   The first electronic computer, built by the British from fifteen hundred radio tubes during World War II. Colossus and nine similar machines running in parallel cracked increasingly complex German codes on military-intelligence and contributed to the Allied forces' winning of World War II.

Combinatorial explosion   The rapid—exponential—growth in the number of possible ways of choosing distinct combinations of elements from a set as the number of elements in that set grows. In an algorithm, the rapid growth in the number of alternatives to be explored while performing a search for a solution to a problem.

Common sense   The ability to analyze a situation based on its context, using millions of integrated pieces of common knowledge. Currently, computers lack common sense. To quote Marvin Minsky: "Deep Blue might be able to win at chess, but it wouldn't know to come in from the rain."

Compact disc read-only memory (CD-ROM)   A laser-read disc that contains up to a half billion bytes of information. "Read only" refers to the fact that information can be read, but not deleted or recorded, on the disc.

Complicated-minded school   The use of sophisticated procedures to evaluate the terminal leaves in a recursive algorithm.

Computation   The process of calculating a result by use of an algorithm (e.g., a computer program) and related data. The ability to remember and solve problems.

Computer   A machine that implements an algorithm. A computer transforms data according to the specifications of an algorithm. A programmable computer allows the algorithm to be changed.

Computer language   A set of rules and specifications for describing an algorithm or process on a computer.

Computing medium   Computing circuitry capable of implementing one or more algorithms. Examples include human neurons and silicon chips.

Connectionism   An approach to studying intelligence and to creating intelligent solutions to problems. Connectionism is based on storing problem-solving knowledge as a pattern of connections among a very large number of simple processing units operating in parallel.

Consciousness   The ability to have subjective experience. The ability of a being, animal, or entity to have self-perception and self-awareness. The ability to feel. A key question in the twenty-first century is whether computers will achieve consciousness (which their human creators are considered to have).

Continuous speech recognition (CSR)   A software program that recognizes and records natural language.

Crystalline computing   A system in which data is stored in a crystal as a hologram, conceived by Stanford professor Lambertus Hesselink. This three-dimensional storage method requires a million atoms for each bit and could achieve a trillion bits of storage for each cubic centimeter. Crystalline computing also refers to the possibility of growing computers as crystals.

CSR   *See* Continuous speech recognition.

Cybernetic artist   A computer program that is able to create original artwork in poetry, visual art, or music. Cybernetic artists will become increasingly commonplace starting in 2009.

Cybernetic chauffeur   Self-driving cars that use special sensors in the roads. Self-driving cars are being experimented with in the late 1990s, with implementation on major highways feasible during the first decade of the twenty-first century.

Cybernetic poet   A computer program that is able to create original poetry.

Cybernetics   A term coined by Norbert Wiener to describe the "science of control and communication in animals and machines." Cybernetics is based on the theory that intelligent living beings adapt to their environments and accomplish objectives primarily by reacting to feedback from their surroundings.

Database   The structured collection of data that is designed in connection with an information retrieval system. A database management system (DBMS) allows monitoring, updating, and interacting with the database.

Debugging   The process of discovering and correcting errors in computer hardware and software. The issue of bugs or errors in a program will become increasingly important as computers are integrated into the human brain and physiology throughout the twenty-first century. The first "bug" was an actual moth, discovered by Grace Murray Hopper, the first programmer of the Mark I computer.

Deep Blue   The computer program, created by IBM, that defeated Gary Kasparov, the world's chess champion, in 1997.

Destroy-all-copies movement   In 2099, a movement to permit an individual to terminate her mind file and to destroy all backup copies of that file.

Destructive scan   The process of scanning one's brain and neural system while destroying it, with a view to replacing it with electronic circuits of far greater capacity, speed, and reliability.

Digital   Varying in discrete steps. The use of combinations of bits to represent data in computation. Contrasted with analog.

Digital video disc (DVD)   A high-density compact disc system that uses a more focused laser than the conventional CD-ROM, with storage capacities of up to 9.4 gigabytes on a double-sided disc. A DVD has sufficient capacity to hold a full-length movie.

Direct neural pathway   Direct electronic communication to the brain. In 2029, direct neural pathways, combined with wireless communication technology, will connect humans directly to the worldwide computing network (the Web).

Diversity   Variety of choices, in which evolution thrives. A key resource for an evolutionary process. The other resource for evolution is its own increasing order.

DNA   Deoxyribonucleic acid; the building blocks of all organic life-forms. In the twenty-first century, intelligent life-forms will be based on new computational technologies and nanoengineering.

DNA computing   A form of computing, pioneered by Leonard Adleman, in which DNA molecules are used to solve complex mathematical problems. DNA computers allow trillions of computations to be performed simultaneously.

DVD   *See* Digital video disc.

Einstein's theory of relativity   Refers to two of Einstein's theories. Einstein's Special Theory of Relativity postulates the speed of light as the fastest speed at which we can transmit information. Einstein's General Theory of Relativity deals with the effects of gravity on the geometry of space. Includes the formula $E=mc^2$ (energy equals mass times the speed of light squared), which is the basis of nuclear power.

EMI   *See* Experiments in Musical Intelligence.

Encryption   Encoding information so that only the intended recipient can understand the message by decoding it. PGP (Pretty Good Privacy) is an example of encryption.

Entropy   In thermodynamics, a measure of the chaos (unpredictable movement) of particles and unavailable energy in a physical system of many components. In other contexts, a term used to describe the extent of randomness and disorder of a system.

Evolution   A process in which diverse entities (sometimes called organisms) compete for limited resources in an environment, with the more successful organisms able to survive and reproduce (to a greater extent) into subsequent generations. Over many such generations, the organisms become better adapted at survival. Over generations, the order (suitability of information for a purpose) of the design of the organisms increases, with the purpose being survival. In an "evolutionary algorithm" (see below), the purpose may be defined to be the discovery of a solution to a complex problem. Evolution also refers to a theory in which each life-form on Earth has its origin in an earlier form.

Evolutionary algorithm   Computer-based problem-solving systems that use computational models of the mechanisms of evolution as key elements in their design.

Experiments in Musical Intelligence (EMI)   A computer program that composes musical scores. Created by the composer David Cope.

Expert system   A computer program, based on various artificial intelligence techniques, that solves a problem using a database of expert knowledge on a topic. Also a system that enables such a database to become available to the nonexpert user. A branch of the artificial intelligence field.

Exponential growth   Characterized by growth in which size increases by a fixed multiple over time.

Exponential trend   Any trend that exhibits exponential growth (such as an exponential trend in population growth).

Femtoengineering   In 2099, a proposed computing technology on the femtometer (one thousandth of a trillionth of a meter) scale. Femtoengineering requires harnessing mechanisms inside a quark. Molly discusses femtoengineering proposals with the author in 2099.

Florence Manifesto Brigade   In 2029, a neo-Luddite group that is based on the "Florence Manifesto" written by Theodore Kaczynski from prison. Members of the brigade protest technology primarily through nonviolent means.

Fog swarm projection   In the mid- and late-twenty-first century, a technology that allows projections of physical objects and entities through the behavior of trillions of foglets. Molly's physical appearance to the author in 2099 is created by a fog swarm projection. *See* Foglet; Utility fog.

Foglet   A hypothetical robot that consists of a human-cell-sized device with twelve arms pointing in all directions. At the end of the arms are grippers so that the Foglets can grasp one another to form larger structures. These nanobots are intelligent and can merge their computational capacities with one another to create a distributed intelligence. Foglets are the brainchild of J. Storrs Hall, a Rutgers University computer scientist.

Free will   Purposeful behavior and decision making. Since the time of Plato, philosophers have explored the paradox of free will, particularly as it applies to machines. During the next century, a key issue will be whether machines will evolve into beings with consciousness and free will. A primary philosophical issue is how free will is possible if events are the result of the predictable—or unpredictable—interaction of particles. Considering the interaction of particles to be unpredictable does not resolve the paradox of free will because there is nothing purposeful in random behavior.

General Problem Solver (GPS)   A procedure and program developed by Allen Newell, J. C. Shaw, and Herbert Simon. GPS attains an objective by using recursive search and by applying rules to generate the alternatives at each branch in the recursive expansion of possible sequences. GPS uses a procedure to measure the "distance" from the goal.

Genetic algorithm   A model of machine learning that derives its behavior from a metaphor of the mechanisms of evolution in nature. Within a program, a population of simulated "individuals" are created and undergo a process of evolution in a simulated competitive environment.

Genetic programming   The method of creating a computer program using genetic or evolutionary algorithms. *See* Evolutionary algorithm; Genetic algorithm.

God spot A tiny locus of nerve cells in the frontal lobe of the brain that appears to be activated during religious experiences. Neuroscientists from the University of California discovered the God spot while studying epileptic patients who have intense mystical experiences during seizures.

Gödel's incompleteness theorem A theorem postulated by Kurt Gödel, a Czech mathematician, that states that in a mathematical system powerful enough to generate the natural numbers, there inevitably exist propositions that can be neither proved nor disproved.

Gordian knot An intricate, practically unsolvable problem. A reference to the knot tied by Gordius, to be untied only by the future ruler of Asia. Alexander the Great circumvented the dilemma of untying the knot by slashing it with his sword.

GPS *See* General Problem Solver.

Grandfather legislation As of 2099, legislation that protects the rights of MOSHs (mostly original substrate humans) and acknowledges the roots of twenty-first-century beings. *See* MOSH.

Haptic interface In virtual reality systems, the physical actuators that provide the user with a sense of touch (including the sensing of pressure and temperature).

Haptics The development of systems that allow one to experience the sense of touch in virtual reality. *See* Haptic interface.

Hologram An interference pattern, often using photographic media, that is encoded by laser beams and read by means of low-power laser beams. This interference pattern can reconstruct a three-dimensional image. An important property of a hologram is that the information is distributed throughout the hologram. Cut a hologram in half, and both halves will have the full picture, only at half the resolution. Scratching a hologram has no noticeable effect on the image. Human memory is regarded to be distributed in a similar way.

Holy Grail Any objective of a long and difficult quest. In medieval lore, the Grail refers to the plate used by Christ at the Last Supper. The Holy Grail subsequently became the object of knights' quests.

Homo erectus "Upright man." *Homo erectus* emerged in Africa about 1.6 million years ago and developed fire, clothing, language, and weapon use.

Homo habilis "Handy human." A direct ancestor leading to *Homo erectus* and eventually to *Homo sapiens*. *Homo habilis* lived approximately 1.6 to 2 million years ago. *Homo habilis* hominids were different from previous hominids in their bigger brain size, diet of both meat and plants, and creation and use of rudimentary tools.

Homo sapiens Human species that emerged perhaps 400,000 years ago. *Homo sapiens* are similar to advanced primates in terms of their genetic heritage and are distinguished by their creation of technology, including art and language.

Homo sapiens neanderthal (neanderthalensis) A subspecies of *Homo sapiens*. *Homo sapiens neanderthalensis* is thought to have evolved from *Homo erectus* about 100,000 years ago in Europe and the Middle East. This highly intelligent subspecies cultivated an involved culture that included elaborate funeral rituals, burying their dead with ornaments, caring for the sick, and making tools for domestic use and for protection. *Homo sapiens neanderthalensis* disappeared about 35,000 to 40,000 years ago, in all likelihood as a result of violent conflict with *Homo sapiens sapiens* (the subspecies of contemporary humans).

Homo sapiens sapiens Another subspecies of *Homo sapiens* that emerged in Africa about 90,000 years ago. Contemporary humans are the direct descendants of this subspecies.

Human Genome Project An international research program with the goal of gathering a resource of genomic maps and DNA sequence information that will provide detailed information about the structure, organization, and characteristics of the DNA of humans and other animals. The project began in the mid-1980s and is expected to be completed by around the year 2005.

Idiot savant   A system or person who is highly skilled in a narrow task area but who lacks context and is otherwise impaired in more general areas of intelligent functioning. The term is taken from psychiatry, where it refers to a person who exhibits brilliance in one very limited domain but is underdeveloped in common sense, knowledge, and competence. For example, some human idiot savants are capable of multiplying very large numbers in their heads, or memorizing a phone book. Deep Blue is an example of an idiot savant system.

Image processing   The manipulation of data representing images, or pictorial representation on a screen, composed of pixels. The use of a computer program to enhance or modify an image.

Improvisor   A computer program that creates original music, written by Paul Hodgson, a British jazz saxophone player. Improvisor can emulate styles ranging from Bach to jazz greats Louis Armstrong and Charlie Parker.

Industrial Revolution   The period in history in the late eighteenth and nineteenth centuries marked by accelerating developments in technology that enabled the mass production of goods and materials.

Information   A sequence of data that is meaningful in a process, such as the DNA code of an organism or the bits in a computer program. Information is contrasted with "noise," which is a random sequence. However, neither noise nor information is predictable. Noise is inherently unpredictable but carries no information. Information is also unpredictable; that is, we cannot predict future information from past information. If we can fully predict future data from past data, then that future data stops being information.

Information Theory   A mathematical theory concerning the difference between information and noise, and the ability of a communications channel to carry information.

Intelligence   The ability to use optimally limited resources—including time—to achieve a set of goals (which may include survival, communication, solving problems, recognizing patterns, performing skills). The products of intelligence may be clever, ingenious, insightful, or elegant. R. W. Young defines intelligence as "that faculty of mind by which order is perceived in a situation previously considered disordered."

Intelligent agent   An autonomous software program that performs a function on its own, such as searching the Web for information of interest to a person based on certain criteria.

Intelligent function   A function that requires increasing intelligence to compute for increasing arguments. The busy beaver is an example of an intelligent function.

Internet computation harvesting proposal   A proposal to harvest the unused computational resources of personal computers on the Internet and thereby create virtual parallel supercomputers. There are sufficient unused "computes" on the Internet in 1998 to create human brain capacity supercomputers, at least in terms of hardware capability.

Knee of the curve   The period in which the exponential nature of the curve of time begins to explode. Exponential growth lingers with no apparent growth for a long period of time and then appears to erupt suddenly. This is now occurring in the capability of computers.

Knowledge engineering   The art of designing and building expert systems. In particular, collecting knowledge and heuristic rules from human experts in their area of specialty and assembling them into a knowledge base or expert system.

Knowledge principle   A principle that emphasizes the important role played by knowledge in many forms of intelligent activity. It states that a system exhibits intelligence in part due to the specific knowledge relevant to the task that it contains.

Knowledge representation A system for organizing human knowledge in a domain into a data structure flexible enough to allow the expression of facts, rules, and relationships.

Law of Accelerating Returns As order exponentially increases, time exponentially speeds up (i.e., the time interval between salient events grows shorter as time passes).

Law of Increasing Chaos As chaos exponentially increases, time exponentially slows down (i.e., the time interval between salient events grows longer as time passes).

Law of Time and Chaos In a process, the time interval between salient events (i.e., events that change the nature of the process, or significantly affect the future of the process) expands or contracts along with the amount of chaos.

Laws of thermodynamics The laws of thermodynamics govern how and why energy is transferred.

The first law of thermodynamics (postulated by Hermann von Helmholtz in 1847), also called the Law of Conservation of Energy, states that the total amount of energy in the Universe is constant. A process may modify the form of energy, but a closed system does not lose energy. We can use this knowledge to determine the amount of energy in a system, the amount lost as waste heat, and the efficiency of the system.

The second law of thermodynamics (articulated by Rudolf Clausias in 1850), also known as the Law of Increasing Entropy, states that the entropy (disorder of particles) in the Universe never decreases. As the disorder in the Universe increases, the energy is transformed into less usable forms. Thus, the efficiency of any process will always be less than 100 percent.

The third law of thermodynamics (described by Walter Hermann Nernst in 1906, based on the idea of a temperature of absolute zero first articulated by Baron Kelvin in 1848), also known as the Law of Absolute Zero, tells us that all molecular movement stops at a temperature called absolute zero, or 0 Kelvin ($-273°C$). Since temperature is a measure of molecular movement, the temperature of absolute zero can be approached, but it can never be reached.

Life The ability of entities (usually organisms) to reproduce into future generations. Patterns of matter and energy that can perpetuate themselves and survive.

LISP (list processing) An interpretive computer language developed in the late 1950s at MIT by John McCarthy used to manipulate symbolic strings of instructions and data. The principal data structure is the list, a finite ordered sequence of symbols. Because a program written in LISP is itself expressed as a list of lists, LISP lends itself to sophisticated recursion, symbol manipulation, and self-modifying code. It has been widely used for AI programming, although it is less popular today than it was in the 1970s and 1980s.

Logical positivism A twentieth-century philosophical school of thought that was inspired by Ludwig Wittgenstein's *Tractatus Logico-Philosophicus*. According to logical positivism, all meaningful statements may be confirmed by observation and experiment or are "analytic" (deducible from observations).

Luddite One of a group of early-nineteenth-century English workmen who destroyed labor-saving machinery in protest. The Luddites were the first organized movement to oppose the mechanized technology of the Industrial Revolution. Today, the Luddites are a symbol of opposition to technology.

Magnetic resonance imaging (MRI) A noninvasive diagnostic technique that produces computerized images of body tissues and is based on nuclear magnetic resonance of atoms within the body produced by the application of radio waves. A person is placed in a magnetic field thirty thousand times stronger than the normal magnetic field on Earth. The person's body is stimulated with radio waves, and the body responds with its own electromagnetic transmissions. These are detected and processed by computer to generate a three-dimensional map of high-resolution internal features such as blood vessels.

Massively parallel neural nets   A neural net built from many parallel processing units. Generally, a separate, specialized computer implements each neuron model.

Microprocessor   An integrated circuit built on a single chip containing the entire central processing unit (CPU) of a computer.

Millions of Instructions per Second   A method of measuring the speed of a computer in terms of the number of millions of instructions performed by the computer in one second. An instruction is a single step in a computer program as represented in the computer's machine language.

Mind-body problem   The philosophical question: How does the nonphysical entity of the mind emerge from the physical entity of the brain? How do feelings and other subjective experiences result from the processing of the physical brain? By extension, will machines emulating the processes of the human brain have subjective experiences? Also, how does the nonphysical entity of the mind exert control over the physical reality of the body?

Mind trigger   A stimulation of an area of the brain that evokes a feeling usually (i.e., otherwise) gained from actual physical or mental experience.

Minimax procedure or theorem   A basic technique used in game-playing programs. An expanding tree of possible moves and countermoves (moves from the opponent) is constructed. An evaluation of the final "leaves" of the tree that minimizes the opponent's ability to win and maximizes the program's ability to win is then passed back down the branches of the tree.

MIPS   *See* Millions of Instructions per Second.

Mission critical system   A software program that controls a process on which people are heavily dependent. Examples of mission critical software include life-support systems in hospitals, automated surgical equipment, autopilot flying and landing systems, and other software-based systems that affect the well-being of a person or organization.

Molecular computer   A computer based on logic gates that is constructed on principles of molecular mechanics (as opposed to principles of electronics) by appropriate arrangements of molecules. Since the size of each logic gate (device that can perform a logical operation) is only one or a few molecules, the resultant computer can be microscopic in size. Limitations on molecular computers arise only from the physics of atoms. Molecular computers can be massively parallel by having parallel computations performed by trillions of molecules simultaneously. Molecular computers have been demonstrated using the DNA molecule.

Moore's Law   First postulated by former Intel CEO Gordon Moore in the mid-1960s, Moore's Law is the prediction that the size of each transistor on an integrated circuit chip will be reduced by 50 percent every twenty-four months. The result is the exponentially growing power of integrated circuit-based computation over time. Moore's Law doubles the number of components on a chip as well as the speed of each component. Both of these aspects double the power of computing, for an effective quadrupling of the power of computation every twenty-four months.

MOSH   In 2099, an acronym for Mostly Original Substrate Humans. In the last half of the twenty-first century, a human being still using native carbon-based neurons and unenhanced by neural implants is referred to as a MOSH. In 2099, Molly refers to the author as being a MOSH.

MOSH art   In 2099, art (that is usually created by enhanced humans) that a MOSH is theoretically capable of appreciating, although MOSH art is not always shared with a MOSH.

MOSH music   In 2099, MOSH art in the form of music.

Moshism   In 2099, an archaic term that is rooted in the MOSH way of life, before the advent of enhanced humans through neural implants and the porting of human brains to new computational substrates. An example of a Moshism: the word *papers* to refer to knowledge structures representing a body of intellectual work.

MRI   *See* Magnetic resonance imaging.

MYCIN   A successful expert system, developed at Stanford University in the mid-1970s, designed to aid medical practitioners in prescribing an appropriate antibiotic by determining the exact identity of a blood infection.

Nanobot   A nanorobot (robot built using nanotechnology). A self-replicating nanobot requires mobility, intelligence, and the ability to manipulate its environment. It also needs to know when to stop its own replication. In 2029, nanobots will circulate through the bloodstream of the human body to diagnose illnesses.

Nanobot swarm   In the last half of the twenty-first century, a swarm comprised of trillions of nanobots. The nanobot swarms can rapidly take on any form. A nanobot swarm can project the visual images, sounds, and pressure contours of any set of objects, including people. The swarms of nanobots can also combine their computational abilities to emulate the intelligence of people and other intelligent entities and processes. A nanobot swarm effectively brings the ability to create virtual environments into the real environment.

Nanoengineering   The design and manufacturing of products and other objects based on the manipulation of atoms and molecules; building machines atom by atom. "Nano" refers to a billionth of a meter, which is the width of five carbon atoms. *See* Pi coengineering; Femtoengineering.

Nanopathogen   A self-replicating nanobot that replicates excessively, possibly without limit, causing destruction to both organic and inorganic matter.

Nanopatrol   In 2029, a nanobot in the bloodstream that checks the body for biological pathogens and other disease processes.

Nanotechnology   A body of technology in which products and other objects are created through the manipulation of atoms and molecules. "Nano" refers to a billionth of a meter, which is the width of five carbon atoms.

Nanotubes   Elongated carbon molecules that resemble long tubes and are formed of the same pentagonal patterns of carbon atoms as buckyballs. Nanotubes can perform the electronic functions of silicon-based components. Nanotubes are extremely small, thereby providing very high densities of computation. Nanotubes are a likely technology to continue to provide the exponential growth of computing when Moore's Law on integrated circuits dies by the year 2020. Nanotubes are also extremely strong and heat resistant, thereby permitting the creation of three-dimensional circuits.

Natural language   Language as ordinarily spoken or written by humans using a human language such as English (as contrasted with the rigid syntax of a computer language). Natural language is governed by rules and conventions sufficiently complex and subtle for there to be frequent ambiguity in syntax and meaning.

Neanderthal   *See* Homo sapiens neanderthal (*neanderthalensis*).

Neural computer   A computer with hardware optimized for using the neural network paradigm. A neural computer is designed to simulate a massive number of models of human neurons.

Neural connection calculation   In a neural network, a term that refers to the primary calculation of multiplying the "strength" of a neural connection by the input to that connection (which is either the output of another neuron or an initial input to the system) and then adding this product to the accumulated sum of such products from other connections to this neuron. This operation is highly repetitive, so neural computers are optimized for performing it.

Neural implant   A brain implant that enhances one's sensory ability, memory, or intelligence. Neural implants will become ubiquitous in the twenty-first century.

Neural network   A computer simulation of human neurons. A system (implemented in software or hardware) that is intended to emulate the computing structure of neurons in the human brain.

Neuron   Information-processing cell of the central nervous system. There are an estimated 100 billion neurons in the human brain.

Noise   A random sequence of data. Because the sequence is random and without meaning, noise carries no information. Contrasted with information.

Objective experience   The experience of an entity as observed by another entity, or measurement apparatus.

OCR   *See* Optical character recognition.

Operating system   A software program that manages and provides a variety of services to application programs, including user interface facilities and management of input-output and memory devices.

Optical character recognition (OCR)   A process in which a machine scans, recognizes, and encodes printed (and possibly handwritten) characters into digital form.

Optical computer   A computer that processes information encoded in patterns of light beams; different from today's conventional computers, in which information is represented in electronic circuitry or encoded on magnetic surfaces. Each stream of photons can represent an independent sequence of data, thereby providing extremely massive parallel computation.

Optical imaging   A brain-imaging technique similar to MRI but potentially providing higher resolution imaging. Optical imaging is based on the interaction between electrical activity in the neurons and blood circulation in the capillaries feeding the neurons.

Order   Information that fits a purpose. The measure of order is the measure of how well the information fits the purpose. In the evolution of life-forms, the purpose is to survive. In an evolutionary algorithm (a computer program that simulates evolution to solve a problem), the purpose is to solve the problem. Having more information, or more complexity, does not necessarily result in a better fit. A superior solution for a purpose—greater order—may require either more or less information, and either more or less complexity. Evolution has shown, however, that the general trend toward greater order does generally result in greater complexity.

Paradigm   A pattern, model, or general approach to solving a problem.

Parallel processing   Refers to computers that use multiple processors operating simultaneously as opposed to a single processing unit. (*Compare with* Serial computer.)

Pattern recognition   Recognition of patterns with the goal of identifying, classifying, or categorizing complex inputs. Examples of inputs include images such as printed characters and faces, and sounds such as spoken language.

Perceptron   In the late 1960s and 1970s, a machine constructed from mathematical models of human neurons. Early Perceptrons were modestly successful in such pattern-recognition tasks as identifying printed letters and speech sounds. The Perceptron was a forerunner of contemporary neural nets.

Personal computer   A generic term for a single-user computer using a microprocessor, and including the computing hardware and software needed for an individual to work autonomously.

PGP   *See* Pretty Good Privacy.

Picoengineering   Technology on the picometer (one trillionth of a meter) scale. Pico-engineering will involve engineering at the level of subatomic particles.

Picture portal   In 2009, a visual display for viewing people and other real-time images. In later years, the portals project three-dimensional, real-time scenes. Molly's son, Jeremy, uses a picture portal to view the Stanford University campus.

Pixel   An abbreviation for picture element. The smallest element on a computer screen that holds information to represent a picture. Pixels contain data giving brightness and possibly color at particular points in the picture.

Pretty Good Privacy (PGP)   A system of encryption (designed by Phil Zimmerman) distributed on the Internet and widely used. PGP uses a public key that can be freely disseminated and used by anyone to encode a message and a private key that is kept only by the intended recipient of the encoded messages. The private key is used by the recipient to decode messages encrypted using the public key. Converting the public key into a private key requires factoring large numbers. If the number of bits in the public key is large enough, then the factors cannot be computed in a reasonable amount of time using conventional computation (and thus the encoded information remains secure). Quantum computing (with a sufficient number of qu-bits) would destroy this type of encryption.

Price-performance   A measure of the performance of a product per unit cost.

Program   A set of computer instructions that enables a computer to perform a specific task. Programs are usually written in a high-level language such as "C" or "FORTRAN" that can be understood by human programmers and then translated into machine language using a special program called a compiler. Machine language is a special set of codes that directly controls a computer.

Punch card   A rectangular card that typically records up to eighty characters of data in a binary coded format as a pattern of holes punched in it.

Quantum computing   A revolutionary method of computing, based on quantum physics, that uses the ability of particles such as electrons to exist in more than one state at the same time. *See* Qu-bit.

Quantum decoherence   A process in which the ambiguous quantum state of a particle (such as the nuclear spin of an electron representing a qu-bit in a quantum computer) is resolved into an unambiguous state as the result of direct or indirect observation by a conscious observer.

Quantum encryption   A possible form of encryption using streams of quantum entangled particles such as photons. *See* Quantum entanglement.

Quantum entanglement   A relationship between two physically separated particles under special circumstances. Two photons may be "quantum entangled" if produced by the same particle interaction and emerging in opposite directions. The two photons remain quantum entangled with each other even when separated by very large distances (even when light-years apart). In such a circumstance, the two quantum entangled photons, if each forced to make a decision to choose among two equally probable pathways, will make the identical decision and will do so at the same instant in time. Since there is no possible communication link between two quantum entangled photons, classical physics would predict that their decisions would be independent. But two quantum entangled photons make the same decision and do so at the same instant in time. Experiments have demonstrated that even if there were an unknown communication path between them, there is not enough time for a message to travel from one photon to the other at the speed of light.

Quantum mechanics   A theory that describes the interactions of subatomic particles, combining several basic discoveries. These include Max Planck's 1900 observation that energy is absorbed or radiated in discrete quantities, called quanta. Also Werner

Heisenberg's 1927 uncertainty principle stating that we cannot know both the exact position and momentum of an electron or other particle at the same time. Interpretations of quantum theory imply that photons simultaneously take all possible paths (e.g., when bouncing off a mirror). Some paths cancel each other out. Remaining ambiguity in the path actually taken is resolved based on the conscious observation of an observer.

Qu-bit   A "quantum bit," used in quantum computing, that is both zero and one at the same time, until quantum decoherence (direct or indirect observation by a conscious observer) causes each quantum bit to disambiguate into a state of zero or one. One qubit stores two possible numbers (zero and one) at the same time. N qu-bits stores 2^N possible numbers at the same time. Thus an N qu-bit quantum computer would try 2^N possible solutions to a problem simultaneously, which gives the quantum computer its enormous potential power.

RAM   *See* Random Access Memory.

Random Access Memory (RAM)   Memory that can be both read and written with random access of memory locations. Random access means that locations can be accessed in any order and do not need to be accessed sequentially. RAM can be used as the working memory of a computer into which applications and programs can be loaded and run.

Ray Kurzweil's Cybernetic Poet   A computer program designed by Ray Kurzweil that uses a recursive approach to create poetry. The Cybernetic Poet analyzes word sequence patterns of poems it has "read" using markov models (a mathematical cousin of neural nets) and creates new poetry based on these patterns.

Read-Only Memory (ROM)   A form of computer storage that can be read from but not written to or deleted (e.g., CD-ROM).

Reading machine   A machine that scans text and reads it aloud. Initially developed for those who are visually impaired, reading machines are currently used by anyone who cannot read at their intellectual level, including reading disabled (e.g., dyslexic) persons and children first learning to read.

Recursion   The process of defining or expressing a function or procedure in terms of itself. Typically, each iteration of a recursive-solution procedure produces a simpler (or possibly smaller) version of the problem than the previous iteration. This process continues until a subproblem whose answer is already known (or that can be readily computed without recursion) is obtained. A surprisingly large number of symbolic and numerical problems lend themselves to recursive formulations. Recursion is typically used by game-playing programs, such as the chess-playing program Deep Blue.

Recursive formula   A computer-programming paradigm that uses recursive search to find a solution to a problem. The recursive search is based on a precise definition of the problem (e.g., the rules of a game such as chess).

Relativity   A theory based on two postulates: (1) that the speed of light in a vacuum is constant and independent of the source or the observer, and (2) that the mathematical forms of the laws of physics are invariant in all inertial systems. Implications of the theory of relativity include the equivalence of mass and energy and of change in mass, dimension, and time with increased velocity. *See also* Einstein's theory of relativity.

Relaxation Response   A neurological mechanism discovered by Dr. Herbert Benson and other researchers at the Harvard Medical School and Boston's Beth Israel Hospital. The opposite of the "fight or flight" or stress response, the Relaxation Response is associated with reduced levels of epinephrine (adrenaline) and norepinephrine (noradrenaline), blood pressure, blood sugar, breathing, and heart rates.

Remember York movement   In the second decade of the twenty-first century, a neo-

Luddite web discussion group. The group is named to commemorate the 1813 trial in York, England, during which a number of the Luddites who destroyed industrial machinery were hanged, jailed, or exiled.

Reverse engineering Examining a product, program, or process to understand it and to determine its methods and algorithms. Scanning and copying a human brain's salient computational methods into a neural computer of sufficient capacity is a future example of reverse engineering.

RKCP *See* Ray Kurzweil's Cybernetic Poet.

Robinson The world's first operational computer, constructed from telephone relays and named after a popular cartoonist who drew "Rube Goldberg" machines (very ornate machinery with many interacting mechanisms). During World War II, Robinson provided the British with a transcription of nearly all significant Nazi coded messages, until it was replaced by Colossus. *See* Colossus.

Robot A programmable device, linked to a computer, consisting of mechanical manipulators and sensors. A robot may perform a physical task normally done by human beings, possibly with greater speed, strength, and/or precision.

Robotics The science and technology of designing and manufacturing robots. Robotics combines artificial intelligence and mechanical engineering.

ROM *See* Read-Only Memory.

Russell's Paradox The ambiguity created by the following question: Does a set that is defined as "all sets that do not include themselves" include itself as a member? Russell's paradox motivated Bertrand Russell to create a new theory of sets.

Search A recursive procedure in which an automatic problem solver seeks a solution by iteratively exploring sequences of possible alternatives.

Second Industrial Revolution The automation of mental rather than physical tasks.

Second law of thermodynamics Also known as the Law of Increasing Entropy, this law states that the disorder (amount of random movement) of particles in the Universe may increase but never decreases. As the disorder in the Universe increases, the energy is transformed into less usable forms. Thus, the efficiency of any process will always be less than 100 percent (hence the impossibility of perpetual motion machines).

Self-replication A process or device that is capable of creating an additional copy of itself. Nanobots are self-replicating if they can create copies of themselves. Self-replication is regarded as a necessary means of manufacturing nanobots due to the very large number (i.e., trillions) of such devices needed to perform useful functions.

Semiconductor A material commonly based on silicon or germanium with a conductivity midway between that of a good conductor and an insulator. Semiconductors are used to manufacture transistors. Semiconductors rely on the phenomenon of tunneling. *See* Tunneling.

Sensorium In 2019, the product name for a total touch virtual reality environment, which provides an all-encompassing tactile environment.

Serial computer A computer that performs a single computation at a time. Thus two or more computations are performed one after the other, not simultaneously (even if the computations are independent). The opposite of a parallel processing computer.

Silicon Valley The area in California, south of San Francisco, that is a key center of high-technology innovation, including the development of software, communication, integrated circuits and related technologies.

Simple-minded school The use of simple procedures to evaluate the terminal leaves in a recursive algorithm. For example, in the context of a chess program, adding up piece values.

Simulated person A realistic, animated personality incorporating a convincing visual appearance and capable of communicating using natural language. By 2019, a simulated person can interact with real persons using visual, auditory, and tactile means in a virtual reality environment.

Simulator   A program that models and represents an activity or environment on a computer system. Examples include the simulation of chemical interaction and fluid flow. Other examples include a flight simulator used to train pilots and a simulated patient to train physicians. Simulators are also often used for entertainment.

Society of mind   A theory of the mind proposed by Marvin Minsky in which intelligence is seen to be the result of proper organization of a large number (a society) of other minds, which are in turn comprised of yet simpler minds. At the bottom of this hierarchy are simple mechanisms, each of which is by itself unintelligent.

Software   Information and knowledge used to perform useful functions by computers and computerized devices. Includes computer programs and their data, but more generally also includes such knowledge products as books, music, pictures, movies, and videos.

Software-based evolution   Software simulation of the evolutionary process. One example of software-based evolution is Network Tierra, designed by Thomas Ray. Ray's "creatures" are software simulations of organisms in which each "cell" has its own DNA-like genetic code. The organisms compete with one another for the limited simulated space and energy resources of their simulated environment.

Speaker independence   Refers to the ability of a speech-recognition system to understand any speaker, regardless of whether or not the system has previously sampled that speaker's speech.

Stored-program computer   A computer in which the program is stored in memory along with the data to be operated on. A stored-program capacity is an important capability for systems of artificial intelligence in that recursion and self-modifying code are not possible without it.

Subjective experience   The experience of an entity as experienced by the entity, as opposed to observations of that entity (including its internal processes) by another entity, or by a measurement apparatus.

Substrate   Computing medium or circuitry. *See* Computing medium.

Supercomputer   The fastest and most powerful computer available at any given time. Supercomputers are used for computations demanding high speed and storage (e.g., analyzing weather data).

Superconductivity   The physical phenomenon whereby some materials exhibit zero electrical resistance at low temperatures. Superconductivity points to the possibility of great computational power with little or no heat dissipation (a limiting factor today). Heat dissipation is a major reason that three-dimensional circuits are difficult to create.

Synthesizer   A device that computes signals in real time. In the context of music, a (usually computer based) device that creates and generates sounds and music electronically.

Tactile virtualism   By 2029, a technology that allows one to use a virtual body to enjoy virtual reality experiences without virtual reality equipment other than the use of neural implants (which include high-bandwidth wireless communication). The neural implants create the pattern of nerve signals that corresponds to a comparable "real" experience.

Technology   An evolving process of tool creation to shape and control the environment. Technology goes beyond the mere fashioning and use of tools. It involves a record of tool making and a progression in the sophistication of tools. It requires invention and is itself a continuation of evolution by other means. The "genetic code" of the evolutionary process of technology is the knowledge base maintained by the tool-making species.

Three-dimensional chip   A chip that is constructed in three dimensions, thus allowing for hundreds or thousands of layers of circuitry. Three-dimensional chips are currently being researched and engineered by a variety of companies.

Total touch environment   In 2019, a virtual-reality environment that provides an all-encompassing tactile environment.

Transistor   A switching and/or amplifying device using semiconductors, first created in 1948 by John Bardeen, Walter Brattain, and William Shockley of Bell Labs.

Translating telephone   A telephone that provides real-time speech translation from one human language to another.

Tunneling   In quantum mechanics, the ability of electrons (negatively charged particles orbiting the nucleus of an atom) to exist in two places at once, in particular on both sides of a barrier. Tunneling allows some of the electrons to effectively move through the barrier and accounts for the "semi" conductor properties of a transistor.

Turing machine   A simple abstract model of a computing machine, designed by Alan Turing in his 1936 paper "On Computable Numbers." The Turing machine is a fundamental concept in the theory of computation.

Turing Test   A procedure proposed by Alan Turing in 1950 for determining whether or not a system (generally a computer) has achieved human-level intelligence, based on whether it can deceive a human interrogator into believing that it is human. A human "judge" interviews the (computer) system, and one or more human "foils" over terminal lines (by typing messages). Both the computer and the human foil(s) try to convince the human judge of their humanness. If the human judge is unable to distinguish the computer from the human foil(s), then the computer is considered to have demonstrated human-level intelligence. Turing did not specify many key details, such as the duration of the interrogation and the sophistication of the human judge and foils. By 2029, computers are passing the test, although the validity of the test remains a point of controversy and philosophical debate.

Utility fog   A space filled with Foglets. At the end of the twenty-first century, utility fog can be used to simulate any environment, essentially providing "real" reality with the environment-transforming capabilities of virtual reality. *See* Fog swarm projection; Foglet.

Vacuum tube   The earliest form of an electronic switch (or amplifier) based on vacuum-filled glass containers. Used in radios and other communication equipment and early computers; replaced by the transistor.

Venture Capital   Refers to funds available for investment by organizations that have raised pools of capital specifically to invest in companies, primarily new ventures.

Virtual body   In virtual reality, one's own body potentially transformed to appear (and ultimately to feel) different than it does in "real" reality.

Virtual reality   A simulated environment in which you can immerse yourself. A virtual reality environment provides a convincing replacement for the visual and auditory senses, and (by 2019) the tactile sense. In later decades, the olfactory sense will be included as well. The key to a realistic visual experience in virtual reality is that when you move your head, the scene instantly repositions itself so that you are now looking at a different region of a three-dimensional scene. The intention is to simulate what happens when you turn your real head in the real world: The images captured by your retinas rapidly change. Your brain nonetheless understands that the world has remained stationary and that the image is sliding across your retinas only because your head is rotating. Initially, virtual reality (including crude contemporary systems) requires the use of special helmets to provide the visual and auditory environments. By 2019, virtual reality will be provided by ubiquitous contact-lens-based systems and implanted retinal-imaging devices (as well as comparable devices for auditory "imaging"). Later in the twenty-first century, virtual reality (which will include all the senses) will be provided by direct stimulation of nerve pathways using neural implants.

Virtual reality auditory lenses In 2019, sonic devices that project high-resolution sounds precisely placed in the three-dimensional virtual environment. These can be built into eyeglasses, worn as body jewelry, or implanted.

Virtual reality blocking display In 2019, a display technology using virtual reality optical lenses (see below) and virtual reality auditory lenses (see above) that creates highly realistic virtual visual environments. The display blocks out the real environment, so you see and hear only the projected virtual environment.

Virtual reality head-directed display In 2019, a display technology using virtual reality optical lenses (see below) and virtual reality auditory lenses (see above) that projects a virtual environment stationary with respect to the position and orientation of your head. When you move your head, the display moves relative to the real environment. This mode is often used to interact with virtual documents.

Virtual reality optical lenses In 2009, three-dimensional displays built into glasses or contact lenses. These "direct eye" displays create highly realistic virtual visual environments overlaying the "real" environment. This display technology projects images directly onto the human retina, exceeds the resolution of human vision, and is widely used regardless of visual impairment. In 1998, the Microvision Virtual Retina Display provides a similar capability for military pilots, with consumer versions anticipated.

Virtual reality overlay display In 2019, a display technology using virtual reality optical lenses (see above) and virtual reality auditory lenses (see above) that integrates real and virtual environments. The displayed images slide when you move or turn your head so that the virtual people, objects, and environment appear to remain stationary in relation to the real environment (which you can still see). Thus if the direct eye display is displaying the image of a person (who could be a geographically remote real person engaging in a three-dimensional visual phone call with you, or a computer-generated simulated person), that projected person will appear to be in a particular place relative to the real environment that you also see. When you move your head, that projected person will appear to remain in the same place relative to the real environment.

Virtual sex Sex in virtual reality incorporating a visual, auditory, and tactile environment. The sex partner can be a real or simulated person.

Virtual tactile environment A virtual reality system that allows the user to experience a realistic and all-encompassing tactile environment.

Vision chip A silicon emulation of the human retina that captures the algorithm of early mammalian visual processing, an algorithm called center surround filtering.

World Wide Web (WWW) A highly distributed (not centralized) communications network allowing individuals and organizations around the world to communicate with one another. Communication includes the sharing of text, images, sounds, video, software, and other forms of information. The primary user interface paradigm of the "web" is based on hypertext, which consists of documents (which can contain any type of data) connected by "links," which the user selects by a pointing device such as a mouse. The Web is a system of data-and-message servers linked by high-capacity communication links that can be accessed by any computer user with a "web browser" and Internet access. With the introduction of Windows98, access to the Web is built into the operating system. By the late twenty-first century, the Web will provide the distributed computing medium for software-based humans.

Y2K (year 2000 problem) Refers to anticipated difficulties caused by software (usually developed several decades prior to the year 2000) in which date fields used only two digits. Unless the software is adjusted, this will cause computer programs to behave erratically when the year becomes "00." These programs will mistake the year 2000 for 1900.

NOTES

PROLOGUE: AN INEXORABLE EMERGENCE

1. My recollections of *The Twilight Zone* episode are essentially accurate, although the gambler is actually a small-time crook named Rocky Valentine. Episode 28, "A Nice Place to Visit" (I learned the name of the episode after writing the prologue), aired during the first season of *The Twilight Zone,* on April 15, 1960.

   The episode begins with a voice-over: "Portrait of a man at work, the only work he's ever done, the only work he knows. His name is Henry Francis Valentine, but he calls himself Rocky, because that's the way his life has been—rocky and perilous and uphill at a dead run all the way. . . ."

   While robbing a pawnbroker's shop, Valentine is shot and killed by a policeman. When he awakens, he is met by his afterlife guide, Pip. Pip explains that he will provide Valentine with whatever he wants. Valentine is suspicious, but he asks for and receives a million dollars and a beautiful girl. He then goes on a gambling spree, winning at the roulette table, at the slot machines, and later, at pool. He is also surrounded by beautiful women, who shower him with attention.

   Eventually Valentine tires of the gambling, the winning, and the beautiful women. He tells Pip that it is boring to win all the time and that he doesn't belong in Heaven. He begs Pip to take him to "the Other Place." With a malicious gleam in his eye, Pip replies, "This is the Other Place!" Episode synopsis adapted from Marc Scott Zicree, *The Twilight Zone Companion* (Toronto: Bantam Books, 1982, 113–115).

2. What were the primary political and philosophical issues of the twentieth century? One was ideological—totalitarian systems of the right (fascism) and left (communism) were confronted and largely defeated by capitalism (albeit with a large public sector) and democracy. Another was the rise of technology, which began to be felt in the nineteenth century and became a major force in the twentieth century. But the issue of "what constitutes a human being" is not yet a primary issue (except as it affects the abortion debate), although the past century did witness the continuation of earlier struggles to include all members of the species as deserving of certain rights.

3. For an excellent overview and technical details on neural-network pattern recognition, see the "Neural Network Frequently Asked Questions" web site, edited by W. S.

Sarle, at <ftp://ftp.sas.com/pub/neural/FAQ.html>. In addition, an article by Charles Arthur, "Computers Learn to See and Smell Us," from *Independent,* January 16, 1996, describes the ability of neural nets to differentiate between unique characteristics.

4. As will be discussed in chapter 6, "Building New Brains," destructive scanning will be feasible early in the twenty-first century. Noninvasive scanning with sufficient resolution and bandwidth will take longer but will be feasible by the end of the first half of the twenty-first century.

CHAPTER 1: THE LAW OF TIME AND CHAOS

1. For a comprehensive overview and detailed references on the big bang theory and the origin of the Universe, see "Introduction to Big Bang Theory," Bowdoin College Department of Physics and Astronomy at <http://www.bowdoin.edu/dept/physics/astro. 1997/astro4/bigbang.html>.

    Print sources on the big bang theory include: Joseph Silk, *A Short History of the Universe* (New York: Scientific American Library, 1994); Joseph Silk, *The Big Bang* (San Francisco: W. H. Freeman and Company, 1980); Robert M. Wald, *Space, Time & Gravity* (Chicago: The University of Chicago Press, 1977); and Stephen W. Hawking, *A Brief History of Time* (New York: Bantam Books, 1988).

2. The strong force holds an atomic nucleus together. It is called "strong" because it needs to overcome the powerful repulsion between the protons in a nucleus with more than one proton.

3. The electroweak force combines electromagnetism and the weak force responsible for beta decay. In 1968, American physicist Steven Weinberg and Pakistani physicist Abdus Salam were successful in their unification of the weak force and the electromagnetic force using a mathematical method called gauge symmetry.

4. The weak force is responsible for beta decay and other slow nuclear processes that occur gradually.

5. Albert Einstein, *Relativity: The Special and the General Theory* (New York: Crown Publishers, 1961).

6. The laws of thermodynamics govern how and why energy is transferred.

    The first law of thermodynamics (postulated by Hermann von Helmholtz in 1847), also called the Law of Conservation of Energy, states that the total amount of energy in the universe is constant.

    The second law of thermodynamics (articulated by Rudolf Clausias in 1850), also known as the Law of Increasing Entropy, states that entropy, or disorder, in the Universe never decreases (and, therefore, usually increases). As the disorder in the Universe increases, the energy is transformed into less usable forms. Thus, the efficiency of any process will always be less than 100 percent.

    The third law of thermodynamics (described by Walter Hermann Nernst in 1906, based on the idea of a temperature of absolute zero first articulated by Baron Kelvin in 1848), also known as the Law of Absolute Zero, tells us that all molecular movement stops at a temperature called absolute zero, or 0 Kelvin ($-273°C$). Since temperature is a measure of molecular movement, the temperature of absolute zero can be approached, but it can never be reached.

7. "Evolution and Behavior" at <http://ccp.uchicago.edu/~jyin/evolution.html> contains an excellent collection of articles and links exploring the theories of evolution. Print sources include Edward O. Wilson, *The Diversity of Life* (New York: W. W. Norton & Company, 1993); and Stephen Jay Gould, *The Book of Life* (New York: W. W. Norton & Company, 1993).

8. Four hundred million years ago, vegetation spread from lowland swamps to create

the first land-based plants. This development permitted vertebrate herbivorous animals to step onto land, creating the first amphibians. Along with the amphibians, arthropods also stepped onto land, some of which evolved into insects. About 200 million years ago, dinosaurs and mammals began sharing the same environment. The dinosaurs were far more noticeable. Mostly the mammals stayed out of the dinosaurs' way, with many mammals being nocturnal.

9. Mammals became dominant in the niche of land-based animals after the demise of the dinosaurs 65 million years ago. Mammals are the more intellectual animal class, distinguished by warm blood, the nourishment of their children with maternal milk, hairy skin, sexual reproduction, four appendages (in most cases) and, most notably, a highly developed nervous system.

10. Primates, the most advanced mammalian order, were distinguished by forward-facing eyes, binocular vision, large brains with a convoluted cortex, which permitted more advanced reasoning faculties, and complicated social patterns. Primates were not the only intelligent animals, but they had one additional characteristic that would hasten the age of computation: the opposable thumb. The two qualities needed for the subsequent emergence of technology were now coming into place: intelligence and the ability to manipulate the environment. It's no coincidence that fingers are called digits. The origin of the word digit, as used in Modern English and appearing first in Middle English, is from the Latin word *digitus,* for "finger" or "toe"; perhaps akin to Greek *deiknynai,* "to show."

11. About 50 million years ago, the anthropoid suborder of primates split off. Unlike their prosimian cousins, the anthropoids underwent rapid evolution, giving rise to advanced primates such as monkeys and apes about 30 million years ago. These sophisticated primates were noted for subtle communication abilities using sounds, gestures, and facial expressions, thereby allowing the development of intricate social groups. About 15 million years ago, the first humanoids emerged. Although they initially walked on their hind legs, they used the knuckles of their front legs for balance.

12. Although it is worth pointing out that a 2 percent change in a computer program can be very significant.

13. *Homo sapiens* are the only technology-creating species on Earth today, but were not the first such species. Emerging about five million years ago was *Homo habilis* (i.e., "handy" human being), known for his erect posture and large brain. He was called handy because he fashioned and used tools. Our most direct ancestor, *Homo erectus,* showed up in Africa about two million years ago. *Homo erectus* was also responsible for advancing technology, including the domestication of fire, the development of language, and the use of weapons.

14. Technology emerged from the mists of humanoid history and has accelerated ever since. Technologies invented by other human species and subspecies included the domestication of fire, tools of stone, pottery, clothing, and other means of providing for basic human needs. Early humanoids also initiated the development of language, visual art, music, and other means for human communication.

About ten thousand years ago, humans began domesticating plants, and soon thereafter, animals. Nomadic hunting tribes began settling down, allowing for more stable forms of social organization. Buildings were constructed to protect both humans and their farming products. More effective means of transportation emerged, facilitating the emergence of trade and large-scale human societies.

The wheel appears to be a relatively recent innovation, with the oldest excavated wheels dating from about 5,500 years ago in Mesopotamia. Emerging around the same time in the same region were rafts, boats, and a system of "cuneiform" inscriptions, the first form of written language that we are aware of.

These technologies enabled humans to congregate in large groups, allowing the emergence of civilization. The first cities emerged in Mesopotamia around 6,000 years ago. Emerging about a millennium later were the ancient Egyptian cities, including Memphis and Thebes, culminating in the reigns of the great Egyptian kings. These cities were constructed as war machines with defensive walls protected by armies utilizing weapons drawn from the most advanced technologies of their time, including chariots, spears, armor, and bows and arrows. Civilization in turn allowed for human specialization of labor through a caste system and organized efforts at advancing technology. An intellectual class including teachers, engineers, physicians, and scribes emerged. Other contributions by the early Egyptian civilization included a paperlike material manufactured from papyrus plants, standardization of measurement, sophisticated metalworking, water management, and a calendar.

More than 2,000 years ago, the Greeks invented elaborate machinery with multiple internal states. Archimedes, Ptolemy, and others described levers, cams, pulleys, valves, cogs, and other intricate mechanisms that revolutionized the measurement of time, navigation, mapmaking, and the construction of buildings and ships. The Greeks are perhaps best known for their contributions to the arts, particularly literature, theater, and sculpture.

The Greeks were superseded by the superior military technology of the Romans. The Roman empire was so successful that it produced the first urban civilization to experience long-term peace and stability. Roman engineers constructed tens of thousands of kilometers of roads and thousands of public constructions such as administrative buildings, bridges, sports stadiums, baths, and sewers. The Romans made particularly notable advances in military technology, including advanced chariots and armor, the catapult and javelin, and other effective tools of war.

The fall of the Roman empire around 500 A.D. ushered in the misnamed Dark Ages. While progress during the next thousand years was slow by contemporary standards, the ever tightening spiral that is technological progress continued to accelerate. Science, technology, religion, art, literature, and philosophy all continued to evolve in Byzantine, Islamic, Chinese, and other societies. Worldwide trade enabled a cross-fertilization in technologies. In Europe, for example, the crossbow and gunpowder were borrowed from China. The spinning wheel was borrowed from India. Paper and printing were developed in China about 2,000 years ago and migrated to Europe many centuries later. Windmills emerged in several parts of the world, facilitating expertise with elaborate gearing machines that would subsequently support the first calculating machines.

The invention in the thirteenth century of a weight-driven clock using the cam technology perfected for windmills and waterwheels freed society from structuring their lives around the sun. Perhaps the most significant invention of the late Middle Ages was Johannes Gutenberg's invention of the movable-type printing press, which opened intellectual life beyond an elite controlled by church and state.

By the seventeenth century, technology had created the means for empires to span the globe. Several European countries, including England, France, and Spain, were developing economies based on far-flung colonies. This colonization spawned the emergence of a merchant class, a worldwide banking system, and early forms of intellectual property protection, including the patent.

On May 26, 1733, the English Patent Office issued a patent to John Kay for his "New Engine for Opening and Dressing Wool." This was good news, for he had plans to manufacture his "flying shuttle" and market it to the burgeoning English textile industry. Kay's invention was a quick success, but he spent all of his profits on litigation, attempting in vain to enforce his patent. He died in poverty, never realizing that his innovation in the weaving of cloth represented the launching of the Industrial Revolution.

The widespread adoption of Kay's innovation created pressure for a more efficient way to spin yarn, which resulted in Sir Richard Arkwright's Cotton Jenny, patented in 1770. In the 1780s, machines were invented to card and comb the wool to feed the new automated spinning machines. By the end of the eighteenth century, the English cottage industry of textiles was replaced with increasingly efficient centralized machines. The birth of the Industrial Revolution led to the founding of the Luddite movement in the early 1800s, the first organized movement opposing technology.

15. Primatologist Carl Van Schaik observed that the orangutans of Sumatra's Suaq Balimbing swamp all make and use tools to reach insects, honey, and fruit. Though captive orangutans are easily taught to use tools, the Suaq primates are the first wild population observed using tools. The use of tools may be a result of necessity. Orangutans in other parts of the world have not been observed to use tools, basically because their food supply is more easily accessible.

Carl Zimmer, "Tooling Through the Trees." *Discover* 16, no. 11 (November 1995): 46–47.

Crows fashion tools from sticks and leaves. The tools are used for different purposes, are highly predictable in their construction, and even have hooks and other mechanisms for finding and manipulating insect prey. They often carry these devices when flying and store them next to their nests.

Tina Adler, "Crows Rely on Tools to Get Their Work Done." *Science News* 149, no. 3 (January 20, 1996): 37.

Crocodiles can't grip prey, so they sometimes trap prey between rocks and/or roots. The tree root acts to anchor the dead prey while the crocodile eats its meal. Some people have attributed the crocodiles' use of stones and roots as using tools.

From the "Animal Diversity Web Site" at the University of Michigan's Museum of Zoology, <http://www.oit.itd.umich.edu/projects/ADW/>.

16. An animal communicates for a variety of reasons: defense (to signal approaching danger to other members of its species), food gathering (to alert other members to a food source), courtship and mating (to alert members of its desirability and to warn potential competitors away), and maintenance of territory. The basic motivation for communication is survival of the species. Some animals use communication not only for survival, but also to express emotion.

There are many fascinating examples of animal communication:

- A female tree frog found in Malaysia uses its toes to tap on vegetation, alerting potential mates to her availability. Lori Oliwenstein, Fenella Saunders, and Rachel Preiser, "Animals 1995." *Discover* 17, no. 1 (January 1996): 54–57.
- Male meadow voles (a small rodent) groom themselves in order to produce body odors that will attract their mates. Tina Adler, "Voles Appreciate the Value of Good Grooming." *Science News* 149, no. 16 (April 20, 1996): 247.
- Whales communicate through a series of calls and cries. Mark Higgins, "Deep Sea Dialogue." *Nature Canada* 26, no. 3 (Summer 1997): 29–34.
- Primates, of course, vocalize to communicate a variety of messages. One group of researchers studied capuchin monkeys, squirrel monkeys, and golden-lion tamarins in Central and South America. Often these animals are unable to see each other through the forest, so they developed a series of calls or trills that would alert members to move toward food sources.

Bruce Bower, "Monkeys Sound Off, Move Out." *Science News* 149, no. 17 (April 27, 1996): 269.

17. Washoe and Koko (male and female gorillas, respectively) are credited with acquiring American Sign Language (ASL). They are the most famous of the communicating primates. Viki, a chimpanzee, was taught to vocalize three words (*mama, papa,* and *cup*). Lana and Kanzi (female chimpanzees) were taught to press buttons with symbols.

Steven Pinker reflects upon researchers' claims that apes fully comprehend sign language. In *The Language Instinct: How the Mind Creates Language* (New York: Morrow, 1994), he notes that the apes learned a very crude form of ASL, not the full nuances of this language. The signs they learned were crude mimics of the "real thing." In addition, according to Pinker, the researchers often misinterpreted apes' hand motions as actual signs. One researcher on Washoe's team who was deaf noted that other researchers would keep a log of long lists of signs, whereas the deaf researcher's log was short.

18. David E. Kalish. "Chip Makers and U.S. Unveil Project." *New York Times,* September 12, 1997.

19. The chart "The Exponential Growth of Computing, 1900–1998" is based on the following data:

Date	Device	Add Time (sec)	Calculations per Second (cps)	Cost (then dollars)	Cost 1998 Dollars	CPS/$1000
1900	Analytical Engine	9.00E−00	1.11E−01	$1,000,000	$19,087,000	5.821E−06
1908	Hollerith Tabulator	5.00E+01	2.00E−02	$9,000	$154,000	1.299E−04
1911	Monroe Calculator	3.00E+01	3.33E−02	$35,000	$576,000	5.787E−05
1919	IBM Tabulator	5.00E−00	2.00E−01	$20,000	$188,000	1.064E−03
1928	National Ellis 3000	1.00E+01	1.00E−01	$15,000	$143,000	6.993E−04
1939	Zuse 2	1.00E−00	1.00E−00	$10,000	$117,000	8.547E−03
1940	Bell Calculator Model 1	3.00E−01	3.33E−00	$20,000	$233,000	1.431E−02
1941	Zuse 3	3.00E−01	3.33E−00	$6,500	$72,000	4.630E−02
1943	Colossus	2.00E−04	5.00E+03	$100,000	$942,000	5.308E−00
1946	ENIAC	2.00E−04	5.00E+03	$750,000	$6,265,000	7.981E−01
1948	IBM SSEC	8.00E−04	1.25E+03	$500,000	$3,380,000	3.698E−01
1949	BINAC	2.86E−04	3.50E+03	$278,000	$1,903,000	1.837E−00
1949	EDSAC	1.40E−03	7.14E+02	$100,000	$684,000	1.044E−00
1951	Univac I	1.20E−04	8.33E+03	$930,000	$5,827,000	1.430E−00
1953	Univac 1103	3.00E−05	3.33E+04	$895,000	$5,461,000	6.104E−00
1953	IBM 701	6.00E−05	1.67E+04	$230,000	$1,403,000	1.188E+01
1954	EDVAC	9.00E−04	1.11E+03	$500,000	$3,028,000	3.669E−01
1955	Whirlwind	5.00E−05	2.00E+04	$200,000	$1,216,000	1.645E+01
1955	IBM 704	2.40E−05	4.17E+04	$1,994,000	$12,120,000	3.438E−00
1958	Datamatic 1000	2.50E−04	4.00E+03	$2,179,100	$12,283,000	3.257E−01
1958	Univac II	2.00E−04	5.00E+03	$970,000	$5,468,000	9.144E−01
1959	Mobidic	1.60E−05	6.25E+04	$1,340,000	$7,501,000	8.332E−00
1959	IBM 7090	4.00E−06	2.50E+05	$3,000,000	$16,794,000	1.489E+01
1960	IBM 1620	6.00E−04	1.67E+03	$200,000	$1,101,000	1.514E−00
1960	DEC PDP-1	1.00E−05	1.00E+05	$120,000	$660,000	1.515E+02
1961	DEC PDP-4	1.00E−05	1.00E+05	$65,000	$354,000	2.825E+02
1962	Univac III	9.00E−06	1.11E+05	$700,000	$3,776,000	2.943E+01

1964	CDC 6600	2.00E−07	5.00E+06	$6,000,000	$31,529,000	1.586E+02
1965	IBM 1130	8.00E−06	1.25E+05	$50,000	$259,000	4.826E+02
1965	DEC PDP-8	6.00E−06	1.67E+05	$18,000	$93,000	1.792E+03
1966	IBM 360 Model 75	8.00E−07	1.25E+06	$5,000,000	$25,139,000	4.972E+01
1968	DEC PDP-10	2.00E−06	5.00E+05	$500,000	$2,341,000	2.136E+02
1973	Intellec-8	1.56E−04	6.41E+03	$2,398	$8,798	7.286E+02
1973	Data General Nova	2.00E−05	5.00E+04	$4,000	$14,700	3.401E+03
1975	Altair 8800	1.56E−05	6.41E+04	$2,000	$6,056	1.058E+04
1976	DEC PDP-11 Model 70	3.00E−06	3.33E+05	$150,000	$429,000	7.770E+02
1977	Cray 1	1.00E−08	1.00E+08	$10,000,000	$26,881,000	3.720E+03
1977	Apple II	1.00E−05	1.00E+05	$1,300	$3,722	2.687E+04
1979	DEC VAX 11 Model 780	2.00E−06	5.00E+05	$200,000	$449,000	1.114E+03
1980	Sun-1	3.00E−06	3.33E+05	$30,000	$59,300	5.621E+03
1982	IBM PC	1.56E−06	6.41E+05	$3,000	$5,064	1.266E+05
1982	Compaq Portable	1.56E−06	6.41E+05	$3,000	$5,064	1.266E+05
1983	IBM AT-80286	1.25E−06	8.00E+05	$5,669	$9,272	8.628E+04
1984	Apple Macintosh	3.00E−06	3.33E+05	$2,500	$3,920	8.503E+04
1986	Compaq Deskpro 386	2.50E−07	4.00E+06	$5,000	$7,432	5.382E+05
1987	Apple Mac II	1.00E−06	1.00E+06	$3,000	$4,300	2.326E+05
1993	Pentium PC	1.00E−07	1.00E+07	$2,500	$2,818	3.549E+06
1996	Pentium PC	1.00E−08	1.00E+08	$2,000	$2,080	4.808E+07
1998	Pentium II PC	5.00E−09	2.00E+08	$1,500	$1,500	1.333E+08

Cost conversions from dollars in each year to 1998 dollars are based on the ratio of the consumer price indices (CPI) for the respective years, based on CPI data as recorded by the Woodrow Federal Reserve Bank of Minneapolis. See their web site, <http://woodrow.mpls.frb.fed.us/economy/calc/cpihome.html>.

Charles Babbage designed the Analytical Engine in the 1830s and continued to refine the concept until his death in 1871 Babbage never completed his invention. I have estimated a date of 1900 for the Analytical Engine as an estimated date for when its mechanical technology became feasible, based on the availability of other mechanical computing technology available in that time period.

Sources for the chart "The Exponential Growth of Computing, 1900–1998" include the following:
25 Years of Computer History
<http://www.compros.com/timeline.html>
BYTE Magazine "Birth of a Chip"
<http://www.byte.com/art/9612/sec6/art2.htm>
cdc.html@www.citybeach.wa.edu (Stretch)
<http://www.citybeach.wa.edu.au/lessons/history/video/sunedu/computer/cdc.html>
Chronology of Digital Computing Machines
<http://www.best.com/~wilson/faq/chrono.html>
Chronology of Events in the History of Microcomputers
<http://www3.islandnet.com/~kpolsson/comphist/comp1977.htm>
The Computer Museum History Center
<http://www.tcm.org/html/history/index.html>

delan at infopad.eecs.berkeley.edu
<http://infopad.eecs.berkeley.edu/CIC/summary/delan>
Electronic Computers Within the Ordnance Corps
<http://ftp.arl.mil/~mike/comphist/61ordnance/index.html>
General Processor Information
<http://infopad.eecs.berkeley.edu/CIC/summary/local/>
The History of Computing at Los Alamos
<http://bang.lanl.gov/video/sunedu/computer/comphist.html>
The Machine Room
<http://www.tardis.ed.ac.uk/~alexios/MACHINE-ROOM/>
Mind Machine Web Museum
<http://userwww.sfsu.edu/~hl/mmm.html>
Hans Moravec at Carnegie Mellon University: Computer Data
<http://www.frc.ri.cmu.edu/~hpm/book97/ch3/processor.list>
PC Magazine Online: Fifteen Years of PC Magazine
<http://www.zdnet.com/pcmag/special/anniversary/>
PC Museum
<http://www.microtec.net/~dlessard/index.html>
PDP-8 Emulation
<http://csbh.mhv.net/~mgraffam/emu/pdp8.html>
Silicon Graphics Webpage press release
<http://www.pathfinder.com/money/latest/press/PW/1998Jun16/270.html>
Stan Augarten, *Bit by Bit: An Illustrated History of Computers* (New York: Ticknor & Fields, 1984).
International Association of Electrical and Electronics Engineers (IEEE), "Annals of the History of the Computer," vol. 9, no. 2, pp. 150–153 (1987).
IEEE, vol. 16, no. 3, p. 20 (1994).
Hans Moravec, *Mind Children: The Future of Robot and Human Intelligence* (Cambridge, MA: Harvard University Press, 1988).
René Moreau, *The Computer Comes of Age* (Cambridge, MA: MIT Press, 1984).

20. For additional views on the future of computer capacity, see: Hans Moravec, *Mind Children: The Future of Robot and Human Intelligence* (Cambridge, MA: Harvard University Press, 1988); and "An Interview with David Waltz, Vice President, Computer Science Research, NEC Research Institute" at Think Quest's web page <http://tqd.advanced.org/2705/waltz.html>. I also discuss this subject in my book *The Age of Intelligent Machines* (Cambridge, MA: MIT Press, 1990), 401–419. These three sources discuss the exponential growth of computing.

21. A mathematical theory concerning the difference between information and noise and the ability of a communications channel to carry information.

22. The Santa Fe Institute has played a pioneering role in developing concepts and technology related to complexity and emergent systems. One of the principal developers of paradigms associated with chaos and complexity has been Stuart Kauffman. Kauffman's *At Home in the Universe: The Search for the Laws of Self-Organization and Complexity* (Oxford: Oxford University Press, 1995) looks "at the forces for order that lie at the edge of chaos" (from the card catalog description).

In his book *Evolution of Complexity by Means of Natural Selection* (Princeton, NJ: Princeton University Press, 1988), John Tyler Bonner asks the question: "How is it that an egg turns into an elaborate adult? How is it that a bacterium, given many millions of years, could have evolved into an elephant?"

John Holland is another leading thinker from the Sante Fe Institute in the emerging field of complexity. His book *Hidden Order: How Adaptation Builds Complexity* (Reading, MA: Addison-Wesley, 1996) presents a series of lectures that Holland presented at the Santa Fe Institute in 1994.

Also see John H. Holland, *Emergence: From Chaos to Order* (Reading, MA: Addison-Wesley, 1998) and M. Mitchell Waldrop, *Complexity: The Emerging Science at the Edge of Order and Chaos* (New York: Simon and Schuster, 1992).

CHAPTER 2: THE INTELLIGENCE OF EVOLUTION

1. In the early 1950s, the chemical composition of DNA was already known. At that time, the important questions were: How is the DNA molecule constructed? How does DNA accomplish its work? These questions would be answered in 1953 by James D. Watson and Francis H. C. Crick.

   Watson and Crick wrote "The Molecular Structure of Nucleic Acid: A Structure for Deoxyribose Nucleic Acid" published in the April 25, 1953 issue of *Nature*. For more information on the race by various research groups to discover the molecular structure of DNA, read Watson's book, *The Double Helix* (New York: Atheneum Publishers, 1968).

2. Translation starts by unwinding a region of DNA to expose its code. A strand of messenger RNA (mRNA) is created by copying the exposed DNA base-pair codes. The appropriately named messenger RNA records a copy of a portion of the DNA letter sequence and travels out of the nucleus into the cell body. There the mRNA encounters a ribosome molecule, which reads the letters encoded in the mRNA molecules and then, using another set of molecules called transfer RNA (tRNA), actually builds protein chains one amino acid at a time. These proteins are the worker molecules that perform the cell's functions. For example, hemoglobin, which is responsible for carrying oxygen in the blood from the lungs to the body's tissues, is a sequence of 500 amino acids. With each amino acid requiring three nucleotide letters, the coding for hemoglobin requires 1,500 positions on the DNA molecule. Molecules of hemoglobin, incidentally, are created 500 trillion times a second in the human body, so the machinery is quite efficient.

3. The goal of the Human Genome Project is to construct detailed genetic sequence maps of the 50,000 to 100,000 genes in the human genome, and to provide information about the overall structure and sequence of the DNA of humans and of other animals. The project began in the mid-1980s. The web site of the Human Genome Project, <http://www.nhgri.nih.gov/HGP/>, contains information on the background of the project, current and future goals, and detailed explanations on the structure of DNA.

4. Thomas Ray's work is described in an article by Joe Flower, "A Life in Silicon." *New Scientist* 150, no. 2034 (June 15, 1996): 32–36. Dr. Ray also has a web site with updates on his software-based evolution at <http://www.hip.atr.co.jp/~ray/>.

5. A selection of books exploring the nature of intelligence includes: H. Gardner, *Frames of Mind* (New York: Basic Books, 1983); Stephen Jay Gould, *The Mismeasure of Man* (New York: Basic Books, 1983); R. J. Herrnstein and C. Murray, *The Bell Curve* (New York: The Free Press, 1994); R. Jacoby and N. Glauberman, eds., *The Bell Curve Debate* (New York: Times Books, 1995).

6. To further explore the theories of expansion and contraction of the Universe, see: Stephen W. Hawking, *A Brief History of Time* (New York: Bantam Books, 1988); and Eric L. Lerner, *The Big Bang Never Happened* (New York: Random House, 1991). For the latest updates, see the International Astronomical Union (IAU) web site at <http://www.intastun.org/>, as well as the above noted "Introduction to Big Bang Theory" at <http://www.bowdoin.edu/dept/physics/astro.1997/astro4/bigbang.html >.

7. See chapter 3, "Of Mind and Machines," including the box "The View from Quantum Mechanics."

8. Peter Lewis, "Can Intelligent Life Be Found? Gorilla Will Go Looking." *New York Times*, April 16, 1998.

9. Voice Xpress Plus from the Dictation Division of Lernout & Hauspie Speech Products (formerly Kurzweil Applied Intelligence) allows users to give "natural language" commands to Microsoft Word. It also provides large-vocabulary continuous-speech dictation. The program is "mode-less," so users do not need to indicate when they are giving commands. For example, if the user says: "I enjoyed my trip to Belgium last week. Make this paragraph four points bigger. Change its font to Arial. I hope to go back to Belgium soon." Voice Xpress Plus automatically determines that the second and third sentences are commands and will carry them out (rather than transcribing them). It also determines that the first and fourth sentences are not commands, and will transcribe them into the document.

CHAPTER 3: OF MIND AND MACHINES

1. To learn more about the current state of brain-scanning research, the article "Brains at Work: Researchers Use New Techniques to Study How Humans Think" by Vincent Kiernan is a good place to begin. This article, in the *Chronicle of Higher Education* (January 23, 1998, vol. 44, no. 20, pp. A16–17), discusses uses of MRI to map brain activity during complex thinking processes.

   "Visualizing the Mind" by Marcus E. Raichle in the April 1994 *Scientific American* provides background on various brain-imaging technologies: MRI, positron emission tomography (PET), magnetoencephalography (MEG), and electroencephalography (EEG).

   "Unlocking the Secrets of the Brain" by Tabitha M. Powledge is a two-part article in the July–August issue of *Bioscience* 47 (pp. 330–334 and 403–409), 1997.

2. Blood-forming cells of the bone marrow and certain layers of the skin grow and reproduce frequently, replenishing themselves in a period of months. In contrast, muscle cells do not reproduce for several years. Neurons have not been considered to reproduce at all after one's birth, but recent findings indicate the possibility of primate neuron reproduction. Dr. Elizabeth Gould of Princeton University and Dr. Bruce S. McEwen of Rockefeller University in New York found that adult marmoset monkeys are able to manufacture brain cells in the hippocampus, a brain region that is connected to learning and memory. Conversely, when the animals are under stress, the ability to manufacture new brain cells in the hippocampus diminishes. This research is described in an article by Gina Kolata, "Studies Find Brain Grows New Cells," *The New York Times,* March 17, 1998.

   Other types of cells will grow and reproduce if necessary. For example, if seven-eighths of the liver cells are removed, the remaining cells will grow and reproduce until most of the cells are replenished. Arthur Guyton, *Physiology of the Human Body*, fifth edition (Phila., PA: W. B. Saunders, 1979): 42–43.

3. Oppression of human races, nationalities, and other groups has often been justified in the same way.

4. Plato's works are available in Greek and English in the Loeb Classical Library editions. A detailed account of Plato's philosophy is presented in J. N. Findlay, *Plato and Platonism: An Introduction.* On the dialogues as Plato's chosen form, see D. Hyland's "Why Plato Wrote Dialogues." *Philosophy and Rhetoric* 1 (1968): 38–50.

5. A brief history of logical positivism can be found in A. J. Ayer, *Logical Positivism* (New York: Macmillan, 1959): 3–28.

6. David J. Chalmers distinguishes "between the easy problems and the hard problem of consciousness," and argues that "the hard problem eludes conventional methods of explanation entirely" in an essay entitled "Facing Up to the Problem of Consciousness." Stuart R. Hameroff, ed., *Toward a Science of Consciousness: The First Tucson Discussions and Debates (Complex Adaptive Systems)* (Cambridge, MA: MIT Press, 1996).

7. This objective view was systematically defined early in the twentieth century by Ludwig Wittgenstein in an analysis of language called logical positivism. This philosophical school, which would subsequently influence the emergence of computational theory and linguistics, drew its inspiration from Wittgenstein's first major work, the *Tractatus Logico-Philosophicus.* The book was not an immediate hit and it took the influence of his former instructor, Bertrand Russell, to secure a publisher.

   In a foreshadowing of early computer-programming languages, Wittgenstein numbered all of the statements in his *Tractatus* indicating their position in the hierarchy of his thinking. He starts out with statement 1: "The world is all that is the case," indicating his ambitious agenda for the book. A typical statement is number 4.0.0.3.1: "All philosophy is a critique of language." His last statement, number 7, is "What we cannot speak about we must pass over in silence." Those who trace their philosophical roots to the early Wittgenstein still regard this short work as the most influential work of philosophy of the past century. Ludwig Wittgenstein, *Tractatus Logico-Philosophicus,* translated by D. F. Pears and B. F. McGuiness, Germany, 1921.

8. In the preface to *Philosophical Investigations,* translated by G. E. M. Anscombe, Wittgenstein "acknowledges" that he made "grave mistakes" in his earlier work, the *Tractatus.*

9. For a useful overview of Descartes's life and work, see *The Dictionary of Scientific Biography,* vol. 1, pp. 55–65. Also, Jonathan Rée's *Descartes* presents a unified view of Descartes's philosophy and its relation to other systems of thought.

10. Quoted from Douglas R. Hofstadter, *Gödel, Escher, Bach: An Eternal Golden Braid* (New York: Basic Books, 1979).

11. "Computing Machinery and Intelligence," *Mind* 59 (1950): 433–460, reprinted in E. Feigenbaum and J. Feldman, eds., *Computers and Thought* (New York: McGraw-Hill, 1963).

12. For a description of quantum mechanics, read George Johnson, "Quantum Theorists Try to Surpass Digital Computing," *New York Times,* February 18, 1997.

CHAPTER 4: A NEW FORM OF INTELLIGENCE ON EARTH

1. Simple calculating devices had been perfected almost two centuries before Babbage, starting with Pascal's Pascaline in 1642, which could add numbers, and a multiplying machine developed by Gottfried Wilhelm Leibniz a couple of decades later. But automating the computing of logarithms was far more ambitious than anything that had been previously attempted.

   Babbage didn't get very far—he exhausted his financial resources, got into a dispute with the British government over ownership, had problems getting the unusual precision parts fabricated, and saw his chief engineer fire all of his workmen and then quit himself. He was also beset with personal tragedies, including the death of his father, his wife, and two of his children.

   The only obvious thing to do now, Babbage figured, was to abandon his "Difference Engine" and embark on something yet more ambitious: the world's first fully programmable computer. Babbage's new conception—the "Analytical Engine"—could be programmed to solve any possible logical or computational problem.

   The Analytical Engine had a random-access memory (RAM) consisting of 1,000 "words" of 50 decimal digits each, equivalent to about 175,000 bits. A number could be retrieved from any location, modified, and stored in any other location. It had a punched-card reader and even included a printer, even though it would be another half century before either typesetting machines or typewriters were to be invented. It had a central processing unit (CPU) that could perform the types of logical and arithmetic operations that CPUs do today. Most important, it had a special storage unit for the software with a machine language very similar to those of today's computers. One decimal field specified the type of operation and another specified the address in memory of the operand. Stan Augarten, *Bit by Bit: An Illustrated History of Computers* (New York: Ticknor and Fields, 1984): 63–64.

Babbage describes the features of his machine in "On the Mathematical Powers of the Calculating Engine," written in 1837 and reprinted as appendix B in Anthony Hyman's *Charles Babbage: Pioneer of the Computer* (Oxford: Oxford University Press, 1982). For biographical information on Charles Babbage and Ada Lovelace, see Hyman's biography, and Dorothy Stein's book *Ada: A Life and a Legacy* (Cambridge, MA: MIT Press, 1985).

2. Stan Augarten, *Bit by Bit,* 63–64. Babbage's description of the Analytical Engine in "On the Mathematical Powers of the Calculating Engine," written in 1837, is reprinted as appendix B in Anthony Hyman's *Charles Babbage: Pioneer of the Computer* (Oxford: Oxford University Press, 1982).

3. Joel Shurkin, in *Engines of the Mind*, p. 104, describes Aiken's machine as "an electro-mechanical Analytical Engine with IBM card handling." For a concise history of the development of the Mark I, see Augarten's *Bit by Bit*, 103–107. I. Bernard Cohen provides a new perspective on Aiken's relation to Babbage in his article "Babbage and Aiken," *Annals of the History of Computing* 10 (1988): 171–193.

4. The idea of the punched card, which Babbage borrowed from the Jacquard looms (automatic weaving machines controlled by punched metal cards), also survived and formed the basis for automating the increasingly popular calculators of the nineteenth century. This culminated in the 1890 U.S. census, which was the first time that electricity was used for a major data-processing project. The punched card itself survived as a mainstay of computing until the 1970s.

5. Turing's Robinson was not a programmable computer. It didn't have to be—it had only one job to do. The first programmable computer was developed by the Germans. Konrad Zuse, a German civil engineer and tinkerer, was motivated to ease what he later called those "awful calculations required of civil engineers." Like Babbage's, his first device, the Z-1, was entirely mechanical—built from an erector set in his parents' living room. The Z-2 used electromechanical relays and was capable of solving complex simultaneous equations. It was his third version—the Z-3—that is the most historic. It stands as the world's first *programmable* computer. As one would retroactively predict from the Law of Accelerating Returns as applied to computation, Zuse's Z-3 was rather slow—a multiplication took more than three seconds.

While Zuse received some incidental support from the German government and his machines played a minor military role, there was little, if any, awareness of computation and its military significance by the German leadership. This explains their apparent confidence in the security of their Enigma code. Instead the German military gave immensely high priority to several other advanced technologies, such as rocketry and atomic weapons.

It would be Zuse's fate that no one would pay much attention to him or his inventions; even the Allies ignored him after the end of the war. Credit for the world's first programmable computer is often given to Howard Aiken, despite the fact that his Mark I was not operational until nearly three years after the Z-3. When Zuse's funding was withdrawn in the middle of the war by the Third Reich, a German officer explained to him that "the German aircraft is the best in the world. I cannot see what we could possibly calculate to improve on."

Zuse's claim to having built the world's first operational fully programmable digital computer is supported by the patent application he filed. See, for instance, K. Zuse, "Verfahren zur Selbst Atigen Durchfurung von Rechnungen mit Hilfe von Rechenmaschinen," German Patent Application Z23624, April 11, 1936. Translated extracts, titled "Methods for Automatic Execution of Calculations with the Aid of Computers," appear in Brian Randell, ed., *The Origins of Digital Computers*, pp. 159–166.

6. "Computing Machinery and Intelligence," *Mind* 59 (1950): 433–460, reprinted in E. Feigenbaum and J. Feldman, eds., *Computers and Thought* (New York: McGraw-Hill, 1963).

7. See A. Newell, J. C. Shaw, and H. A. Simon, "Programming the Logic Theory Machine," *Proceedings of the Western Joint Computer Conference*, 1957, pp. 230–240.

8. Russell and Whitehead's *Principia Mathematica* (see reference at the end of this endnote), first published in 1910–1913, was a seminal work that reformulated mathematics based on Russell's new conception of set theory. Russell's breakthrough in set theory set the stage for Turing's subsequent development of computational theory based on the Turing machine (see note below). Following is my version of "Russell's paradox," which stimulated Russell's discovery:

Before ending up in "the Other Place," our friend the gambler had lived a rough life. He was short of temper and not fond of losing. In our story, he is also a bit of a logician. This time he has picked the wrong man to dispatch. If only he had known that the fellow was the judge's nephew.

Known anyway as a hanging judge, the magistrate is furious and wishes to mete out the most severe sentence he can think of. So he tells the gambler that not only is he sentenced to die but the sentence is to be carried out in a unique way. "First off, we're gonna dispense with you quickly, just like you done with the victim. This punishment must be carried out no later than Saturday. Furthermore, I don't want you preparing yourself for the judgment day. On the morning of your execution, you won't know for certain that the day is at hand. When we come for you, it'll be a surprise."

To which the gambler replies, "Well, that's great, judge, I am greatly relieved."

To which the judge exclaims, "I don't understand, how can you be relieved? I have condemned you to be executed. I have ordered that the sentence be carried out soon, but you'll be unable to prepare yourself because on the morning that we carry it out, you won't know for certain that you'll be dying that day."

"Well, Your Honor," the gambler points out, "in order for your sentence to be carried out, I cannot be executed on Saturday."

"Why is that?" asks the judge.

"Because since the sentence must be carried out by Saturday, if we actually get to Saturday, I will know for certain that I am to be executed on that day, and thus it would not be a surprise."

"I suppose you are right," replies the judge. "You cannot be executed on Saturday. But I still don't see why you're relieved."

"Well, if we have definitely ruled out Saturday, then I can't be executed on Friday either."

"Why is that?" asks the judge, being a little slow.

"We have agreed that I can't be executed on Saturday. Therefore Friday is the last day I can be executed. But if Friday rolls around, I will definitely know that I am to be executed on that day and therefore it would not be a surprise. So I can't be executed on Friday."

"I see," says the judge.

"Thus the last day I can be executed would be Thursday. But if Thursday rolls around, I would know I had to be executed on that day, and thus it would not be a surprise. So Thursday is out. By the same reasoning, we can eliminate Wednesday, Tuesday, Monday, and today."

The judge scratches his head as the confident gambler is led back to his prison cell.

There is an epilogue to the story. On Thursday, the gambler is taken to be executed. And he is very surprised. So the judge's orders are successfully carried out.

This is my version of what has become known as "Russell's paradox" after Bertrand Russell, perhaps the last person to secure major achievements in both mathematics and philosophy. If we analyze this story, we see that the conditions that the judge has set up result in a conclusion that none of the days comply, because, as the prisoner so adroitly points out, each one of them in turn would not be a surprise. But the *conclusion itself* changes the situation, and now surprise *is* possible again. This brings us back to the original situation in which the prisoner could (in theory) demonstrate that each day in turn would be impossible, and so on, ad infinitum. The judge applies "Alexander's solution" in which King Alexander slashed the hopelessly tied Gordian knot.

A simpler example, and the one that Russell actually struggled with, is the following question about sets. A set is a mathematical construct that, as its name implies, is a collection of things. A set may include chairs, books, authors, gamblers, numbers, other sets, themselves, whatever. Now consider set A, which is defined to contain all sets that are not members of themselves. Does set A contain itself?

As we consider this famous problem, we realize there are only two possible answers: Yes and No. We can, therefore, try them all (this is not the case for most problems in mathematics). So let's consider Yes. If the answer is Yes, then set A does contain itself. But if set A contains itself, then according to its defining condition, set A would not belong to set A, and thus it does not belong to itself. Since the answer of Yes led to a contradiction, it must be wrong.

So let's try No. If the answer is No, then set A does not contain itself. But again according to the defining condition, if set A does not belong to itself, then it would belong to set A, another contradiction. As with the story about the prisoner, we have incompatible propositions that imply one another. Yes implies No, which yields Yes, and so on.

This may not seem like a big deal, but to Russell it threatened the foundation of mathematics. Mathematics is based on the concept of sets, and the issue of inclusion (i.e., what belongs to a set) is fundamental to the idea. The definition of set A appears to be a reasonable one. The question of whether set A belongs to itself also appears reasonable. Yet we have difficulty coming up with a reasonable answer to this reasonable question. Mathematics was in big trouble.

Russell pondered this dilemma for more than a decade, nearly exhausting himself and wrecking at least one marriage. But he came up with an answer. To do so, he invented the equivalent of a theoretical computer (although not by name). Russell's "computer" is a logic machine and it implements one logical transformation at a time, each one requiring a quantum of time—so things don't happen all at once. Our question about set A is examined in an orderly fashion. Russell turns on his theoretical computer (which, lacking a real computer, ran only in his head) and the logical operations are "executed" in turn. So at one point, our answer is Yes, but the program keeps running, and a few quantums of time later the answer becomes No. The program runs in an infinite loop, constantly alternating between Yes and No.

But the answer is never Yes and No at the same time!

Impressed? Well Russell was very pleased. Eliminating the possibility of the answer being Yes and No *at the same time* was enough to save mathematics. With the help of his friend and former tutor Alfred North Whitehead, Russell recast all of mathematics in terms of his new theory of sets and logic, which they published in their *Principia Mathematica* in 1910–1913. It is worth pointing out that the concept of a computer, theoretical or otherwise, was not widely understood at the time. The nineteenth-century efforts of Charles Babbage, which are discussed in chapter 4, were largely unknown at the time. It is not clear if Russell was aware of Babbage's efforts. Russell's highly influential and revolutionary work invented a logical theory of computation and recast mathematics as one of its branches. Mathematics was now part of computation.

Russell and Whitehead did not explicitly talk about computers but cast their ideas in the mathematical terminology of set theory. It was left to Alan Turing to create the first theoretical computer in 1936, in his Turing machine (see note 16 below).

Alfred N. Whitehead and Bertrand Russell, *Principia Mathematica*, 3 vols., second edition (Cambridge: Cambridge University Press, 1925–1927). (The first edition was published in 1910, 1912, and 1913.)

Russell's paradox was first introduced in Bertrand Russell, *Principles of Mathematics* (Reprint, New York: W. W.. Norton & Company, 1996), 2nd ed., 79–81. Russell's paradox is a subtle variant of the Liar Paradox. See E. W. Beth, *Foundations of Mathematics* (Amsterdam: North Holland, 1959), p. 485.

 9. "Heuristic Problem Solving: The Next Advance in Operations Research," *Journal of the Operations Research Society of America* 6, no. 1 (1958), reprinted in Herbert Simon, *Models of Bounded Rationality*, vol. 1, Economic Analysis and Public Policy (Cambridge, MA: MIT Press, 1982).

10. "A Mean Chess-Playing Computer Tears at the Meaning of Thought," *New York Times*, February 19, 1996, contains the reactions of Gary Kasparov and a number of noted thinkers concerning the ramifications of Deep Blue beating the world chess champion.

11. Daniel Bobrow, "Natural Language Input for a Computer Problem Solving System," in Marvin Minsky, *Semantic Information Processing*, pp. 146–226.

12. Thomas Evans, "A Program for the Solution of Geometric-Analogy Intelligence Test Questions," in Marvin Minsky, ed., *Semantic Information Processing* (Cambridge, MA: MIT Press, 1968), pp. 271–353.

13. Robert Lindsay, Bruce Buchanan, Edward Feigenbaum, and Joshua Lederberg describe DENDRAL in *Applications of Artificial Intelligence for Chemical Inference: The DENDRAL Project* (New York: McGraw-Hill, 1980). For a brief and clear explanation of the essential mechanisms behind DENDRAL, see Patrick Winston, *Artificial Intelligence* (1984), pp. 163–164, 195–197.

14. For many years SHRDLU was cited as a prominent accomplishment of artificial intelligence. Winograd describes his research in his thesis *Understanding Natural Language* (New York: Academic Press, 1972). A brief version appears as "A Procedural Model of Thought and Language," in Roger Schank and Kenneth Colby, eds., *Computer Models of Thought and Language* (San Francisco: W. H. Freeman, 1973).

15. Haneef A. Fatmi and R. W. Young, "A Definition of Intelligence," *Nature* 228 (1970): 97.
16. Alan Turing showed that the essential basis of computation could be modeled with a
    very simple theoretical machine. He created the first theoretical computer in 1936
    (first introduced in Alan M. Turing, "On Computable Numbers with an Application
    to the Entscheinungs Problem," *Proc. London Math. Soc.* 42 [1936]: 230–265) in an
    eponymous conception called the Turing machine. As with a number of Turing's
    breakthroughs, he would have both the first and last word. The Turing machine rep-
    resented the founding of modern computational theory. It has also persisted as our
    primary theoretical model of a computer because of its combination of simplicity and
    power.

    The Turing machine is one example of the simplicity of the foundations of intelli-
    gence. A Turing machine consists of two primary (theoretical) units: a tape drive and
    a computation unit. The tape drive has a tape of infinite length on which it can write,
    and (subsequently) read, a series of two symbols: zero and one. The computation unit
    contains a program consisting of a sequence of commands, drawing from only seven
    possible operations:

    - Read the tape
    - Move the tape left one symbol
    - Move the tape right one symbol
    - Write 0 on the tape
    - Write 1 on the tape
    - Jump to another command
    - Halt

    Turing was able to show that this extremely simple machine can compute any-
    thing that any machine can compute, no matter how complex. If a problem cannot be
    solved by a Turing machine, then it cannot be solved by any machine. Occasionally
    there are challenges to this position, but in large measure it has stood the test of time.

    In the same paper, Turing reports another unexpected discovery, that of unsolv-
    able problems. These are problems that are well defined with unique answers that can
    be shown to exist, but that we can also prove can never be computed by a Turing
    machine—that is to say by any machine, yet another reversal of what had been a
    nineteenth-century confidence that problems that could be defined would ultimately
    be solved. Turing showed that there are as many unsolvable problems as solvable ones.

    Turing and Alonzo Church, his former professor, went on to assert what has be-
    come known as the Church-Turing thesis: If a problem that can be presented to a Tur-
    ing machine is not solvable by one, then it is also not solvable by human thought.
    "Strong" interpretations of the Church-Turing thesis propose an essential equivalence
    between what a human can think or know and what is computable by a machine. The
    Church-Turing thesis can be viewed as a restatement in mathematical terms of one of
    Wittgenstein's primary theses in his *Tractatus*. The basic idea is that the human brain
    is subject to natural law, and thus its information-processing ability cannot exceed
    that of a machine. We are thus left with the perplexing situation of being able to de-
    fine a problem, to prove that a unique answer exists, and yet know that the answer
    can never be known.

    Perhaps the most interesting unsolvable problem is called the Busy Beaver, which
    may be stated as follows: Each Turing machine has a certain number of commands in
    its program. Given a positive integer n, we construct all of the Turing machines that
    have n states (i.e., n commands). Next we eliminate those n-state Turing machines
    that get into an infinite loop (i.e., never halt). Finally, we select the machine (one that

halts) that writes the largest number of 1s on its tape. The number of 1s that this Turing machine writes is called busy beaver of n.

Tibor Rado, a mathematician and admirer of Turing, showed that there is no algorithm (that is, no Turing machine) that can compute the busy beaver function for all n's. The crux of the problem is sorting out those n-state Turing machines that get into infinite loops. If we program a Turing machine to generate and simulate every possible n-state Turing machine, this simulator itself goes into an infinite loop when it attempts to simulate one of the n-state Turing machines that gets into an infinite loop. Busy beaver can be computed for some ns, and interestingly it is also an unsolvable problem to separate those ns for which we can determine busy beaver of n from those for which we cannot.

Busy beaver is an "intelligent function." More precisely stated, it is a function that requires increasing intelligence to compute for increasing arguments. As we increase n, the complexity of the processes needed to compute busy beaver of n increases.

With $n = 6$, we are dealing with addition and busy beaver of 6 equals 35. In other words, addition is the most complex operation that a Turing machine with only 6 steps in its program is capable of performing. At 7, busy beaver learns to multiply and busy beaver of 7 equals 22,961. At 8, busy beaver can exponentiate, and the number of 1s that our eighth busy beaver writes on its tape is approximately 10^{43}. Note that this is even faster growth than Moore's Law. By the time we get to 10 we need an exotic notation in which we have a stack of exponents (10 to the 10 to the 10, etc.), the height of which is determined by another stack of exponents, the height of which is determined by another stack of exponents, and so on. For the twelfth busy beaver we need an even more exotic notation. Human intelligence (in terms of the complexity of mathematical operations that we can understand) is surpassed well before the busy beaver gets to 100. The computers of the twenty-first century will do a bit better.

The busy beaver problem is one example of a large class of noncomputable functions, as one can see from Tibor Rado, "On Noncomputable Functions," *Bell System Technical Journal* 41, no. 3 (1962): 877–884.

17. Raymond Kurzweil, *The Age of Intelligent Machines* (Cambridge, MA: MIT Press, 1990), pp. 132–133.

18. H. J. Berliner, "Backgammon Computer Program Beats World Champion," *Artificial Intelligence* 14, no. 1 (1980). Also see Hans Berliner, "Computer Backgammon," *Scientific American*, June 1980.

19. To download Ray Kurzweil's Cybernetic Poet (RKCP), go to: <http://www.kurzweil tech.com>. RKCP is further discussed in the section *The Creative Machine* in chapter 8, "1999."

20. See the discussion on these music composition programs in the section *The Creative Machine* in chapter 8, "1999."

21. See W. S. Sarle, ed., "Neural Network Frequently Asked Questions," <ftp://ftp.sas. com/pub/neural/FAQ.html>. This web site has numerous resources on past and current research on neural nets. G. E. Hinton's "How Neural Networks Learn from Experience," in the September 1992 issue of *Scientific American* (144–151), also provides a good introduction to neural networks.

22. Researchers at the Productivity from Information Technology (PROFIT) Initiative at MIT have studied the effectiveness of neural networks in understanding handwriting.

The PROFIT Initiative is based at MIT's Sloan School of Management. The mission of the initiative is to study how the private and public sectors use information technology. Abstracts of working papers on this and other research on neural networks and data mining can be found at <http://scanner-group.mit.edu/papers.html>.

23. "Miros, Inc. is located in Wellesley, Massachusetts, and specializes in providing face

recognition software. Miros' products include TrueFace PC, the first face recognition solution for computer, network and data security; and TrueFace GateWatch, a complete hardware/software security solution that allows or denies access to buildings and rooms by automatically recognizing a person's face taken by a video camera." From Miros Company Information at <http://www.miros.com/About_Miros.htm>.

24. For more information on BrainMaker's aptitude to diagnose illnesses, and to predict the Standard and Poor 500 for LBS Management, see California Scientific's home page at <http://www.calsci.com/>.

25. The reset time stated here is an estimated average for neural connection calculations. For example, Vadim Gerasimov estimates the peak firing frequency of neurons (which significantly exceeds the average rate) to be 250-2,000 Hz (0.5-4 ms intervals) in "Information Processing in the Human Body" at <http://vadim.www.media.mit.edu/MAS862/Project.html>. The firing time is affected by a number of variables, including, for example, the level and duration of a sound, as discussed in Jos. J. Eggermont, "Firing Rate and Firing Synchrony Distinguish Dynamic from Steady State Sound," *NeuroReport* 8, issue 12, 2709–2713.

26. Hugo de Garis maintains a web site on his research for ATR's Brain Builder Group at <http://www.hip.atr.co.jp/~degaris/>.

27. For an intriguing account of this research, read Carver Mead, *Analog VSLI and Neural Systems* (Reading, MA: Addison-Wesley, 1989), 257–278. Synaptics is briefly highlighted in Carol Levin, "Here's Looking at You," *PC Magazine* (December 20, 1994): 31. Carver Mead's web site also provides detailed information on this research at the "Physics of Computation–Carver Mead's Group" at <http://www.pcmp.caltech.edu/>.

28. The SETI (Search for Extraterrestrial Intelligence) Institute conducts research on other signs of life in the Universe, its primary goal being the search for extraterrestrial intelligence. The institute is a nonprofit research organization, funded by government agencies, private foundations, and individuals, which in turn provides funding for several dozen projects. For more information, see the SETI Institute web site, <http://www.seti.org>.

29. The author is dictating portions of this book to his computer through the continuous speech recognition program called Voice Xpress Plus from the dictation division of Lernout & Hauspie (formerly Kurzweil Applied Intelligence). See note 9 on Voice Xpress Plus in chapter 2 for more information.

30. To find out more on State Street Global Advisor's purchase in a majority stake in Advanced Investment Technology, read Frank Byrt, "State Street Global Invests in Artificial Intelligence." *Dow Jones Newswires*, October 29, 1997. The genetic algorithm system used by the AIT Vision mutual fund is described in S. Mahfoud and G. Mani, "Financial Forecasting Using Genetic Algorithms." *Applied Artificial Intelligence* 10 (1996): 543–565. The AIT Vision mutual fund opened at the beginning of 1996 and has publicly available performance numbers. In its first full calendar year (1996), the mutual fund increased 27.2 percent in net asset value, compared to 21.2 percent for its benchmark, the Russell 3000 index.

It should be noted that outperforming its benchmark index does not in itself prove a superior level of decision making. The algorithm may have been making higher-risk investments (on average) than the average in the index.

31. There are many online resources on evolutionary computation and evolutionary and genetic algorithms. One of the best is "The Hitchhiker's Guide to Evolutionary Computation: A List of Frequently Asked Questions (FAQ)," edited by Jörg Heitkötter and David Beasley at <http://www.cs.purdue.edu/coast/archive/clife/FAQ/www/>. This guide includes everything from a glossary to links to various research groups.

Another helpful online resource is the web site for the Santa Fe Institute. The institute's web site can be accessed at <http://www.santafe.edu>.

For an offline introduction to genetic algorithms, read John Holland's article "Genetic Algorithms," *Scientific American* 267, no. 1 (1992): 66–72. As mentioned in note 22 in chapter 1, Holland and his colleagues at the University of Michigan developed genetic algorithms in the 1970s.

For more information on the use of genetic algorithm technology to manage the development and manufacturing of Volvo trucks, read Srikumar S. Rao, "Evolution at Warp Speed," *Forbes* 161, no. 1 (January 12, 1998): 82–83.

See also note 22 on complexity in chapter 1.

32. See "Information Processing in the Human Body," by Vadim Gerasimov, at <http://vadim.www.media.mit.edu/MAS862/Project.html>.
33. See "Information Processing in the Human Body," by Vadim Gerasimov, at <http://vadim.www.media.mit.edu/MAS862/Project.html>.
34. I founded Kurzweil Applied Intelligence (Kurzweil AI) in 1982. The company is now a subsidiary of Lernout & Hauspie Speech Products (L&H), an international leader in the development of speech and language technologies and related applications and products. For more information about these speech recognition products, see <http://www.lhs.com/dictation/>.

CHAPTER 5: CONTEXT AND KNOWLEDGE

1. Victor L. Yu, Lawrence M. Fagan, S. M. Wraith, William Clancey, A. Carlisle Scott, John Hannigan, Robert Blum, Bruce Buchanan, and Stanley Cohen, "Antimicrobial Selection by Computer: A Blinded Evaluation by Infectious Disease Experts," *Journal of the American Medical Association* 242, no. 12 (1979): 1279–1282.
2. For an introduction to the development of expert systems and their use in various companies, read: Edward Feigenbaum, Pamela McCorduck, and Penny Nii, *The Rise of the Expert Company* (Reading, MA: Addison-Wesley, 1983).
3. William Martin, Kenneth Church, and Ramesh Patil, "Preliminary Analysis of a Breadth-First Parsing Algorithm: Theoretical and Experiential Results." MIT Laboratory for Computer Science, Cambridge MA, 1981. In this document, Church cites the synthetic sentence:

"It was the number of products of products of products of products of products of products of products of products?" as having 1,430 syntactically correct interpretations.

He cites the following sentence:

"What number of products of products of products of products of products of products of products of products was the number of products of products of products of products of products of products of products of products of products?" as having 1,430 × 1,430 = 2,044,900 interpretations.

4. These and other theoretical aspects of computational linguistics are covered in Mary D. Harris, *Introduction to Natural Language Processing* (Reston, VA: Reston Publishing Co., 1985).

CHAPTER 6: BUILDING NEW BRAINS . . .

1. Hans Moravec is likely to make this argument in his 1998 book *Robot: Mere Machine to Transcendent Mind* (Oxford University Press; not yet available as of this writing).

2. One hundred fifty million calculations per second for a 1998 personal computer doubling twenty-seven times by the year 2025 (this assumes doubling both the number of components, and the speed of each component every two years) equals about 20 million billion calculations per second. In 1998, it takes multiple calculations on a conventional personal computer to simulate a neural-connection calculation. However, computers by 2020 will be optimized for the neural-connection calculation (and other highly repetitive calculations needed to simulate neuron functions). Note that neural-connection calculations are simpler and more regular than the general-purpose calculations of a personal computer.

3. Five billion bits per $1,000 in 1998 will be doubled seventeen times by 2023, which is about a million billion bits for $1,000 in 2023.

4. NEC's goals to build a supercomputer with a maximum performance of more than 32 teraflops is chronicled in "NEC Begins Designing World's Fastest Computer," *Newsbytes News Network*, January 21, 1998, located online at <http://www.nb-pacifica.com/headline/necbeginsdesigningwo_1208.shtml >.

   In 1998, IBM was one of four companies chosen to participate in PathForward, an initiative from the Department of Energy to develop supercomputers for the twenty-first century. Other companies involved in the project are Digital Equipment Corporation; Sun Microsystems, Inc.; and Silicon Graphics/Cray Computer Systems (SGI/Cray). PathForward is part of the Accelerated Strategic Computing Initiative (ASCI). For more information on this initiative, see <http://www.llnl.gov/asci>.

5. By harnessing the accelerating improvement in both density of components and speed of components, computer power will double every twelve months, or a factor of one thousand every ten years. Based on the projection of $1,000 of computing being equal to the estimated processing power of the human brain (20 million billion calculations per second) by the year 2020, we get a projection of $1,000 of computing being equal to a million human brains in 2040, a billion human brains in 2050, and a trillion human brains in 2060.

6. By 2099, $1,000 of computing will equal 10^{24} times the processing power of the human brain. Based on an estimate of 10 billion persons, that is 10^{14} times the processing power of all human brains. Thus one penny of computing will equal 10^9 (one billion) times the processing power of all human brains.

7. In the Punctuated Equilibrium theories, evolution is seen to progress in sudden leaps followed by periods of relative stability. Interestingly, we often see similar behavior in the performance of evolutionary algorithms (see chapter 4).

8. Dean Takahashi, "Small Firms Jockeying for Position in 3D Chip Market," *Knight-Ridder/Tribune News Service*, September 21, 1994, p. 0921K4365.

9. The entire February 1998 issue of *Computer* (vol. 31, no. 2) explores the status of optical computing and optical storage methods.

   Sunny Bains writes of companies using optical computing for fingerprint recognition and other applications in "Small, Hybrid Digital/Electronic Optical Correlators Ready to Power Commercial Products: Optical Computing Comes into Focus." *EE Times*, January 26, 1998, issue 990. This article is online at <http://www.techweb.com/se/directlink.cgi?EET19980126S0019>.

10. For a nontechnical introduction to DNA computing, read Vincent Kiernan, "DNA-Based Computers Could Race Past Supercomputers, Researchers Predict," in

the *Chronicle of Higher Education* (November 28, 1997). Kiernan discusses the research of Dr. Robert Corn from the University of Wisconsin as well as the research of Dr. Leonard Adleman. The article can be accessed online at <http://chronicle.com/data/articles.dir/art-44.dir/issue-14.dir/14a02301.htm>.

Research at the University of Wisconsin can be accessed online at <http://corninfo.chem.wisc.edu/writings/DNAcomputing.html> .

Leonard Adleman's "Molecular Computation of Solutions to Combinatorial Problems" from the November 11, 1994, issue of *Science* (vol. 266, p. 1021) provides a technical overview of his design of DNA programming for computers.

11. Lambertus Hesselink's research is reported by Phillip F. Schewe and Ben Stein in *Physics News Update* (no. 219, March 28, 1995). The description is available online at <http://www.aip.org/enews/physnews/1995/split/pnu219-2.htm>.

12. For information on nanotubes and buckyballs, read Janet Rae-Dupree's article "Nanotechnology Could Be Foundation for Next Mechanical Revolution," *Knight-Ridder/Tribune News Service*, December 17, 1997, p. 1217K1133.

13. Dr. Sumio Iijima's research on nanotubes is summarized in the following article at the NEC site, <http://www.labs.nec.co.jp/rdletter/letter01/index1.html>.

14. The research of Isaac Chuang and Neil Gershenfeld is reported in "Cue the Qubits: Quantum Computing," *The Economist* 342, no. 8005 (February 22, 1997): 91–92; and in an article by Dan Vergano, "Brewing a Quantum Computer in a Coffee Cup," *Science News* 151, no. 3 (January 18, 1997): 37. More technical details and a list of Chuang and Gershenfeld's publications can be found at the Physics and Media Group/MIT Media Lab <http://physics.www.media.mit.edu/publications/> and at the Los Alamos National Laboratory <http://qso.lanl.gov/qc/>.

Other groups working on quantum computation include the Information Mechanics group at MIT's Lab for Computer Science <http://www-im.lcs.mit.edu/> and the Quantum Computation Group at IBM <http://www.research.ibm.com/quantuminfo/>.

15. "Student Cracks Encryption Code," *USA Today Tech Report*, September 2, 1997.

16. Mark Buchanan, "Light's Spooky Connections Set Distance Record," *New Scientist*, June 28, 1997.

17. Roger Penrose, *The Emperor's New Mind* (New York: Penguin USA, 1990).

18. To understand the concept of tunneling, it is important to understand how transistors on an integrated circuit chip work. An integrated chip is engraved with circuits comprised of thousands or millions of transistors, which electronic devices use to control the flow of electricity. Transistors are made up of a small block of a semiconductor, a material that acts as both an insulator and a conductor of electricity. The first transistors were comprised of germanium and were later replaced with silicon.

Transistors work by holding a pattern of electric charge, allowing that pattern of charge to change millions of times every second. Tunneling refers to the ability of electrons (small particles that circle around the nucleus of an atom) to move or "tunnel" through the silicon. Electrons are said to tunnel through the barrier as a result of the quantum uncertainty as to which side of the barrier they are actually on.

19. Knowledge chunks would be greater than the number of distinct words because words are used in more than one way and with more than one meaning. Each different word meaning or usage is often referred to as a word "sense." It is likely that Shakespeare used more than 100,000 word senses.

20. Quoted from Douglas R. Hofstadter, *Gödel, Escher, Bach: An Eternal Golden Braid* (New York: Basic Books, 1979).

21. Michael Winerip, "Schizophrenia's Most Zealous Foe," *New York Sunday Times*, February 22, 1998.

22. The goal of the Visible Human Project is to create highly detailed, three-dimensional views of the male and female human body. The project is collecting transverse CT, MRI, and cryosection images. The web site is located at <http://www..nlm.nih.gov/research/visible/visible_human.html>.

23. Researchers Mark Hübener, Doron Shoham, Amiram Grinvald, and Tobias Bonhoeffer published their experiments on optical imaging in "Spatial Relationships among Three Columnar Systems in Cat Area 17," *Journal of Neuroscience* 17 (1997): 9270–9284.

   More information on this and other brain-imaging research is located at the Weizmann Institute's web site <http://www.weizmann.ac.il/> and at Amiram Grinvald's web site <http://www.weizmann.ac.il/brain/grinvald/grinvald.htm>.

24. The work of Dr. Benebid and other researchers is summarized in an online article, "Neural Prosthetics Come of Age as Research Continues," by Robert Finn, *The Scientist* 11, no. 19 (September 29, 1997): 13, 16. This article may be found at <http://www.the-scientist.library.upenn.edu/yr1997/sept/research_970929.html>.

25. From an April 1998 phone interview by the author with Dr. Trosch.

26. Dr. Rizzo's research is also reviewed in Finn's article, "Neural Prosthetics Come of Age as Research Continues."

27. To read more about the "neuron transistor," visit the web site of the Membrane and Neurophysics Department at the Max Planck Institute for Biochemistry <http://mnphys.biochem.mpg.de/>.

28. Robert Finn, "Neural Prosthetics Come of Age as Research Continues."

29. Carver Mead's research is described at <http://www.pcmp.caltech.edu/>.

30. W. B. Yeats, "Sailing to Byzantium," from *Selected Poems and Two Plays of William Butler Yeats*, edited by M. L. Rosenthal (New York: Macmillan, 1966).

CHAPTER 7: . . . AND BODIES

1. Herbert Dreyfus is well known for his critique of artificial intelligence in his book *What Computers Can't Do: The Limits of Artificial Intelligence* (New York: Harper and Row, 1979). Other theorists who may be considered to support the mind-beyond-machine perspective include J. R. Lucas and John Searle. See J. R. Lucas's "Minds, Machines and Gödel," *Philosophy* 36 (1961): 120–124; and John Searle's "Mind, Brains, and Programs," *The Behavioral and Brain Sciences* 3 (1980): 417–424. Also, see Searle's more recent book *The Rediscovery of the Mind* (Cambridge, MA: MIT Press, 1992).

2. "Researchers led by Dr. Clifford Steer at the University of Minnesota Medical School report in the current *Nature Medicine* that they have eliminated the need for viruses by harnessing the body's own genetic repair processes. In a landmark proof-of-concept experiment, the Minnesota team permanently altered a blood-clotting gene in 40 percent of the liver cells in a group of rats. The researchers started by splicing their DNA patch into a slip of RNA. Then they encased the hybrid molecule in a protective coating, laced it with sugars that seek out liver cells, and injected it into lab rats. True to plan, the hybrid molecules zeroed in on the targeted gene and lined up alongside it. An enzyme in the rats' own liver cells did the rest: Whenever it spotted a mismatched DNA, it simply removed the offending DNA and stitched in a replacement. Now the trick is to show that it will work with other tissues—and other species." From "DNA Therapy: The New, Virus-Free Way to Make Genetic Repairs." *Time*, March 16, 1998.

3. Hans Moravec, *Mind Children: The Future of Robot and Human Intelligence* (Cambridge, MA: Harvard University Press, 1988), p. 108.

4. Ralph Merkle's comments on nanotechnology can be found in an overview at his web site at the Xerox Palo Alto Research Center <http://sandbox.xerox.com/nano>. His

site contains links to important publications on nanotechnology, such as Richard Feynman's 1959 talk and Eric Drexler's dissertation, as well as links to various research centers that focus on nanotechnology.

5. Richard Feynman presented these ideas on December 29, 1959, at the annual meeting of the American Physical Society at the California Institute of Technology (Cal Tech). His talk was first published in the February 1960 issue of Cal Tech's *Engineering and Science*. This article is available online at <http://nano.xerox.com/nanotech/feynman.html>.

6. Eric Drexler, *Engines of Creation* (New York: Anchor Press/Doubleday, 1986). The book is also accessible online from the Xerox nanotechnology site <http://sandbox.xerox.com/nano> and also from Drexler's web site at the Foresight Institute <http://www.foresight.org/EOC/index.html>.

7. Eric Drexler, *Nanosystems: Molecular Machinery, Manufacturing, and Computation* (New York: John Wiley and Sons, 1992).

8. According to Nanothinc's web site <http://www.nanothinc.com/>, "Nanotechnology, broadly defined to include a number of nanoscale-related activities and disciplines, is a global industry in which more than 300 companies generate over $5 billion in annual revenues today—and $24 billion in 4 years." Nanothinc includes a list of companies and revenues upon which the figure is based. Some of the nanoapplications generating revenues are micromachines, microelectromechanical systems, autofabrication, nanolithography, nanotechnology tools, scanning probe microscopy, software, nanoscale materials, and nanophase materials.

9. Richard Smalley's publications and work on nanotechnology can be found at the web site for the Center for Nanoscale Science and Technology at Rice University <http://cnst.rice.edu/>.

10. For information on the use of nanotechnology in creating IBM's corporate logo, read Faye Flam, "Tiny Instrument Has Big Implications." *Knight-Ridder/Tribute News Service*, August 11, 1997, p. 811K7204.

11. Dr. Jeffrey Sampsell at Texas Instruments has written a white paper summarizing research on micromirrors, available at <http://www.ti.com/dlp/docs/it/resources/white/overview/over.shtml>.

12. A description of the flying machines can be found at the web site of the MEMS (MicroElectroMechanical Systems) and Fluid Dynamics Research Group at the University of California at Los Angeles (UCLA) <http://ho.seas.ucla.edu/new/main.htm>.

13. Xerox's nanotechnology research is described in Brian Santo, "Smart Matter Program Embeds Intelligence by Combining Sensing, Actuation, Computation—Xerox Builds on Sensor Theory for Smart Materials." *EETimes* (March 23, 1998):129. More information on this research can be found at the web site for the Smart Matter Research Group at Xerox's Palo Alto Research Center at <http://www.parc.xerox.com/spl/projects/smart-matter/> .

14. For information on the use of nanotechnology in creating the nanoguitar, read Faye Flam, "Tiny Instrument Has Big Implications." *Knight-Ridder/Tribune News Service*.

15. Learn more about the Chelyabinsk region by visiting the web site dedicated to helping the people living in that area at <http://www.logtv.com/chelya/chel.html>.

16. For more about the story behind Space War, see "A History of Computer Games," *Computer Gaming World* (November 1991): 16–26; and Eric S. Raymond, ed., *New Hacker's Dictionary* (Cambridge, MA: MIT Press, 1992). Space War was developed by Steve Russell in 1961 and implemented by him on the PDP-1 at MIT a year later.

17. Medical Learning Company is a joint venture between the American Board of Family Practice (an organization that certifies the sixty thousand family practice physicians in the United States) and Kurzweil Technologies. The goal of the company is to develop

educational software for continuing medical education of physicians as well as other markets. A key aspect of the technology will include an interactive simulated patient that can be examined, interviewed, and treated.

18. Hall's Utility Fog concept is described in J. Storrs Hall, "Utility Fog Part 1," *Extropy,* issue no. 13 (vol. 6, no. 2), third quarter 1994; and J. Storrs Hall, "Utility Fog Part 2," *Extropy,* issue no. 14 (vol. 7, no. 1), first quarter 1995. Also see Jim Wilson, "Shrinking Micromachines: A New Generation of Tools Will Make Molecule-Size Machines a Reality." *Popular Mechanics* 174, no. 11 (November 1997): 55–58.

19. Mark Yim, "Locomotion with a Unit-Modular Reconfigurable Robot," Stanford University Technical Report STAN-CS-TR-95-1536.

20. Joseph Michael, UK Patent #94004227.2.

21. For examples of early "prurient" text publications, see *A History of Erotic Literature* by Patrick J. Kearney (Hong Kong, 1982); and *History Laid Bare* by Richard Zachs (New York: HarperCollins, 1994).

22. *Upside Magazine*, April 1998.

23. For example, the "TFUI" (Touch-and-Feel User Interface) from pixis, as used in their Diva and Space Sirens series of CD-ROMs.

24. From "Who Needs Jokes? Brain Has a Ticklish Spot," Malcolme W. Browne, *New York Times*, March 10, 1998. Also see I. Fried (with C. L. Wilson, K. A. MacDonald, and E. J. Behnke), "Electric Current Stimulates Laughter," *Scientific Correspondence* 391: 650, 1998.

25. K. Blum et al., "Reward Deficiency Syndrome," *American Scientist*, March–April, 1996.

26. Brain Generated Music is a patented technology of NeuroSonics, a small company in Baltimore, Maryland. The founder, CEO, and principal developer of the technology is Dr. Geoff Wright, who is head of computer music at Peabody Conservatory.

27. For details about Dr. Benson's work, see his book *The Relaxation Response* (New York: Avon, 1990).

28. " 'God Spot' Is Found in Brain," *Sunday Times* (Britain), November 2, 1997.

CHAPTER 8: 1999

1. The U.S. Federal Government Gateway for Year 2000 Information Directories, at <http://www.itpolicy.gsa.gov/mks/yr2000/y2khome.htm>, contain a number of links to web pages devoted to Y2K issues. There are also many discussion groups on the Web about the Y2K topic. Simply do a search for "Y2K discussion" using a search engine such as Yahoo (www.yahoo.com) to find a number of web pages devoted to this subject.

2. David Cope talks about his EMI program in his book *Experiments in Musical Intelligence* (Madison, WI: A-R Editions, 1996). EMI is also discussed in Margaret Boden "Artificial Genius," *Discover* magazine, October 1996.

3. For more about the Improvisor program, see Margaret Boden, "Artificial Genius," *Discover* magazine, October 1996. The article addresses the question of who is the actual creator of original art produced by computer programs—the developer of the program or the program itself?

4. Laurie Flynn, "Program Proves Bad Puns Not Limited to Humans," *New York Times*, January 3, 1998.

5. "ParaMind copies any text you type or paste into its screen and systematically merges your text with new words. The words are all related, such as adjectives related to sight, or adverbs related to walking. In the text that you type or paste in, a word or two is selected where these new words will fit in, in the way that you want. The result is a new listing of your idea changed in several fascinating ways." From the ParaMind

Brainstorming Software web page at <http://www.paramind.net/>. For more information about other computer writing programs, see Marius Watz's web page called Computer Generated Writing at <http://www.notam.uio.no/~mariusw/c-g.writing/>.

6. More information on BRUTUS.1 Story Generator and its inventors can be found at <http://www.rpi.edu/dept/ppcs/BRUTUS/brutus.html>.

7. Ray Kurzweil's Cybernetic Poet (RKCP) is a software program designed by Ray Kurzweil and developed by Kurzweil Technologies. You can download a copy of the program at <http://www.kurzweiltech.com>.

8. For examples of Mutator's artistic creations, visit the web site of Computer Artworks at <http://www.artworks.co.uk/welcome.htm>.
   Karl Sims has written several articles about his work, including "Artificial Evolution for Computer Graphics," *Computer Graphics* 25, no. 4 (July 1991): 319–328.

9. Drawings and paintings by Aaron, Harold Cohen's cybernetic artist, have hung at London's Tate Gallery, Amsterdam's Stedelijk Museum, the Brooklyn Museum, the San Francisco Museum of Modern Art, the Washington Capitol Children's Museum, and others.

10. Harold Cohen, "How to Draw Three People in a Botanical Garden," AAAI-88, *Proceedings of the Seventh National Conference on Artificial Intelligence*, 1988, pp. 846–855. Some of the implications of Aaron are discussed in Pamela McCorduck, "Artificial Intelligence: An Aperçu," *Daedalus*, Winter 1988, pp. 65–83.

11. A list of sites on Cohen's Aaron can be found at <http://www.umcs.maine.edu/~larry/latour/aaron.html>. Also see Harold Cohen's article in "Constructions of the Mind" at <http://shr.stanford.edu/shreview/4-2/text/cohen.html>.

12. Raymond Kurzweil, *The Age of Intelligent Machines* (Cambridge, MA: MIT Press, 1990). Also see the publications section at the web site for Kurzweil Technologies at <http://www.kurzweiltech.com> and the publications section at the web site for Kurzweil Educational Systems at <http://www.kurzweiledu.com>.

13. Venture capital refers to funds available for investment by organizations that have raised pools of capital specifically to invest in companies, primarily new ventures. Angel capital refers to funds available for investment by networks of wealthy investors who invest in start-up companies. In the United States, both venture and angel capital have emphasized high-technology investments.

14. For a comprehensive list of available speech- and face-recognition products and research projects, go to The Face Recognition Home Page at <http://cherry.kist.re.kr/center/html/sites.html>.

15. For an excellent overview of this subject, see "The Intelligent Vehicle Initiative: Advancing 'Human-Centered' Smart Vehicles," by Cheryl Little of the Volpe National Transportation Systems Center. This article is available through the Turner-Fairbank Highway Research Center web page at <http://www.tfhrc.gov/pubrds/pr97-10/p18.htm>. For details about the tests on Interstate 15 in California, go to National AHS Consortium Home Page at <http://monolith-mis.com/ahs/default.htm>.

16. For example, Voice XpressPlus, from the dictation division of Lernout & Hauspie (formerly Kurzweil Applied Intelligence), combines large-vocabulary, continuous speech recognition for dictation, with natural-language understanding for commands. Continuous speech recognition without natural-language understanding (as of 1998) is also available from Dragon System's Naturally Speaking and IBM's ViaVoice.

17. Examples of translation products include Langenscheidt's T1 Professional from Gesellschaft für Multilinguale Systeme, a division of Lernout & Hauspie Speech Products; Globalink Power Translator; and SYSTRAN Classic for Windows.

18. Duncan Bythell, *The Handloom Weavers: A Study in the English Cotton Industry During*

the Industrial Revolution, p. 70. There are also a number of web sites exploring both the original Luddite history and the contemporary neo-Luddite movement. For one example, see the web page Luddites On-Line at <http://www.luddites.com/index2.html>.

19. Ben J. Wattenberg, ed., *The Statistical History of the United States from Colonial Times to the Present*; U.S. Department of Commerce, Bureau of the Census, *Statistical Abstract of the United States*, 1997.

20. Ben J. Wattenberg, ed., *The Statistical History of the United States from Colonial Times to the Present*.

21. U.S. Department of Commerce, Bureau of the Census, *Statistical Abstract of the United States*, 1997.

22. Ted Kaczynski's Unabomber Manifesto was published in both the *New York Times* and the *Washington Post* in September 1995. The full text of the document is available on numerous web pages, including: <http://www.soci.niu.edu/~critcrim/uni/uni.txt>.

CHAPTER 9: 2009

1. A consortium of eighteen manufacturers of cellular telephones and other portable electronic devices is developing a technology called Bluetooth, which provides wireless communications within a radius of about ten meters, at a data rate of 700 to 900 kilobits per second. Bluetooth is expected to be introduced in late 1999 and will initially have a cost of about $20 per unit. This cost is expected to decline rapidly after introduction. Bluetooth will allow personal communications and electronics devices to communicate with one another.

2. Technology such as Bluetooth (see note 1) will allow computer components such as computing units, keyboards, pointing devices, printers, etc. to communicate with one another without the use of cables.

3. Microvision of Seattle has a product called a Virtual Retina Display (VRD) that projects images directly onto the user's retinas while allowing the user to see the normal environment. The Microvision VRD is currently expensive and is sold primarily to the military for use by pilots. Microvision's CEO Richard Rutkowski projects a consumer version built on a single chip before the year 2000.

4. Projecting from the speed of personal computers, a 1998 personal computer can perform about 150 million instructions per second for about $1,000. By doubling every twelve months, we get a projection of 150 million multiplied by 2^{11} (2,048) = 300 billion instructions per second in 2009. Instructions are less powerful than calculations, so calculations per second will be around 100 billion. However, projecting from the speed of neural computers, a 1997 neural computer provided about 2 billion neural connection calculations per second for around $2,000, which is 1 billion calculations per $1,000. By doubling every twelve months, we get a projection of 1 billion times 2^{12} (4,096) = 4 trillion calculations per second in 2009. By 2009, computers will routinely combine both types of computations, so if even 25 percent of the computations are of the neural connection calculation type, the estimate of 1 trillion calculations per second for $1,000 of computing in 2009 is reasonable.

5. The most powerful supercomputers are twenty thousand times more powerful than a $1,000 personal computer. With $1,000 personal computers providing about 1 trillion calculations per second (particularly of the neural-connection type of calculation) in 2009, the more powerful supercomputers will provide about 20 million billion calculations per second, which is about equal to the estimated processing power of the human brain.

6. As of this writing, there has been much publicity surrounding the work of Dr. Judah Folkman of Children's Hospital in Boston, Massachusetts, and the effects of angiogenesis inhibitors. In particular, the combination of Endostatin and Angiostatin, bioengineered drugs that inhibit the reproduction of capillaries, has been remarkably effective in mice. Although there has been a lot of commentary pointing out that drugs that work in mice often do not work in humans, the degree to which this drug combination worked in these laboratory animals was remarkable. Drugs that work this well in mice often do work in humans.

See "HOPE IN THE LAB: A Special Report. A Cautious Awe Greets Drugs That Eradicate Tumors in Mice," *New York Times*, May 3, 1998.

CHAPTER 10: 2019

1. See note 3 of chapter 9, "2009," on the Microvision Virtual Retina Display.
2. A 1997 neural computer provided about 2 billion neural-connection calculations per second for $2,000. By doubling twenty-two times by the year 2019, that comes to about 8 million billion calculations per second for $2,000 and 16 million billion calculations per second for $4,000. In 2020, we get 16 million billion calculations per second for $2,000.
3. With each human brain providing about 10^{16} calculations per second and an estimated 10 billion (10^{10}) persons, we get an estimated 10^{26} calculations per second for all human brains on Earth. There are about 100 million computers in the world in 1998. A conservative estimate for 2019 would be a billion computers equal to the then state-of-the-art for $1,000 machines. Thus the total computing power of the computers equals one billion (10^9) times $10^{16} = 10^{25}$ calculations per second, which is 10 percent of 10^{26}.

CHAPTER 11: 2029

1. With each human brain providing about 10^{16} calculations per second and an estimated 10 billion (10^{10}) persons, we get an estimated 10^{26} calculations per second for all human brains on Earth. There are about 100 million computers in the world in 1998. A (very) conservative estimate for 2029 would be a billion computers equal to the then state-of-the-art for $1,000 machines. This is actually too conservative, but still sufficient for our purposes. Thus the total computing power of the computers equals one billion (10^9) times $10^{19} = 10^{28}$ calculations per second, which is one hundred times the processing power of all human brains (which is 10^{26} calculations per second).
2. See Raymond Kurzweil, *The 10% Solution for a Healthy Life: How to Eliminate Virtually All Risk of Heart Disease and Cancer* (New York: Crown Publishers, 1993).

CHAPTER 12: 2099

1. As discussed in chapter 6, "Building New Brains," and chapter 10, "2019," human capacity of an estimated 2×10^{16} (neural connection) calculations per second will be achieved in a $1,000 computing device by around the year 2020. Also as noted, the capacity of computing will double every twelve months, or ten times every decade, which is a factor of one thousand (2^{10}) every ten years. Thus by the year 2099, $1,000 of computing will be roughly equivalent to 10^{24} times the computing capacity of the human brain, or 10^{40} calculations per second. Estimating a trillion virtual persons

(hundred times greater than the roughly 10 billion persons in the early twenty-first century), and an estimated \$1 million of computing devoted to each person, we get an estimated 10^{55} calculations per second.

2. One thousand qu-bits would enable $2^{1,000}$ (approximately 10^{300}) calculations to be performed at the same time. If 10^{42} of the calculations each second are such quantum calculations, then that is equivalent to $10^{42} \times 10^{300} = 10^{342}$ calculations per second. $10^{55} + 10^{342}$ still equals about 10^{342}.

3. What happened to picoengineering, you're wondering? Picoengineering refers to engineering at the scale of a picometer, which is one trillionth of a meter. Remember that the author has not spoken to Molly for seventy years. Nanotechnology (technology on the scale of a billionth of a meter) is becoming practical in the decade between 2019 and 2029. Note that in the twentieth century, the Law of Accelerating Returns as applied to computation has been achieved through engineering at ever smaller scales of physical size. Moore's Law is a good example of this, in that the size of a transistor (in two dimensions) has been decreasing by 50 percent every two years. This means that transistors have been shrinking by a factor of $2^5 = 32$ in ten years. Thus the feature size of a transistor in *each* dimension has been shrinking by a factor of the square root of $32 = 5.6$ every ten years. We are shrinking, therefore, the feature size of components by a factor of about 5.6 in each dimension every decade.

If engineering at the nanometer scale (nanotechnology) is practical in the year 2032, then engineering at the picometer scale should be practical about forty years later (because $5.6^4 =$ approximately 1,000), or in the year 2072. Engineering at the femtometer (one thousandth of a trillionth of a meter, also referred to as a quadrillionth of a meter) scale should be feasible, therefore, by around the year 2112. Thus I am being a bit conservative to say that femtoengineering is controversial in 2099.

Nanoengineering involves manipulating individual atoms. Picoengineering will involve engineering at the level of subatomic particles (e.g., electrons). Femtoengineering will involve engineering inside a quark. This should not seem particularly startling, as contemporary theories already postulate intricate mechanisms within quarks.

EPILOGUE: THE REST OF THE UNIVERSE REVISITED

1. We could use the Busy Beaver Function (see note 16 on the Turing machine in chapter 4) as a quantitative measure of the software of intelligence.

TIME LINE

Sources for the timeline include Raymond Kurzweil, *The Age of Intelligent Machines* (Cambridge, MA: MIT Press, 1990).

Introduction to big bang theory at <http://www.bowdoin.edu/dept/physics/astro.1997/astro4/bigbang.html>; Joseph Silk, *A Short History of the Universe* (New York: Scientific American Library, 1994); Joseph Silk, *The Big Bang* (San Francisco: W. H. Freeman and Company, 1980); Robert M. Wald, *Space, Time and Gravity* (Chicago: The University of Chicago Press, 1977); Stephen W. Hawking, *A Brief History of Time* (New York: Bantam Books, 1988).

Evolution and behavior at <http://ccp.uchicago.edu/~jyin/evolution.html>; Edward O. Wilson, *The Diversity of Life* (New York: W. W. Norton and Company, 1993); Stephen Jay Gould, *The Book of Life* (New York: W. W. Norton and Company, 1993); Alexander Hellemans and Bryan Bunch, *The Timetable of Science* (Simon and

Schuster, 1988). "CBN History: Radio/Broadcasting Timeline" at <http://www.wcbn.org/history/wcbntime.html>.

"Chronology of Events in the History of Microcomputers" at <http://www3.islandnet.com/~kpolsson/comphist.htm>.

"The Computer Museum History Center" at <http://www.tcm.org/history/index.html>.

1. Picoengineering involves engineering at the level of subatomic particles (e.g., electrons). See note 3 on picoengineering and femtoengineering in chapter 12.
2. Femtoengineering will involve engineering using mechanisms within a quark. See note 3 on picoengineering and femtoengineering in chapter 12.

HOW TO BUILD AN INTELLIGENT MACHINE IN THREE EASY PARADIGMS

1. See "Information Processing in the Human Body," by Vadim Gerasimov, at <http://vadim.www.media.mit.edu/MAS862/Project.html>.
2. Marvin Minsky and Seymour A. Papert, *Perceptrons: An Introduction to Computational Geometry* (Cambridge, MA: MIT Press, 1988).
3. The quoted text on the "two daughter sciences" is from Seymour Papert, "One AI or Many," *Daedalus*, Winter 1988.

    "Dr. Seymour Papert is a mathematician and one of the early pioneers of Artificial Intelligence. Additionally, he is internationally recognized as the seminal thinker about ways in which computers can change learning. Born and educated in South Africa where he participated actively in the anti-apartheid movement, Dr. Papert pursued mathematical research at Cambridge University from 1954 through 1958. He then worked with Jean Piaget at the University of Geneva from 1958 through 1963. It was this collaboration that led him to consider using mathematics in the service of understanding how children can learn and think. In the early 1960s, Papert came to MIT where, with Marvin Minsky, he founded the Artificial Intelligence Laboratory and coauthored their seminal work *Perceptrons*." From the web page entitled "Seymour Papert" at <http://papert.www.media.mit.edu/people/papert/>.
4. "[Marvin] Minsky was . . . one of the pioneers of intelligence-based mechanical robotics and telepresence. . . . In 1951 he built the first randomly wired neural network learning machine (called SNARC, for Stochastic Neural-Analog Reinforcement Computer), based on the reinforcement of simulated synaptic transmission coefficients. . . . Since the early 1950s, Marvin Minsky has worked on using computational ideas to characterize human psychological processes, as well as working to endow machines with intelligence." From the brief academic biography of Marvin Minsky at <http://minsky.www.media.mit.edu/people/minsky/minskybiog.html>.
5. Dr. Raj Reddy is dean of the School of Computer Science at Carnegie Mellon University and the Herbert A. Simon University Professor of Computer Science and Robotics. Dr. Reddy is a leading AI researcher whose research interests include the study of human-computer interaction and artificial intelligence.

SUGGESTED READINGS

Abbott, E. A. *Flatland: A Romance in Many Dimensions*. Reprint. Oxford: Blackwell, 1962.

Abelson, Harold and Andrea diSessa. *Turtle Geometry: The Computer as a Medium for Exploring Mathematics*. Cambridge, MA: MIT Press, 1980.

Abrams, Malcolm and Harriet Bernstein. *Future Stuff*. New York: Viking Penguin, 1989.

Adams, James L. *Conceptual Blockbusting: A Guide to Better Ideas*. Reading, MA: Addison-Wesley, 1986.

————. *The Care and Feeding of Ideas: A Guide to Encouraging Creativity*. Reading, MA: Addison-Wesley, 1986.

Adams, Scott. *The Dilbert Future: Thriving on Stupidity in the 21st Century*. New York: Harper Business, 1997.

Alexander, S. *Art and Instinct*. Reprint. Oxford: Folcroft Press, 1970.

Allen, Peter K. *Robotic Object Recognition Using Vision and Touch*. Boston: Kluwer Academic, 1987.

Allman, William F. *Apprentices of Wonder: Inside the Neural Network Revolution*. New York: Bantam Books, 1989.

Amit, Daniel J. *Modeling Brain Function: The World of Attractor Neural Networks*. Cambridge: Cambridge University Press, 1989.

Anderson, James A. *An Introduction to Neural Networks*. Cambridge, MA: MIT Press, 1997.

Andriole, Stephen, ed. *The Future of Information Processing Technology*. Princeton, NJ: Petrocelli Books, 1985.

Antébi, Elizabeth and David Fishlock. *Biotechnology: Strategies for Life*. Cambridge, MA: MIT Press, 1986.

Anton, John P. *Science and the Sciences in Plato*. New York: EIDOS, 1980.

Ashby, W. Ross. *Design for a Brain*. New York: John Wiley and Sons, 1960.

————. *An Introduction to Cybernetics*. New York: John Wiley and Sons, 1963.

Asimov, Isaac. *Asimov on Numbers*. New York: Bell Publishing Company, 1977.

————. *I, Robot*. New York: Doubleday, 1950.

————. *Robot Dreams*. New York: Berkley Books, 1986.

————. *Robots of Dawn*. New York: Doubleday and Company, 1983.

Asimov, Isaac and Karen A. Frenkel. *Robots: Machines in Man's Image*. New York: Harmony Books, 1985.

Atkins, P. W. *The Second Law*. New York: Scientific American Books, 1984.

Augarten, Stan. *Bit by Bit: An Illustrated History of Computers*. New York: Ticknor and Fields, 1984.

Austrian, Geoffrey D. *Herman Hollerith: Forgotten Giant of Information Processing*. New York: Columbia University Press, 1982.

Axelrod, Robert. *The Evolution of Cooperation*. New York: Basic Books, 1984.

Ayache, Nicholas and Peter T. Sander. *Artificial Vision for Mobile Robots: Stereo Vision and Multisensory Perception*. Cambridge, MA: MIT Press, 1991.

Ayer, Alfred J. *The Foundations of Empirical Knowledge*. London: Macmillan and Company, 1964.

————. *Language, Truth and Logic*. New York: Dover Publications, 1936.

————, ed. *Logical Positivism*. New York: Macmillan, 1959.

Ayers, M. *The Refutation of Determinism: An Essay in Philosophical Logic*. London: Methuen, 1968.

Ayres, Robert U., et al. *Robotics and Flexible Manufacturing Technologies: Assessment, Impacts, and Forecast*. Park Ridge, NJ: Noyes Publications, 1985.

Babbage, Charles. *Charles Babbage and His Calculating Engines*. Edited by Philip Morrison and Emily Morrison. New York: Dover Publications, 1961.

————. *Ninth Bridgewater Treatise: A Fragment*. London: Murray, 1838.

Babbage, Henry Prevost. *Babbage's Calculating Engines: A Collection of Papers by Henry Prevost Babbage (Editor)*. Vol. 2. Los Angeles: Tomash, 1982.

Bailey, James. *After Thought: The Computer Challenge to Human Intelligence*. New York: Basic Books, 1996.

Bara, Bruno G. and Giovanni Guida. *Computational Models of Natural Language Processing*. Amsterdam: North Holland, 1984.

Barnsley, Michael F. *Fractals Everywhere*. Boston: Academic Press Professional, 1993.

Baron, Jonathan. *Rationality and Intelligence*. Cambridge: Cambridge University Press, 1985.

Barrett, Paul H., ed. *The Collected Papers of Charles Darwin*. Vols. 1 and 2. Chicago: University of Chicago Press, 1977.

Barrow, John. *Theories of Everything*. Oxford: Oxford University Press, 1991.

Barrow, John D. and Frank J. Tipler. *The Anthropic Cosmological Principle*. Oxford: Oxford University Press, 1986.

Bartee, Thomas C., ed. *Digital Communications*. Indianapolis, IN: Howard W. Sams and Company, 1986.

Basalla, George. *The Evolution of Technology*. Cambridge: Cambridge University Press, 1988.

Bashe, Charles J., Lyle R. Johnson, John H. Palmer, and Emerson W. Pugh. *IBM's Early Computers*. Cambridge, MA: MIT Press, 1986.

Bateman, Wayne. *Introduction to Computer Music*. New York: John Wiley and Sons, 1980.

Baxandall, D. *Calculating Machines and Instruments*. Rev. ed. London: Science Museum, 1975. Original, 1926.

Bell, C. Gordon with John E. McNamara. *High-Tech Ventures: The Guide for Entrepreneurial Success*. Reading, MA: Addison-Wesley, 1991.

Bell, Gordon. "Ultracomputers: A Teraflop Before Its Time." *Science* 256 (April 3, 1992).

Benedikt, Michael, ed. *Cyberspace: First Steps*. Cambridge, MA: MIT Press, 1992.

Bernstein, Jeremy. *The Analytical Engine: Computers—Past, Present and Future*. Revised ed. New York: William Morrow, 1981.

Bertin, Jacques. *Semiology of Graphics: Diagrams, Networks, Maps*. Madison: University of Wisconsin Press, 1983.

Beth, E. W. *Foundations of Mathematics*. Amsterdam: North Holland, 1959.

Block, Irving, ed. *Perspectives on the Philosophy of Wittgenstein*. Cambridge, MA: MIT Press, 1981.

Block, Ned, Owen Flanagan, Guven Guzeldere, eds. *The Nature of Consciousness: Philosophical Debates*. Cambridge, MA: MIT Press, 1997.

Bobrow, Daniel G. and A. Collins, eds. *Representation and Understanding*. New York: Academic Press, 1975.

Boden, Margaret. *Artificial Intelligence and Natural Man*. New York: Basic Books, 1977.

———. *The Creative Mind: Myths & Mechanisms*. New York: Basic Books, 1991.

Bolter, J. David. *Turing's Man: Western Culture in the Computer Age*. Chapel Hill: The University of North Carolina Press, 1984.

Boole, George. *An Investigation of the Laws of Thought on Which Are Founded the Mathematical Theories of Logic and Probabilities*. 1854. Reprint. Peru, IL: Open Court Publishing, 1952.

Botvinnik, M. M. *Computers in Chess: Solving Inexact Search Problems*. New York: Springer-Verlag, 1984.

Bowden, B. W., ed. *Faster Than Thought*. London: Pittman, 1953.

Brachman, Ronald J. and Hector J. Levesque. *Readings and Knowledge Representation*. Los Altos, CA: Morgan Kaufmann, 1985.

Brady, M., L. A. Gerhardt, and H. F. Davidson. *Robotics and Artificial Intelligence*. Berlin: Springer-Verlag, 1984.

Brand, Stewart. *The Media Lab: Inventing the Future at MIT*. New York: Viking Penguin, 1987.

Briggs, John. *Fractals: The Patterns of Chaos*. New York: Simon and Schuster, 1992.

Brittan, Gordon G. *Kant's Theory of Science*. Princeton, NJ: Princeton University Press, 1978.

Bronowski, J. *The Ascent of Man*. Boston: Little, Brown and Company, 1973.

Brooks, Rodney A. "Elephants Don't Play Chess." *Robotics and Autonomous Systems* 6 (1990).

———. "Intelligence Without Representation." *Artificial Intelligence* 47 (1991).

———. "New Approaches to Robotics." *Science* 253 (1991).

Brooks, Rodney A. and Anita Flynn. "Fast, Cheap and Out of Control: A Robot Invasion of the Solar System." *Journal of the British Interplanetary Society* 42 (1989).

Brooks, Rodney A., Pattie Maes, Maja J. Mataric, and Grinell More. "Lunar Base Construction Robots." *IROS*, IEEE International Workshop on Intelligence Robots and Systems, 1990.

Brown, John Seeley. *Seeing Differently: Insights on Innovation*. Cambridge, MA: Harvard Business School Press, 1997.

Brown, Kenneth A. *Inventors at Work: Interviews with 16 Notable American Inventors*. Redmond, WA: Tempus Books of Microsoft Press, 1988.

Brumbaugh, R. S. *Plato's Mathematical Imagination*. Bloomington: Indiana University Press, 1954.

Bruner, Jerome S., Jacqueline J. Goodnow, and George A. Austin. *A Study of Thinking*. 1956. Reprint. New York: Science Editions, 1965.

Buderi, Robert. *The Invention That Changed the World: How a Small Group of Radar Pioneers Won the Second World War and Launched a Technological Revolution*. New York: Simon and Schuster, 1996.

Burger, Peter and Duncan Gillies. *Interactive Computer Graphics: Functional, Procedural and Device-Level Methods*. Workingham, UK: Addison-Wesley Publishing Company, 1989.

Burke, James. *The Day the Universe Changed*. Boston: Little, Brown and Company, 1985.

Butler, Samuel. "Darwin Among the Machines." *Canterbury Settlement*. AMS Press, 1923. (Written in 1863 by the author of *Erewhon*.)

Buxton, H. W. *Memoir of the Life and Labours of the Late Charles Babbage, Esq. F.R.S.* Edited by A. Hyman. Los Angeles: Tomash, 1988.

Byrd, Donald. "Music Notation by Computer." Ph.D. dissertation, Indiana University Computer Science Department, 1984.

Bythell, Duncan. *The Handloom Weavers: A Study in the English Cotton Industry During the Industrial Revolution.* Cambridge: Cambridge University Press, 1969.

Cairns-Smith, A. G. *Seven Clues to the Origin of Life.* Cambridge: Cambridge University Press, 1985.

Calvin, William H. *The Cerebral Code: Thinking a Thought in the Mosaics of the Mind.* Cambridge, MA: MIT Press, 1996.

Campbell, Jeremy. *The Improbable Machine.* New York: Simon and Schuster, 1989.

Carpenter, Gail A. and Stephen Grossberg. *Pattern Recognition by Self-Organizing Neural Networks.* Cambridge, MA: MIT Press, 1991.

Carroll, Lewis. *Through the Looking Glass.* London: Macmillan, 1871.

Cassirer, Ernst. *The Philosophy of the Enlightenment.* Princeton, NJ: Princeton University Press, 1951.

Casti, John L. *Complexification: Explaining the Paradoxical World Through the Science of Surprise.* New York: HarperCollins, 1994.

Cater, John P. *Electronically Hearing: Computer Speech Recognition.* Indianapolis, IN: Howard W. Sams and Company, 1984.

———. *Electronically Speaking: Computer Speech Generation.* Indianapolis, IN: Howard W. Sams and Company, 1983.

Caudill, Maureen and Charles Butler. *Naturally Intelligent Systems.* Cambridge, MA: MIT Press, 1990.

Chaitin, Gregory J. *Algorithmic Information Theory.* Cambridge: Cambridge University Press, 1987.

Chalmers, D. J. *The Conscious Mind.* New York: Oxford University Press, 1996.

Chamberlin, Hal. *Musical Applications of Microprocessors.* Indianapolis, IN: Hayden Books, 1985.

Chapuis, Alfred and Edmond Droz. *Automata: A Historical and Technological Study.* New York: Griffon, 1958.

Cherniak, Christopher. *Minimal Rationality.* Cambridge, MA: MIT Press, 1986.

Chomsky, Noam. *Cartesian Linguistics.* New York: Harper and Row, 1966.

———. *Language and Mind.* Enlarged edition. New York: Harcourt Brace Jovanovich, 1972.

———. *Language and Problems of Knowledge: The Managua Lectures.* Cambridge, MA: MIT Press, 1988.

———. *Language and Thought.* Wakefield, RI, and London: Moyer Bell, 1993.

———. *Reflections on Language.* New York: Pantheon, 1975.

———. *Rules and Representation.* Cambridge, MA: MIT Press, 1980.

———. *Syntactic Structures.* The Hague: Mouton, 1957.

Choudhary, Alok N. and Janak H. Pattl. *Parallel Architectures and Parallel Algorithms for Integrated Vision Systems.* Boston: Kluwer Academic, 1990.

Christensen, Clayton. *The Innovator's Dilemma: When New Technologies Cause Great Firms to Fail.* Cambridge, MA: Harvard Business School Press, 1997.

Church, Alonzo. *Introduction to Mathematical Logic.* Vol. 1. Princeton, NJ: Princeton University Press, 1956.

Church, Kenneth W. *Phonological Parsing in Speech Recognition.* Norwell, MA: Kluwer Academic, 1987.

Churchland, P. S. and T. J. Sejnowski. *The Computational Brain.* Cambridge, MA: MIT Press, 1992.

Churchland, Paul. *The Engine of Reason, the Seat of the Soul*. Cambridge, MA: MIT Press, 1995.

————. *Matters and Consciousness: A Contemporary Introduction to the Philosophy of Mind*. Cambridge, MA: MIT Press, 1984.

————. *A Neurocomputational Perspective: The Nature of Mind and the Structure of Science*. Cambridge, MA: MIT Press, 1989.

Clark, Andy. *Being There: Putting Brain, Body, and World Together Again*. Cambridge, MA: MIT Press, 1997.

Clarke, Arthur C. *3001: The Final Odyssey*. New York: Ballantine Books, 1997.

Coates, Joseph F., John B. Mahaffie, and Andy Hines. *2025: Scenarios of U.S. and Global Society Reshaped by Science and Technology*. Greensboro, NC: Oak Hill Press, 1997.

Cohen, I. Bernard. *The Newtonian Revolution*. Cambridge: Cambridge University Press, 1980.

Cohen, John. *Human Robots in Myth and Science*. London: Allen and Unwin, 1966.

Cohen, Paul R. *Empirical Methods for Artificial Intelligence*. Cambridge, MA: MIT Press, 1995.

Cohen, Paul R. and Edward A. Feigenbaum. *The Handbook of Artificial Intelligence*, Vols. 3 and 4. Los Altos, CA: William Kaufmann, 1982.

Connell, Jonathan H. *Minimalist Mobile Robotics: A Colony-Style Architecture for an Artificial Creature*. Boston: Academic Press, 1990.

Conrad, Michael and H. H. Pattee. "Evolution Experiments with an Artificial Ecosystem." *Journal of Theoretical Biology* 28 (1970).

Conrad, Michael et al. "Towards an Artificial Brain." *BioSystems* 23 (1989).

Cornford, Francis M. *Plato's Cosmology*. London: Routledge and Kegan Paul, 1937.

Crandall, B. C., ed. *Nanotechnology: Molecular Speculations on Global Abundance*. Cambridge, MA: MIT Press, 1997.

Crease, Robert P. and Charles C. Mann. *The Second Creation*. New York: Macmillan, 1986.

Crick, Francis. *The Astonishing Hypothesis: The Scientific Search for the Soul*. New York: Charles Scribner's Sons, 1994.

————. *Life Itself*. London: Mcdonald, 1981.

Critchlow, Arthur J. *Introduction to Robotics*. New York: Macmillan Publishing Company, 1985.

Cullinane, John J. *The Entrepreneur's Survival Guide: 101 Tips for Managing in Good Times and Bad*. Homewood, IL: Business One Irwin, 1993.

Daedalus: Journal of the American Academy of Arts and Sciences. Artificial Intelligence. Winter 1998. Vol. 117.

Darwin, Charles. *The Descent of Man, and Selection in Relation to Sex*. Second ed. New York: Hurst and Company, 1874.

————. *The Expression of the Emotions in Man and Animals*. 1872. Reprint. Chicago: University of Chicago Press, 1965.

————. *Origin of Species*. Reprint. London: Penguin, 1859.

Davies, Paul. *Are We Alone? Implications of the Discovery of Extraterrestrial Life*. New York: Basic Books, 1995.

————. *The Cosmic Blueprint*. New York: Simon and Schuster, 1988.

————. *God and the New Physics*. New York: Simon and Schuster, 1983.

————. *The Mind of God*. New York: Simon and Schuster, 1992.

————. "A New Science of Complexity." *New Scientist* 26 (November 1988).

Davis, Philip J. and David Park, eds. *No Way: The Nature of the Impossible*. New York: W. H. Freeman, 1988.

Davis, Philip J. and Reben Hersh. *Descartes' Dream: The World According to Mathematics*. San Diego, CA: Harcourt Brace Jovanovich, 1986.

Davis, R. and D. B. Lenat. *Knowledge-Based Systems in Artificial Intelligence*. New York: McGraw-Hill, 1980.

Dawkins, Richard. *The Blind Watchmaker: Why the Evidence of Evolution Reveals a Universe Without Design*. New York: W. W. Norton and Company, 1986.

————. "The Evolution of Evolvability." *Artificial Life*, edited by Christopher G. Langton. Reading, MA: Addison-Wesley, 1988.

————. *The Extended Phenotype*. San Francisco: Freeman, 1982.

————. *River out of Eden: A Darwinian View of Life*. New York: Basic Books, 1995.

————. "Universal Darwinism." *Evolution from Molecules to Men*, edited by D. S. Bendall. Cambridge: Cambridge University Press, 1983.

————. *The Selfish Gene*. Oxford: Oxford University Press, 1976.

Dechert, Charles R., ed. *The Social Impact of Cybernetics*. New York: Simon and Schuster, 1966.

Denes, Peter B. and Elliot N. Pinson. *The Speech Chain: The Physics and Biology of Spoken Language*. Bell Telephone Laboratories, 1963.

Dennett, Daniel C. *Brainstorms: Philosophical Essays on Mind and Psychology*. Cambridge, MA: MIT Press, 1981.

————. *Consciousness Explained*. Boston: Little, Brown and Company, 1991.

————. *Content and Consciousness*. London: Routledge and Kegan Paul, 1969.

————. *Darwin's Dangerous Idea: Evolution and the Meanings of Life*. New York: Simon and Schuster, 1995.

————. *Elbow Room: The Varieties of Free Will Worth Wanting*. Cambridge, MA: MIT Press, 1985.

————. *The Intentional Stance*. Cambridge, MA: MIT Press, 1987.

————. *Kinds of Minds: Toward an Understanding of Consciousness*. New York: Basic Books, 1996.

Denning, Peter J. and Robert M. Metcalfe. *Beyond Calculation: The Next Fifty Years of Computing*. New York: Copernicus, 1997.

Depew, David J. and Bruce H. Weber, eds. *Evolution at a Crossroads*. Cambridge, MA: MIT Press, 1985.

Dertouzos, Michael. *What Will Be: How the New World of Information Will Change Our Lives*. New York: HarperCollins, 1997.

Dertouzos, Michael L. and Joel Moses Dertouzos. *The Computer Age: A Twenty Year View*. Cambridge, MA: MIT Press, 1979.

Descartes, R. *Discourse on Method, Optics, Geometry, and Meteorology*. 1637. Reprint. Indianapolis, IN: Bobbs-Merrill, 1956.

————. *Meditations on First Philosophy*. Paris: Michel Soly, 1641.

————. *Treatise on Man*. Paris, 1664.

Devlin, Keith. *Mathematics: The Science of Patterns*. New York: Scientific American Library, 1994.

Dewdney, A. K. *The Armchair Universe: An Exploration of Computer Worlds*. New York: W. H. Freeman, 1988.

De Witt, Bryce and R. D. Graham, eds. *The Many-Worlds Interpretation of Quantum Mechanics*. Princeton, NJ: Princeton University Press, 1974.

Diebold, John. *Man and the Computer: Technology as an Agent of Social Change*. New York: Avon Books, 1969.

Dixit, Avinash and Robert S. Pindyck. *Investment Under Uncertainty*. Princeton, NJ: Princeton University Press, 1994.

Dobzhansky, Theodosius. *Mankind Evolving: The Evolution of the Human Species*. New Haven, CT: Yale University Press, 1962.

Dodds, E. R. *Greeks and the Irrational*. Berkeley: University of California Press, 1951.

Downes, Larry, Chunka Mui, and Nicholas Negroponte. *Unleashing the Killer App: Digital Strategies for Market Dominance*. Cambridge, MA: Harvard Business School Press, 1998.

Drachmann, A. G. *The Mechanical Technology of Greek and Roman Antiquity*. Madison: University of Wisconsin Press, 1963.

Drexler, K. Eric. *Engines of Creation*. New York: Doubleday, 1986.

————. "Hypertext Publishing and the Evolution of Knowledge." *Social Intelligence* 1:2 (1991).

Dreyfus, Hubert. "Alchemy and Artificial Intelligence," *Rand Technical Report*, December 1965.

————. *Philosophic Issues in Artificial Intelligence*. Chicago: Quadrangle Books, 1967.

————. *What Computers Can't Do: The Limits of Artificial Intelligence*. New York: Harper and Row, 1979.

————. *What Computers Still Can't Do: A Critique of Artificial Reason*. Cambridge, MA: MIT Press, 1992.

————, ed. *Husserl, Intentionality & Cognitive Science*. Cambridge, MA: MIT Press, 1982.

Dreyfus, Hubert L. and Stuart E. Dreyfus. *Mind over Machine: The Power of Human Intuition and Expertise in the Era of the Computer*. New York: The Free Press, 1986.

Drucker, Peter F. *Innovation and Entrepreneurship: Practice and Principles*. New York: Harper and Row, 1985.

Durrett, H. John, ed. *Color and the Computer*. Boston: Academic Press, 1987.

Dyson, Esther. *Release 2.0: A Design for Living in the Digital Age*. New York: Broadway Books, 1997.

Dyson, Freeman. *Disturbing the Universe*. New York: Harper and Row, 1979.

————. *From Eros to Gaia*. New York: HarperCollins, 1990.

————. *Infinite in All Directions*. New York: Harper and Row, 1988.

————. *Origins of Life*. Cambridge: Cambridge University Press, 1985.

Dyson, George B. *Darwin Among the Machines: The Evolution of Global Intelligence*. Reading, MA: Addison-Wesley, 1997.

Eames, Charles and Ray Eames. *A Computer Perspective*. Cambridge, MA: Harvard University Press, 1973.

Ebeling, Carl. *All the Right Moves: A VLSI Architecture for Chess*. Cambridge, MA: MIT Press, 1987.

Edelman, G. M. *Neural Darwinism: The Theory of Neuronal Group Selection*. New York: Basic Books, 1987.

Einstein, Albert. *Relativity: The Special and the General Theory*. New York: Crown, 1961.

Elithorn, Alick and Ranan Banerji. *Artificial and Human Intelligence*. Amsterdam: North Holland, 1991.

Enderle, G. *Computer Graphics Programming*. Berlin: Springer-Verlag, 1984.

Fadiman, Clifton, ed. *Fantasia Mathematica: Being a Set of Stories, Together with a Group of Oddments and Diversion, All Drawn from the Universe of Mathematics*. New York: Simon and Schuster, 1958.

Fahlman, Scott E. *NETL: A System for Representing and Using Real-World Knowledge*. Cambridge, MA: MIT Press, 1979.

Fant, Gunnar. *Speech Sounds and Features*. Cambridge, MA: MIT Press, 1973.

Feigenbaum, E. and Avron Barr, eds. *The Handbook of Artificial Intelligence*. Vol. 1. Los Altos, CA: William Kaufmann, 1981.

Feigenbaum, Edward A. and Julian Feldman, eds. *Computers and Thought*. New York: McGraw-Hill, 1963.

Feigenbaum, Edward A. and Pamela McCorduck. *The Fifth Generation: Artificial Intelligence and Japan's Computer Challenge to the World*. Reading, MA: Addison-Wesley, 1983.

Feynman, Richard. "There's Plenty of Room at the Bottom." *Miniaturization*, edited by H. D. Gilbert. New York: Reinhold, 1961.

Feynman, Richard P. *Surely You're Joking, Mr. Feynman!* New York: Norton, 1985.

———. *What Do You Care What Other People Think?* New York: Bantam, 1988.

Feynman, Richard P., Robert B. Leighton, and Matthew Sands. *The Feynman Lectures in Physics*. Reading, MA: Addison-Wesley, 1965.

Findlay, J. N. *Plato and Platonism: An Introduction*. New York: Times Books, 1978.

Finkelstein, Joseph, ed. *Windows on a New World: The Third Industrial Revolution*. New York: Greenwood Press, 1989.

Fischler, Martin A. and Oscar Firschein. *Intelligence: The Eye, the Brain and the Computer*. Reading, MA: Addison-Wesley, 1987.

———, eds., *Readings in Computer Vision: Issues, Problems, Principles, and Paradigms*. Los Altos, CA: Morgan Kaufmann, 1987.

Fjermedal, Grant. *The Tomorrow Makers: A Brave New World of Living Brain Machines*. New York: Macmillan Publishing Company, 1986.

Flanagan, Owen. *Consciousness Reconsidered*. Cambridge, MA: MIT Press, 1992.

Flynn, Anita, Rodney A. Brooks, and Lee S. Tavrow. "Twilight Zones and Cornerstones: A Gnat Robot Double Feature." *A.I. Memo 1126*. MIT Artificial Intelligence Laboratory, 1989.

Fodor, Jerry A. *The Language of Thought*. Hassocks, UK: Harvester, 1975.

———. "Methodological Solipsism Considered as a Research Strategy in Cognitive Psychology." *Behavioral and Brain Sciences*. Vol. 3, 1980.

———. *The Modularity of Mind*. Cambridge, MA: MIT Press, 1983.

———. *Psychosemantics*. Cambridge, MA: MIT Press, 1987.

———. *Representations: Philosophical Essays on the Foundations of Cognitive Science*. Cambridge, MA. MIT Press, 1982.

———. *A Theory of Content and Other Essays*. Cambridge, MA: MIT Press, 1990.

Fogel, Lawrence J., Alvin J. Owens and Michael J. Walsh. *Artificial Evolution Through Simulated Evolution*. New York: John Wiley and Sons, 1966.

Foley, James, Andries van Dam, Steven Feiner, and John Hughes. *Computer Graphics: Principles and Practice*. Reading, MA: Addison-Wesley, 1990.

Forbes, R. J. *Studies in Ancient Technology*. 9 vols. Leiden, Netherlands: E. J. Brill, 1955–1965.

Ford, Kenneth M., Clark Glymour, and Patrick J. Hayes. *Android Epistemology*. Cambridge, MA: MIT Press, 1995.

Forester, Tom. *Computers in the Human Context*. Cambridge, MA: MIT Press, 1989.

———. *High-Tech Society: The Story of the Information Technology Revolution*. Cambridge, MA: MIT Press, 1987.

———. *The Information Technology Revolution*. Cambridge, MA: MIT Press, 1985.

———. *The Materials Revolution*. Cambridge, MA: MIT Press, 1988.

Forrest, Stephanie, ed. *Emergent Computation*. Amsterdam: North Holland, 1990.

Foster, Richard. *Innovation: The Attacker's Advantage*. New York: Summit Books, 1986.

Fowler, D. H. *The Mathematics of Plato's Academy*. Oxford: Clarendon Press, 1987.

Franke, Herbert W. *Computer Graphics–Computer Art*. Berlin: Springer-Verlag, 1985.

Franklin, Stan. *Artificial Minds*. Cambridge, MA: MIT Press, 1997.

Frauenfelder, Uli H. and Lorraine Komisarjevsky Tyler. *Spoken Word Recognition*. Cambridge, MA: MIT Press, 1987.

Freedman, David H. *Brainmakers: How Scientists Are Moving Beyond Computers to Create a Rival to the Human Brain*. New York: Simon and Schuster, 1994.

Freeman, Herbert, ed. *Machine Vision for Three-Dimensional Scenes*. Boston: Academic Press, 1990.

Freud, Sigmund. *The Interpretation of Dreams*. Reprint. London: Hogarth Press, 1955.

———. *Jokes and Their Relation to the Unconscious*. Vol. 8 of *Standard Edition of the Complete Psychological Works of Sigmund Freud*. 1905. Reprint. London: Hogarth Press, 1957.

Freudenthal, Hans. *Mathematics Observed*. Trans. Stephen Rudolfer and I. N. Baker. New York: McGraw-Hill, 1967.

Frey, Peter W., ed. *Chess Skill in Man and Machine*. New York: Springer-Verlag, 1983.

Friend, David, Alan R. Pearlman, and Thomas D. Piggott. *Learning Music with Synthesizers*. Lexington, MA: Hal Leonard, 1974.

Gamow, George. *One Two Three . . . Infinity*. Toronto: Bantam Books, 1961.

Gardner, Howard. *The Mind's New Science: A History of the Cognitive Revolution*. New York: Basic Books, 1985.

Gardner, Martin. *Time Travel and Other Mathematical Bewilderments*. New York: W. H. Freeman, 1988.

Garey, Michael R. and David S. Johnson. *Computers and Intractability*. San Francisco: W. H. Freeman, 1979.

Gates, Bill. *The Road Ahead*. New York: Viking Penguin, 1995.

Gay, Peter. *The Enlightenment: An Interpretation*. Vol. 1, *The Rise of Modern Paganism*. New York, W. W. Norton, 1966.

———. *The Enlightenment: An Interpretation*. Vol. 2, *The Science of Freedom*. New York, W. W. Norton, 1969.

Gazzaniga, Michael S. *Mind Matters: How Mind and Brain Interact to Create Our Conscious Lives*. Boston: Houghton-Mifflin Company, 1988.

Geissler, H. G. et al. *Advances in Psychology*. Amsterdam: Elsevier Science, B.V., 1983.

Geissler, Hans-George, et al. *Modern Issues in Perception*. Amsterdam: North Holland, 1983.

Gelernter, David. *Mirror Worlds: Or the Day Software Puts the Universe in a Shoebox . . . How It Will Happen and What It Will Mean*. New York: Oxford University Press, 1991.

———. *The Muse in the Machine: Computerizing the Poetry of Human Thought*. New York: The Free Press, 1994.

Gell-Mann, Murray. *The Quark and the Jaguar: Adventures in the Simple and the Complex*. New York: W. H. Freeman, 1994.

———. "Simplicity and Complexity in the Description of Nature." *Engineering & Science* 3, Spring 1988.

Ghiselin, Brewster. *The Creative Process: A Symposium*. New York: New American Library, 1952.

Gilder, George. *Life After Television*. New York: W. W. Norton and Company, 1994.

———. *The Meaning of Microcosm*. Washington, D.C.: The Progress and Freedom Foundation, 1997.

———. *Microcosm: The Quantum Revolution in Economics and Technology*. New York: Simon and Schuster, 1989.

———. *Telecosm*. New York: American Heritage Custom Publishing, 1996.

Gillispie, Charles. *The Edge of Objectivity*. Princeton, NJ: Princeton University Press, 1960.

Glass, Robert L. *Computing Catastrophes*. Seattle, WA: Computing Trends, 1983.

Gleick, James. *Chaos: Making a New Science*. New York: Viking Penguin, 1987.

Glenn, Jerome Clayton. *Future Mind: Artificial Intelligence: The Merging of the Mystical and the Technological in the 21st Century*. Washington, D.C.: Acropolis Books, 1989.

Gödel, Kurt. *On Formally Undecidable Propositions in "Principia Mathematica" and Related Systems*. New York: Basic Books, 1962.

Goldberg, David E. *Genetic Algorithms in Search, Optimization, and Machine Learning*. Reading, MA: Addison-Wesley, 1989.

Goldstine, Herman. *The Computer from Pascal to von Neumann*. Princeton, NJ: Princeton University Press, 1972.

Goleman, Daniel. *Emotional Intelligence: Why It Can Matter More Than IQ*. New York: Bantam Books, 1995.

Good, I. J. "Speculations Concerning the First Ultraintelligent Machine." *Advances in Computers*. Vol. 6. Edited by Franz L. Alt and Morris Rubinoff. Academic Press, 1965.

Goodman, Cynthia. *Digital Visions: Computers and Art*. New York: Harry N. Abrams, 1987.

Gould, Stephen J. *Ever Since Darwin*. New York: Norton, 1977.

————. *Full House: The Spread of Excellence from Plato to Darwin*. New York: Crown, 1995.

————. *Hen's Teeth and Horse's Toes*. New York: Norton, 1983.

————. *The Mismeasure of Man*. New York: Norton, 1981.

————. *Ontogeny and Phylogeny*. Cambridge, MA: Harvard University Press, 1977.

————. "Opus 200." *Natural History*, August 1991.

————. *The Panda's Thumb*. New York: Norton, 1980.

————. *Wonderful Life: The Burgess Shale and the Nature of History*. New York: Norton, 1989.

Gould, Stephen J. and Elisabeth S. Vrba. "Exaptation—A Missing Term in the Science of Form." *Paleobiology* 8:1, 1982.

Gould, Stephen J. and R. C. Lewontin. "The Spandrels of San Marco and the Panglossian Paradigm: A Critique of the Adaptationist Programme." *Proceedings of the Royal Society of London*, B 205 (1979).

Graubart, Steven R., ed. *The Artificial Intelligence Debate: False Starts, Real Foundations*. Cambridge, MA: MIT Press, 1990.

Greenberg, Donald, Aaron Marcus, Alan H. Schmidt, and Vernon Gorter. *The Computer Image: Applications of Computer Graphics*. Reading, MA: Addison-Wesley, 1982.

Greenberger, Martin, ed. *Computers and the World of the Future*. Cambridge, MA: MIT Press, 1962.

Greenblatt, R. D. et al. *The Greenblatt Chess Program*. Proceedings of the Fall Joint Computer Conference. ACM, 1967.

Gribbin, J. *In Search of Schrödinger's Cat: Quantum Physics and Reality*. New York: Bantam Books, 1984.

Grimson, W. Eric L. *Object Recognition by Computer: The Role of Geometric Constraints*. Cambridge, MA: MIT Press, 1990.

Grimson, W. Eric L. and Ramesh S. Patil, eds. *AI in the 1980s and Beyond: An MIT Survey*. Cambridge, MA: MIT Press, 1987.

Grimson, William Eric Leifur. *From Images to Surfaces: A Computational Study of the Human Early Visual System*. Cambridge, MA: MIT Press, 1981.

Grossberg, Stephen, ed. *Neural Networks and Natural Intelligence*. Cambridge, MA: MIT Press, 1988.

Grossman, Reinhardt. *Phenomenology and Existentialism: An Introduction*. London: Routledge and Kegan Paul, 1984.

Guillen, Michael. *Bridges to Infinity: The Human Side of Mathematics*. Los Angeles: Jeremy P. Tarcher, 1983.

Guthrie, W. K. C. *A History of Greek Philosophy*. 6 vols. Cambridge: Cambridge University Press, 1962–1981.

Hafner, Katie and John Markoff. *Cyberpunk: Outlaws and Hackers on the Computer Frontier*. New York: Simon and Schuster, 1991.

Halberstam, David. *The Next Century*. New York: William Morrow, 1991.

Hameroff, Stuart R., Alwyn W. Kaszniak, and Alwyn C. Scott, eds. *Toward a Science of Consciousness: The First Tucson Discussions and Debates*. Cambridge, MA: MIT Press, 1996.

Hamming, R. W. *Introduction to Applied Numerical Analysis*. New York: McGraw-Hill, 1971.

Hankins, Thomas L. *Science and the Enlightenment*. Cambridge: Cambridge University Press, 1985.

Harel, David. *Algorithmics: The Spirit of Computing*. Menlo Park, CA: Addison-Wesley, 1987.

Harman, Willis. *Global Mind Change: The New Age Revolution in the Way We Think*. New York: Warner Books, 1988.

Harmon, Paul and David King. *Expert Systems: Artificial Intelligence in Business*. New York: John Wiley and Sons, 1985.

Harre, Rom, ed. *American Behaviorial Scientist: Computation and the Mind*. Vol. 40, no. 6, May 1997.

Harrington, Steven. *Computer Graphics: A Programming Approach*. New York: McGraw-Hill, 1987.

Harris, Mary Dee. *Introduction to Natural Language Processing*. Reston, VA: Reston, 1985.

Haugeland, John. *Artificial Intelligence: The Very Idea*. Cambridge, MA: MIT Press, 1985.

——————, ed. *Mind Design: Philosophy, Psychology, Artificial Intelligence*. Cambridge, MA: MIT Press, 1981.

——————, ed. *Mind Design II: Philosophy, Psychology, Artificial Intelligence*. Cambridge, MA: MIT Press, 1997.

Hawking, Stephen W. *A Brief History of Time: From the Big Bang to Black Holes*. Toronto: Bantam Books, 1988.

Hayes-Roth, Frederick, D. A. Waterman, and D. B. Lenat, eds. *Building Expert Systems*. Reading, MA: Addison-Wesley, 1983.

Heisenberg, Werner. *Physics and Beyond: Encounters and Conversations*. New York: Harper and Row, 1971.

Hellemans, Alexander and Bryan Bunch. *The Timetables of Science*. New York: Simon and Schuster, 1988.

Herbert, Nick. *Quantum Reality*. Garden City, NY: Anchor Press, 1985.

Hildebrandt, Stefan and Anthony Tromba. *Mathematics and Optimal Form*. New York: Scientific American Books, 1985.

Hillis, W. Daniel. *The Connection Machine*. Cambridge, MA: MIT Press, 1985.

——————. "Intelligence as an Emergent Behavior; Or: The Songs of Eden," in S. R. Graubard, ed. *The Artificial Debate: False Starts and Real Foundations*. Cambridge, MA: MIT Press, 1988.

Hindle, Brooke and Steven Lubar. *Engines of Change: The American Industrial Revolution, 1790–1860*. Washington, D.C.: Smithsonian Institution Press, 1986.

Hoage, R. J. and Larry Goldman. *Animal Intelligence: Insights into the Animal Mind*. Washington, D.C.: Smithsonian Institution Press, 1986.

Hodges, Andrew. *Alan Turing: The Enigma*. New York: Simon and Schuster, 1983.

Hoel, Paul G., Sidney C. Port, and Charles J. Stone. *Introduction to Stochastic Processes*. Boston: Houghton-Mifflin, 1972.

Hofstadter, Douglas R. *Gödel, Escher, Bach: An Eternal Golden Braid*. New York: Basic Books, 1979.

——————. *Metamagical Themas: Questing for the Essence of Mind and Pattern*. New York: Basic Books, 1985.

Hofstadter, Douglas R. and Daniel C. Dennett. *The Mind's I: Fantasies and Reflections on Self and Soul*. New York: Basic Books, 1981.

Hofstadter, Douglas R., Gray Clossman, and Marsha Meredith. "Shakespeare's Plays Weren't Written by Him, but by Someone Else of the Same Name." Bloomington: Indiana University Computer Science Department Technical Report 96, 1980.

Holland, J. H., K. J. Holyoke, R. E. Nisbett, and P. R. Thagard. *Induction: Processes of Inference, Learning, and Discovery*. Cambridge, MA: MIT Press, 1986.

Hookway, Christopher, ed. *Minds, Machines, and Evolution: Philosophical Studies*. Cambridge: Cambridge University Press, 1984.

Hopper, Grace Murray and Steven L. Mandell. *Understanding Computers*. Second ed. St. Paul, MN: West Publishing Co., 1987.

Horn, Berthold Klaus Paul. *Robot Vision*. Cambridge, MA: MIT Press, 1986.

Horn, Berthold K. P. and Michael J. Brooks. *Shape from Shading*. Cambridge, MA: MIT Press, 1989.

Hsu, F. *Two Designs of Functional Units for VLSI Based Chess Machines*. Technical Report. Computer Science Department, Carnegie Mellon University, 1986.

Hubel, David H. *Eye, Brain, and Vision*. New York: Scientific American Library, 1988.

Hume, D. *Inquiry Concerning Human Understanding*. 1748. Reprint. Indianapolis, IN: Bobbs-Merrill, 1955.

Hunt, V. Daniel. *Understanding Robotics*. San Diego, CA: Academic Press, 1990.

Huxley, Aldous. *Brave New World*. New York: Harper, 1946.

Hyman, Anthony. *Charles Babbage: Pioneer of the Computer*. Oxford: Oxford University Press, 1982.

Inose, Hiroshi and John R. Pierce. *Information Technology and Civilization*. New York: W. H. Freeman, 1984.

Jacobs, François. *The Logic of Life*. New York: Pantheon Books, 1973.

James, Mike. *Pattern Recognition*. New York: John Wiley and Sons, 1988.

James, William. *The Varieties of Religious Experience*. New York: Collier Books, 1961.

Jamieson, Leah H., Dennis Gannon, and Robert J. Douglas. *The Characteristics of Parallel Algorithms*. Cambridge, MA: MIT Press, 1987.

Johnson, Mark and George Lakoff. *Metaphors We Live By*. Chicago: University of Chicago Press, 1980.

Jones, Steve. *The Language of Genes: Solving the Mysteries of Our Genetic Past, Present, and Future*. New York: Anchor Books, 1993.

Jones, W. T. *Kant and the Nineteenth Century*. Vol. 4 of *A History of Western Philosophy*. Second ed. New York: Harcourt Brace Jovanovich, 1975.

————. *The Twentieth Century to Wittgenstein and Sartre*. Vol. 5 of *A History of Western Philosophy*. Second ed. New York: Harcourt Brace Jovanovich, 1975.

Joy, Kenneth I., Charles W. Grant, Nelson L. Max, and Lansing Hatfield. *Tutorial: Computer Graphics: Image Synthesis*. Washington, D.C.: Computer Society Press, 1988.

Judson, Horace F. *The Eighth Day of Creation*. New York: Simon and Schuster, 1979.

Jung, Carl. *Memories, Dreams, Reflections*. Rev. ed. Edited by Aniela Jaffé and translated by Richard and Clara Winston. New York: Pantheon Books, 1961.

Jung, Carl, et al. *Man and His Symbols*. Garden City, NY: Doubleday, 1964.

Kaku, Michio. *Hyperspace: A Scientific Odyssey Through Parallel Universes, Time Warps, and the 10th Dimension*. New York: Anchor Books, 1995.

————. *Visions: How Science Will Revolutionize the 21st Century*. New York: Doubleday, 1997.

Kant, Immanuel. *Prolegomena to Any Future Metaphysics*. Indianapolis, IN: Bobbs-Merrill, 1950.

Kasner, Edward and James Newman. *Mathematics and the Imagination*. New York: Simon and Schuster, 1940.

Kauffman, Stuart A. "Antichaos and Adaptation." *Scientific American*, August 1991.

————. *At Home in the Universe: The Search for the Laws of Self-Organization and Complexity*. New York: Oxford University Press, 1995.

————. *The Origins of Order: Self-Organization and Selection in Evolution*. Oxford: Oxford University Press, 1993.

————. "The Sciences of Complexity and 'Origins of Order.' " Santa Fe Institute, 1991, technical report 91-04-021.

Kaufmann, William J. and Larry L. Smarr. *Supercomputing and the Transformation of Science*. New York: Scientific American Library, 1993.

Kay, Alan C. "Computers, Networks and Education." *Scientific American*, September 1991.

Kelly, Kevin. *Out of Control: The New Biology of Machines, Social Systems and the Economic World*. Reading, MA: Addison-Wesley, 1994.

Kent, Ernest W. *The Brains of Men and Machines*. Peterborough, NH: BYTE/McGraw-Hill, 1981.

Kidder, Tracy. *The Soul of a New Machine*. London: Allen-Lane. 1982.

Kirk, G. S., J. E. Raven, and M. Schofield. *The Presocratic Philosophers*. Cambridge: Cambridge University Press, 1983.

Kleene, Stephen Cole. *Introduction to Metamathematics*. New York: D. Van Nostrand, 1952.

Kline, Morris. *Mathematics and the Search for Knowledge*. Oxford: Oxford University Press, 1985.

Klivington, Kenneth A. *The Science of Mind*. Cambridge, MA: MIT Press, 1989.

Klix, Friedhart, ed. *Human and Artificial Intelligence*. Amsterdam: North Holland, 1979.

Knorr, Wilbur Richard. *The Ancient Tradition of Geometric Problems*. Boston: Birkhäuser, 1986.

Kobayashi, Koji. *Computers and Communications: A Vision of C & C*. Cambridge, MA: MIT Press, 1986.

Kohonen, Teuvo. *Self-Organization and Associative Memory*. Berlin: Springer-Verlag, 1984.

Kosslyn, Stephen M. *Image and Brain: The Resolution of the Imagery Debate*. Cambridge, MA: MIT Press, 1996.

Koza, John R. *Genetic Programming: On the Programming of Computers by Means of Natural Selection*. Cambridge, MA: MIT Press, 1992.

Krauss, Lawrence N. *The Physics of Star Trek*. New York: Basic Books, 1995.

Kullander, Sven and Borje Larsson. *Out of Sight! From Quarks to Living Cells*. Cambridge: Cambridge University Press, 1994.

Kuno, Susumu. *Functional Syntax: Anaphora, Discourse, and Empathy*. Chicago: University of Chicago Press, 1987.

Kurzweil, Raymond. *The Age of Intelligent Machines*. Cambridge, MA: MIT Press, 1990.

————. *The Age of Spiritual Machines: When Computers Exceed Human Intelligence*. New York: Viking Penguin, 1999.

————. "When Will HAL Understand What We Are Saying? Computer Speech Recognition and Understanding." Chapter in *HAL's Legacy: 2001's Computer as Dream & Reality*. Edited by David G. Stork. Cambridge, MA: MIT Press, 1996.

————. *The 10% Solution for a Healthy Life: How to Eliminate Virtually All Risk of Heart Disease and Cancer*. New York: Crown Publishers, 1993.

Lammers, Susan. *Programmers at Work: Interviews*. Redmond, WA: Microsoft Press, 1986.

Landes, David S. *Revolution in Time: Clocks and the Making of the Modern World*. Cambridge, MA: Harvard University Press, 1983.

Landreth, Bill. *Out of the Inner Circle: A Hacker's Guide to Computer Security*. Bellevue, WA: Microsoft Press, 1985.

Langley, Pat, Herbert A. Simon, Gary L. Bradshaw, and Jan M. Zytkow. *Scientific Discovery: Computational Explorations of the Creative Process*. Cambridge, MA: MIT Press, 1987.

Langton, Christopher G., ed. *Artificial Life: An Overview*. Cambridge, MA: MIT Press, 1997.

Lasserre, François. *The Birth of Mathematics in the Age of Plato*. New York: World Publishing Co., 1964.

Latil, Pierre de. *Thinking by Machine: A Study of Cybernetics*. Boston: Houghton-Mifflin, 1956.

Laver, Murray. *Computers and Social Change*. Cambridge: Cambridge University Press, 1980.

Lea, Wayne A., ed. *Trends in Speech Recognition*. Englewood Cliffs, NJ: Prentice-Hall, 1980.

Leavitt, Ruth, ed. *Artist and Computer*. Morristown, NJ: Creative Computing Press, 1976.

Lee, Kai-Fu and Raj Reddy. *Automatic Speech Recognition: The Development of the SPHINX Recognition System*. Boston: Kluwer, 1989.

Lee, Thomas F. *The Human Genome Project: Cracking the Genetic Code of Life*. New York: Plenum Press, 1991.

Leebaert, Derek, ed. *Technology 2001: The Future of Computing and Communications*. Cambridge, MA: MIT Press, 1991.

Leibniz, Gottfried Wilhelm. *Philosophical Writings*. Ed. G. H. R. Parkinson. London and Toronto: J. M. Dent and Sons, 1973.

Leibniz, Gottfried Wilhelm and Samuel Clarke. *The Leibniz-Clarke Correspondence*. Ed. H. G. Alexander. Manchester, UK: Manchester University Press, 1956.

Lenat, Douglas B. "The Heuristics of Nature: The Plausible Mutation of DNA." Stanford Heuristic Programming Project, 1980, technical report HPP-80-27.

Lenat, Douglas B. and R. V. Guha. *Building Large Knowledge-Based Systems: Representation and Inference in the CYC Project*. Reading, MA: Addison-Wesley, 1990.

Leontief, Wassily W. *The Impact of Automation on Employment, 1963–2000*, Institute for Economic Analysis, New York University, 1984.

Leontief, Wassily W. and Faye Duchin, eds. *The Future Impact of Automation on Workers*. Oxford: Oxford University Press, 1986.

Lettvin, J. Y., U. Maturana, W. McCulloch, and W. Pitts. "What the Frog's Eye Tells the Frog's Brain." *Proceedings of the IRE*, 47 (1959).

Levy, Steven. *Artificial Life: The Quest for a New Creation*. New York: Pantheon Books, 1992.

———. *Hackers: Heroes of the Computer Revolution*. Garden City, NY: Anchor Press/ Doubleday, 1968.

Lewin, Roger. *Complexity: Life at the Edge of Chaos*. New York: Macmillan, 1992.

———. *In the Age of Mankind: A Smithsonian Book of Human Evolution*. Washington, D.C.: Smithsonian Books, 1988.

———. *Thread of Life: The Smithsonian Looks at Evolution*. Washington, D.C.: Smithsonian Books, 1982.

Lieff, Jonathan D. (M.D.) *Computer Applications in Psychiatry*. Washington, D.C.: American Psychiatric Press, 1987.

Lloyd, G. E. R. *Aristotle: The Growth and Structure of His Thought*. Cambridge: Cambridge University Press, 1968.

———. *Early Greek Science: Thales to Aristotle*. New York: W. W. Norton, 1970.

Locke, John. *Essay Concerning Human Understanding*. London, 1690.

Lord, Norman W. and Paul A. Guagosian. *Advanced Computers: Parallel and Biochip Processors*. Ann Arbor, MI: Ann Arbor Science, Butterworth Group, 1983.

Lowe, David G. *Perceptual Organization and Visual Recognition*. Boston: Kluwer Academic, 1985.

Lubar, Steven. *InfoCulture: The Smithsonian Book of Information Age Inventions*. Boston: Houghton-Mifflin Company, 1993.

Luce, R. D. and H. Raiffa. *Games and Decisions*. New York: John Wiley and Sons, 1957.

Lucky, Robert W. *Silicon Dreams: Information, Man, and Machine*. New York: St. Martin's Press, 1989.

MacEy, Samuel L. *Clocks and the Cosmos: Time in Western Life and Thought*. Hamden: Archon Books, 1980.

Maes, Pattie. *Designing Autonomous Agents*. Cambridge, MA: MIT Press, 1991.

Magnenat-Thalmann, Nadia and Daniel Thalmann. *Computer Animation: Theory and Practice*. Tokyo: Springer-Verlag, 1985.

Malcolm, Norman. *Ludwig Wittgenstein: A Memoir, with a Biographical Sketch by Georg Henrik Von Wright*. Oxford: Oxford University Press, 1958.

Mamdani, E. H. and B. R. Gaines. *Fuzzy Reasoning and Its Applications*. London: Academic Press, 1981.

Mandelbrot, Benoit B. *The Fractal Geometry of Nature*. New York: W. H. Freeman, 1988.
———. *Fractals: Form, Chance, and Dimension*. San Francisco: W. H. Freeman, 1977.

Mander, Jerry. *In the Absence of the Sacred: The Failure of Technology and the Survival of the Indian Nations*. San Francisco: Sierra Club Books, 1992.

Margulis, Lynn and Dorion Sagan. *Microcosmos: Four Billion Years of Evolution from Our Microbial Ancestors*. New York: Summit Books, 1986.

Markle, Sandra and William Markle. *In Search of Graphics: Adventures in Computer Art*. New York: Lothrop, Lee and Shepard Books, 1985.

Markoff, John. "The Creature That Lives in Pittsburgh." *New York Times*, April 21, 1991.

Markov, A. *The Theory of Algorithms*. Moscow: National Academy of Sciences, USSR, 1954.

Marr, D. *Vision*. New York: W. H. Freeman, 1982.

Martin, James and Steven Oxman. *Building Expert Systems: A Tutorial*. Englewood Cliffs, NJ: Prentice-Hall, 1988.

Martin, William A., K. W. Church, and R. S. Patil. "Preliminary Analysis of a Breadth-First Parsing Algorithm: Theoretical and Experiential Results." Cambridge, MA: MIT Laboratory for Computer Science, 1981.

Marx, Leo. *The Machine in the Garden: Technology and the Pastoral Ideal in America*. London: Oxford University Press, 1964.

Mason, Matthew T. and Kenneth Salisbury, Jr. *Robot Hands and the Mechanics of Manipulation*. Cambridge, MA: MIT Press, 1985.

Massaro, D. W., et al. *Letter and Word Perception: Orthographic Structure and Visual Processing in Reading*. Amsterdam: North Holland, 1980.

Mathews, Max V. *The Technology of Computer Music*. Cambridge, MA: MIT Press, 1969.

Mayr, Ernst. *Animal Species and Evolution*. Cambridge, MA: Harvard University Press, 1963.
———. *Toward a New Philosophy of Biology*. Cambridge, MA: Harvard University Press, 1988.

Mayr, Otto. *Authority, Liberty, and Automatic Machinery in Early Modern Europe*. Baltimore, MD: Johns Hopkins University Press, 1986.

Mazlish, Bruce. *The Fourth Discontinuity: The Co-Evolution of Humans and Machines*. New Haven, CT: Yale University Press, 1993.

McClelland, James L. and David E. Rumelhart. *Parallel Distributed Processing: Explorations in the Microstructure of Cognition Volume 1*. Cambridge, MA: MIT Press, 1986.
———. *Parallel Distributed Processing: Explorations in the Microstructure of Cognition Volume 2*. Cambridge, MA: MIT Press, 1986.

McCorduck, Pamela. *Aaron's Code: MetaArt, Artificial Intelligence, and the Work of Harold Cohen*. New York: W. H. Freeman, 1991.
———. *Machines Who Think: A Personal Inquiry into the History and Prospects of Artificial Intelligence*. San Francisco: W. H. Freeman, 1979.

McCulloch, Warren S. *An Account of the First Three Conferences of Teleological Mechanisms*. Josiah Macy, Jr. Foundation, 1947.

———. *Embodiments of Mind*. Cambridge, MA: MIT Press, 1965.

McLuhan, Marshall. *The Medium Is the Message*. New York: Bantam Books, 1967.

———. *Understanding Media: The Extension of Man*. New York: McGraw-Hill, 1964.

McRae, Hamish. *The World in 2020: Power, Culture, and Prosperity*. Cambridge, MA: Harvard Business School Press, 1994.

Mead, Carver. *Analog VLSI Implementation of Neural Systems*. Reading, MA: Addison-Wesley, 1989.

Mead, Carver and Lynn Conway. *Introduction to VLSI Systems*. Reading, MA: Addison-Wesley, 1980.

Meisel, William S. *Computer-Oriented Approaches to Pattern Recognition*. New York: Academic Press, 1972.

Mel, Bartlett W. *Connectionist Robot Motion Planning: A Neurally-Inspired Approach to Visually-Guided Reaching*. Boston: Academic Press, 1990.

Metropolis, N., J. Howlett, and Gian-Carlo Rota, eds. *A History of Computing in the Twentieth Century*. New York: Academic Press, 1980.

Miller, Eric, ed. *Future Vision: The 189 Most Important Trends of the 1990s*. Naperville, IL: Sourcebooks Trade, 1991.

Minsky, Marvin. *Computation: Finite and Infinite Machines*. Englewood Cliffs, NJ: Prentice-Hall, 1967.

———. "A Framework for Representing Knowledge." In *The Psychology of Computer Vision*, edited by P. H. Winston. New York: McGraw-Hill, 1975.

———. *The Society of Mind*. New York: Simon and Schuster, 1985.

———, ed. *Robotics*. New York: Doubleday, 1985.

———, ed. *Semantic Information Processing*. Cambridge, MA: MIT Press, 1968.

Minsky, Marvin and Seymour A. Papert. *Perceptrons: An Introduction to Computational Geometry*. Cambridge, MA: MIT Press, 1969 (revised edition, 1988).

Mitchell, Melanie. *An Introduction to Genetic Algorithms*. Cambridge, MA: MIT Press, 1996.

Mohr, Richard R. *The Platonic Cosmology*. Leiden, Netherlands: E. J. Brill, 1985.

Moore, Thomas J. *Lifespan: New Perspectives on Extending Human Longevity*. New York: Simon and Schuster, 1993.

Moore, Walter. *Schrödinger: Life and Thought*. Cambridge: Cambridge University Press, 1989.

Moravec, Hans. *Mind Children: The Future of Robot and Human Intelligence*. Cambridge, MA: Harvard University Press, 1988.

Morgan, Christopher P., ed. *The "Byte" Book of Computer Music*. Peterborough, NH: Byte Books, 1979.

Morowitz, Harold J. and Jerome L. Singer. *The Mind, the Brain, and Complex Adaptive Systems*. Reading, MA: Addison-Wesley, 1995.

Morris, Desmond. *The Naked Ape: A Zoologist's Study of the Human Animal*. New York: McGraw-Hill, 1967.

Morse, Stephen S., ed. *Emerging Viruses*. Oxford: Oxford University Press, 1997.

Mumford, Lewis. *The Myth of the Machine: Technics and Human Development*. New York: Harcourt Brace and World, 1967.

Murphy, Pat. *By Nature's Design*. San Francisco: Chronicle Books, 1993.

Murray, David W. and Bernard F. Buxton. *Experiments in the Machine Interpretation of Visual Motion*. Cambridge, MA: MIT Press, 1990.

Myers, Terry, John Laver, and John Anderson, eds. *The Cognitive Representation of Speech*. Amsterdam: North Holland, 1981.

Naisbitt, John. *Global Paradox: The Bigger the World Economy, the More Powerful Its Smallest Players*. New York: William Morrow, 1994.

Naisbitt, John and Patricia Aburdene. *Megatrends 2000: Ten New Directions for the 1990's*. New York: William Morrow, 1990.

———. *Re-Inventing the Corporation: Transforming Your Job and Your Company for the New Information Society*. New York: Warner Books, 1985.

Nayak, P. Ranganath and John M. Ketteringham. *Breakthroughs! How the Vision and Drive of Innovators in Sixteen Companies Created Commercial Breakthroughs That Swept the World*. New York: Arthur D. Little, 1986.

Negroponte, Nicholas. *Being Digital*. New York: Alfred A. Knopf, 1995.

———. "Products and Services for Computer Networks." *Scientific American*, September 1991.

Neuberger, A. P. *The Technical Arts and Sciences of the Ancients*. London: Methuen, 1930.

Newell, Allen. *Intellectual Issues in the History of Artificial Intelligence*. Pittsburgh, PA: Carnegie Mellon University, 1982.

———. *The Unified Theories of Cognition*. Cambridge, MA: Harvard University Press, 1990.

Newell, Allen and Herbert A. Simon. *Human Problem Solving*. Englewood Cliffs, NJ: Prentice-Hall, 1972.

Newell, Allen, et al. "Speech Understanding Systems: Final Report of a Study Group." Computer Science Department, Carnegie Mellon University, Pittsburgh, May 1971.

Newmeyer, Frederick J. *Linguistic Theory in America*. Second ed. Orlando, FL: Academic Press, 1986.

Newton, Isaac. *Philosophiae Naturalis Principia Mathematica*. Third ed. Cambridge, MA: Harvard University Press, 1972. Original, 1726.

Nierenberg, Gerard. *The Art of Creative Thinking*. New York: Simon and Schuster, 1982.

Nilsson, Lennart. *The Body Victorious: The Illustrated Story of Our Immune System and Other Defenses of the Human Body*. Trans. Clare James. New York: Delacorte Press, 1985.

Nilsson, Nils J. *Principles of Artificial Intelligence*. Los Altos, CA: Morgan Kaufmann, 1980.

Nilsson, Nils J. and Bonnie Lynn Webber. *Readings in Artificial Intelligence*. Los Altos, CA: Morgan Kaufmann, 1985.

Nocera, Joseph. *A Piece of the Action: How the Middle Class Joined the Money Class*. New York: Simon and Schuster, 1994.

Norretranders, Tor. *The User Illusion: Cutting Consciousness Down to Size*. New York: Viking, 1998.

O'Keefe, Bernard J. *Nuclear Hostages*. Boston: Houghton-Mifflin Company, 1983.

Oakley, D. A., ed. *Brain and Mind*. London and New York: Methuen, 1985.

Oliver, Dick. *FractalVision: Put Fractals to Work for You*. Carmel, IN: Sams Publishing, 1992.

Ornstein, Robert. *The Evolution of Consciousness: Of Darwin, Freud, and Cranial Fire; the Origins of the Way We Think*. New York: Prentice-Hall Press, 1991.

———. *The Mind Field*. London: Octagon Press, 1976.

———. *Multimind: A New Way of Looking at Human Behavior*. Boston: Houghton-Mifflin, 1986.

———. *On the Experience of Time*. London: Penguin Books, 1969.

———. *The Psychology of Consciousness*. Second ed. New York: Harcourt Brace Jovanovich, 1972.

———, ed. *The Nature of Human Consciousness: A Book of Readings*. New York: Viking, 1973.

Ornstein, Robert and Paul Ehrlich. *New World, New Mind: Moving Toward Conscious Evolution*. New York: Doubleday, 1989.

Ornstein, Robert and D. S. Sobel. *The Healing Brain*. New York: Simon and Schuster, 1987.

Ornstein, Robert and Richard F. Thompson. *The Amazing Brain*. Boston: Houghton-Mifflin, 1984.

Osherson, Daniel N., Michael Stob, and Scott Weinstein. *Systems That Learn: An Introduction to Learning Theory for Cognitive and Computer Scientists*. Cambridge, MA: MIT Press, 1986.

Ouellette, Pierre. *The Deus Machine*. New York: Villard Books, 1994.

Owen, G. *The Universe of the Mind*. Baltimore, MD: Johns Hopkins University Press, 1971.

Pagels, Heinz R. *The Cosmic Code: Quantum Physics as the Language of Nature*. New York: Bantam Books, 1983.

———. *The Dreams of Reason: The Computer and the Rise of the Sciences of Complexity*. New York: Bantam Books, 1988.

———. *Perfect Symmetry: The Search for the Beginning of Time*. New York: Bantam Books, 1986.

Papert, Seymour. *The Children's Machine: Rethinking School in the Age of the Computer*. New York: Basic Books, 1993.

———. *Mindstorms: Children, Computers, and Powerful Ideas*. New York: Basic Books, 1980.

Pascal, Blaise. *Pensées*. New York: E. P. Dutton, 1932. Original, 1670.

Paul, Gregory S. and Earl D. Cox. *Beyond Humanity: CyberEvolution and Future Minds*. Rockland, MA: Charles River Media, 1996.

Paul, Richard P. *Robot Manipulators: Mathematics, Programming, and Control*. Cambridge, MA: MIT Press, 1981.

Paulos, John Allen. *Beyond Numeracy: Ruminations of a Number Man*. New York: Alfred A. Knopf, 1991.

Pavlov, I. P. *Conditioned Reflexes*. London: Oxford University Press, 1927.

Peat, F. David. *Artificial Intelligence: How Machines Think*. New York: Baen Enterprises, 1985.

———. *Synchronicity: The Bridge Between Matter and Mind*. Toronto: Bantam Books, 1987.

Peitgen, H. O., D. Saupe, et al. *The Science of Fractal Images*. New York: Springer-Verlag, 1988.

Peitgen, H. O. and P. H. Richter. *The Beauty of Fractals: Images of Complex Dynamical Systems*. Berlin: Springer-Verlag, 1986.

Penfield, W. *The Mystery of the Mind*. Princeton, NJ: Princeton University Press, 1975.

Penrose, R. and C. J. Isham, eds. *Quantum Concepts in Space and Time*. Oxford: Oxford University Press: 1986.

Penrose, Roger. *The Emperor's New Mind: Concerning Computers, Minds, and the Laws of Physics*. New York: Oxford University Press, 1989.

———. *Shadows of the Mind*. Oxford: Oxford University Press, 1994.

Pentland, Alex P., ed. *From Pixels to Predicates: Recent Advances in Computational and Robotic Vision*. Norwood, NJ: Ablex Publishing Corporation, 1986.

Peterson, Dale. *Genesis II: Creation and Recreation with Computers*. Reston, VA: Reston Publishing Co., 1983.

Petroski, Henry. *To Engineer Is Human: The Role of Failure in Successful Design*. New York: St. Martin's Press, 1985.

Piaget, Jean. *The Psychology of Intelligence*. London: Routledge and Kegan Paul, 1967.

Pickover, Clifford A. *Computers and the Imagination: Visual Adventures Beyond the Edge*. New York: St. Martin's Press, 1991.

Pierce, John R. *The Science of Musical Sound*. New York: Scientific American Books, 1983.

Pines, David, ed. *Emerging Syntheses in Science*. Reading, MA: Addison-Wesley, 1988.

Pinker, Steven. *How the Mind Works*. New York: W. W. Norton and Company, 1997.

———. *The Language Instinct*. New York: William Morrow, 1994.

———. *Language Learnability and Language Development*. Cambridge, MA: Harvard University Press, 1984.

———. *Learnability and Cognition: The Acquisition of Argument Structure*. Cambridge, MA: MIT Press, 1989.

———, ed. *Visual Cognition*. Cambridge, MA: MIT Press, 1984.

Pinker, Steven and J. Mehler, eds. *Connections and Symbols*. Cambridge, MA: MIT Press, 1988.

Plato. *Epinomis*. The Loeb Classical Library. Ed. W. R. M. Lamb. Vol. 8. New York: G. P. Putnam's Sons, 1927.

———. *Protagoras and Meno*. Baltimore, MD: Penguin Books, 1956.

———. *Timaeus*. Indianapolis, IN: Bobbs-Merrill, 1959.

Pollock, John. *How to Build a Person: A Prolegomenon*. Cambridge, MA: MIT Press, 1989.

Poole, Robert M. *The Incredible Machine*. Washington, D.C.: The National Geographic Society, 1986.

Poppel, Ernst. *Mindworks: Time and Conscious Experience*. Boston: Harcourt Brace Jovanovich, 1988.

Popper, Karl and John Eccles. *The Self and Its Brain*. Berlin, London: Springer-Verlag, 1977.

Posner, Michael I. and Marcus E. Raichle. *Images of Mind*. New York: Scientific American Library, 1994.

Potter, Jerry L., ed. *The Massively Parallel Processor*. Cambridge, MA: MIT Press, 1985.

Poundstone, William. *Prisoner's Dilemma*. New York: Doubleday, 1992.

———. *The Recursive Universe: Cosmic Complexity and the Limits of Scientific Knowledge*. New York: William Morrow, 1985.

Pratt, Vernon. *Thinking Machines: The Evolution of Artificial Intelligence*. New York: Basil Blackwell, 1987.

Pratt, William K. *Digital Image Processing*. New York: John Wiley and Sons, 1978.

Price, Derek J. de Solla. *Gears from the Greeks: The Antikythera Mechanism—A Calendar Computer from Circa 80 B.C.* New York: Science History Publications, 1975.

Prigogine, Ilya. *The End of Certainty: Time's Flow and the Laws of Nature*. New York: Simon and Schuster, 1997.

Prueitt, Melvin L. *Art and the Computer*. New York: McGraw-Hill, 1984.

Prusinkiewicz, Przemyslaw and Aristid Lindenmayer. *The Algorithmic Beauty of Plants*. New York: Springer-Verlag, 1990.

Rabiner, Lawrence R. and Ronald W. Schafer. *Digital Processing of Speech Signals*. Englewood Cliffs, NJ: Prentice-Hall, 1978.

RACTER. *The Policeman's Beard Is Half Constructed: Computer Prose and Poetry by RACTER*. [William Chamberlain and Joan Hall.] New York: Warner Books, 1984.

Radford, Andrew. *Transformational Syntax: A Student's Guide to Chomsky's Extended Standard Theory*. Cambridge: Cambridge University Press, 1981.

Raibert, Marc H. *Legged Robots That Balance*. Cambridge, MA: MIT Press, 1986.

Randell, Brian, ed. *The Origins of Digital Computers: Selected Papers*. New York: Springer-Verlag, 1975.

Raphael, Bertram. *The Thinking Computer: Mind Inside Matter*. San Francisco: W. H. Freeman, 1976.

Rasmussen, S., et al. "Computational Connectionism Within Neurons: A Model of Cytoskeletal Automata Subserving Neural Networks," *Emergent Computation*. Edited by Stephanie Forrest. Cambridge, MA: MIT Press, 1991.

Raup, David M. *Extinction: Bad Genes or Bad Luck?* New York: W. W. Norton, 1991.

Rawlings, Gregory J. E. *Moths to the Flame: The Seductions of Computer Technology*. Cambridge, MA: MIT Press, 1996.

Rée, Jonathan. *Descartes*. New York: Pica Press, 1974.

Reichardt, Jasia. *Robots: Fact, Fiction and Prediction*. Middlesex, UK: Penguin Books, 1978.

Reid, Robert H. *Architects of the Web: 1,000 Days That Built the Future of Business*. New York: John Wiley and Sons, 1997.

Restak, Richard M. (M.D.) *The Brain*. Toronto: Bantam Books, 1984.

Rheingold, Howard. *Virtual Reality*. New York: Summit Books, 1991.

Rich, Elaine. *Artificial Intelligence*. New York: McGraw-Hill, 1983.

Rich, Elaine and Kevin Knight. *Artificial Intelligence*. Second ed. New York: McGraw-Hill, 1991.

Ringle, Martin D., ed. *Philosophical Perspectives in Artificial Intelligence*. Brighton, Sussex: Harvester Press, 1979.

Roads, Curtis, ed. *Composers and the Computer*. Los Altos, CA: William Kaufmann, 1985.

————, ed. *The Music Machine: Selected Readings from "Computer Music Journal."* Cambridge, MA: MIT Press, 1988.

Roads, Curtis and John Strawn. *Foundations of Computer Music*. Cambridge, MA: MIT Press, 1989.

Robin, Harry and Daniel J. Kevles. *The Scientific Image: From Cave to Computer*. New York: Harry N. Abrams, 1992.

Rock, Irvin. *Perception*. New York: Scientific American Books, 1984.

Rogers, David F. and Rae A. Ernshaw, eds. *Computer Graphics Techniques: Theory and Practice*. New York: Springer-Verlag, 1990.

Rose, Frank. *Into the Heart of the Mind: An American Quest for Artificial Intelligence*. New York: Vintage Books, 1984.

Rosenberg, Jerry M. *Dictionary of Artificial Intelligence and Robotics*. New York: John Wiley and Sons, 1986.

Rosenblatt, Frank. *Principles of Neurodynamics*. New York: Spartan, 1962.

Rosenfield, Israel. *The Invention of Memory: A New View of the Brain*. New York: Basic Books, 1988.

Rothchild, Joan, ed. *Machina ex Dea: Feminist Perspectives on Technology*. New York: Pergamon Press, 1982.

Rothschild, Michael. *Bionomics: The Inevitability of Capitalism*. New York: Henry Holt and Company, 1990.

Rucker, Rudy. *Infinity and the Mind*. Boston: Birkhäuser, 1982.

————. *Mind Tools: The Five Levels of Mathematical Reality*. Boston: Houghton-Mifflin Company, 1987.

————. *Software*. Middlesex, UK: Penguin Books, 1983.

Rumelhart, D. E., J. L. McClelland, and the PDP Research Group. *Parallel Distributed Processing*. Vols. 1 and 2. Cambridge, MA: MIT Press, 1982.

Russell, Bertrand. *The ABC of Relativity*. Fourth ed. 1925. Reprint. London: Allen and Unwin, 1985.

————. *The Autobiography of Bertrand Russell: 1872–1914*. Toronto: Bantam Books, 1967.

————. *The Autobiography of Bertrand Russell: 1914–1944*. Toronto: Bantam Books, 1968.

————. *A History of Western Philosophy*. New York: Simon and Schuster, 1945.

————. *Introduction to Mathematical Philosophy*. New York: Macmillan, 1919.

————. *Mysticism and Logic*. New York: Doubleday Anchor Books, 1957.

————. *The Principles of Mathematics*. Reprint. New York: W. W. Norton & Company, 1996.

————. *The Problems of Philosophy*. New York: Oxford University Press, 1959.

Russell, Peter. *The Global Brain: Speculations on the Evolutionary Leap to Planetary Consciousness*. Los Angeles: J. P. Tarcher, 1976.

Sabbagh, Karl. *The Living Body*. London: Macdonald & Company, 1984.

Sacks, Oliver. *The Man Who Mistook His Wife for a Hat and Other Clinical Tales*. New York: Harper and Row, 1985.

Sagan, Carl. *Contact*. New York: Simon and Schuster, 1985.

————. *The Dragons of Eden: Speculations on the Evolution of Human Intelligence*. New York: Ballantine Books, 1977.

————, ed. *Communication with Extraterrestrial Intelligence*. Cambridge, MA: MIT Press, 1973.

Sambursky, S. *The Physical World of the Greeks*. London: Routledge and Kegan Paul, 1963. Original, 1956.

Sanderson, George and Frank Mcdonald, eds. *Marshall McLuhan: The Man and His Message*. Golden, CO: Fulcrum, 1989.

Saunders, Peter T. "The Complexity of Organisms." *Evolutionary Theory: Paths into the Future*, edited by J. W. Pollard. New York: John Wiley and Sons, 1984.

Savage, John E., Susan Magidson, and Alex M. Stein. *The Mystical Machine: Issues and Ideas in Computing*. Reading, MA: Addison-Wesley, 1986.

Saxby, Graham. *Holograms: How to Make and Display Them*. London: Focal Press, 1980.

Sayre, Kenneth M. and Frederick J. Crosson. *The Modeling of Mind: Computers and Intelligence*. New York: Simon and Schuster, 1963.

Schank, Roger. *The Creative Attitude: Learning to Ask and Answer the Right Questions*. New York: Macmillan Publishing Company, 1988.

————. *Dynamic Memory: A Theory of Reminding and Learning in Computers and People*. Cambridge: Cambridge University Press, 1982.

————. *Tell Me a Story: A New Look at Real and Artificial Memory*. New York: Charles Scribner's Sons, 1990.

Schank, Roger C. and Kenneth Mark Colby., eds. *Computer Models of Thought and Language*. San Francisco: W. H. Freeman, 1973.

Schank, Roger [with Peter G. Childers]. *The Cognitive Computer: On Language, Learning, and Artificial Intelligence*. Reading, MA: Addison-Wesley, 1984.

Schilpp, P. A., ed. *The Philosophy of Bertrand Russell*. Chicago: Chicago University Press, 1944.

Schön, Donald A. *Educating the Reflective Practitioner: Toward a New Design for Teaching and Learning in the Professions*. San Francisco: Jossey-Bass, 1987.

Schorr, Herbert and Alain Rappaport, eds. *Innovative Applications of Artificial Intelligence*. Menlo Park, CA: AAAI Press, 1989.

Schrödinger, Erwin. *What Is Life?* Cambridge: Cambridge University Press, 1967.

Schull, Jonathan. "Are Species Intelligent?" *Behavioral and Brain Sciences* 13:1 (1990).

Schulmeyer, G. Gordon. *Zero Defect Software*. New York: McGraw-Hill, 1990.

Schwartz, Lillian F. *The Computer Artist's Handbook: Concepts, Techniques, and Applications*. New York: W. W. Norton and Company, 1992.

Searle, John R. "Minds, Brains, and Programs." *The Behavioral and Brain Sciences*. Vol. 3. Cambridge: Cambridge University Press, 1980.

————. *Minds, Brains and Science*. Cambridge, MA: Harvard University Press, 1985.

————. *The Rediscovery of the Mind*. Cambridge, MA: MIT Press, 1992.

Sejnowski, T. and C. Rosenberg. "Parallel Networks That Learn to Pronounce English Text." *Complex Systems* 1 (1987).

Serra, Jean, ed. *Image Analysis and Mathematical Morphology*. Vol. 1. London: Academic Press, 1988.

————, ed. *Image Analysis and Mathematical Morphology*. Vol. 2: Theoretical Advances. London: Academic Press, 1988.

Shapiro, Stuart D., ed. *Encyclopedia of Artificial Intelligence*. 2 vols. New York: John Wiley and Sons, 1987.

Sharples, M. D., et al. *Computers and Thought: A Practical Introduction to Artificial Intelligence*. Cambridge, MA: MIT Press, 1989.

Shear, Jonathan, ed. *Explaining Consciousness—The "Hard" Problem*. Cambridge, MA: MIT Press, 1995–1997.

Shortliffe, E. *MYCIN: Computer-Based Medical Consultations*. New York: American Elsevier, 1976.

Shurkin, Joel. *Engines of the Mind: A History of the Computer*. New York: W. W. Norton, 1984.

Siekmann, Jorg and Graham Wrightson. *Automation of Reasoning 1: Classical Papers on Computational Logic 1957–1966*. Berlin: Springer-Verlag, 1983.

————. *Automation of Reasoning 2: Classical Papers on Computational Logic 1967–1970*. Berlin: Springer-Verlag, 1983.

Simon, Herbert A. *Models of My Life*. New York: Basic Books, 1991.

————. *The Sciences of the Artificial*. Cambridge, MA: MIT Press, 1969.

Simon, Herbert A. and Allen Newell. "Heuristic Problem Solving: The Next Advance in Operations Research." *Operations Research*. Vol. 6. 1958.

Simon, Herbert A. and L. Siklossy, eds. *Representation and Meaning: Experiments with Information Processing Systems*. Englewood Cliffs, NJ: Prentice-Hall, 1972.

Simpson, George Gaylord. *The Meaning of Evolution*. The New American Library of World Literature. New York: A Mentor Book, 1951.

Singer, C., E. J. Holmyard, A. R. Hall, and T. I. Williams, eds. *A History of Technology*. 5 vols. Oxford: Oxford University Press, 1954–1958.

Singer, Michael A. *The Search for Truth*. Alachua, FL: Shanti Publications, 1974.

Slater, Robert. *Portraits in Silicon*. Cambridge, MA: MIT Press, 1987.

Smith, John Maynard. *Did Darwin Get It Right? Essays on Games, Sex and Evolution*. New York: Chapman and Hall, 1989.

Smullyan, Raymond. *Forever Undecided: A Puzzle Guide to Gödel*. New York: Alfred A. Knopf, 1987.

Solso, Robert L. *Mind and Brain Sciences in the 21st Century*. Cambridge, MA: MIT Press, 1997.

Soltzberg, Leonard J. *Sing a Song of Software: Verse and Images for the Computer-Literate*. Los Altos, CA: William Kaufmann, 1984.

Soucek, Branko and Marina Soucek. *Neural and Massively Parallel Computers: The Sixth Generation*. New York: John Wiley and Sons, 1988.

Spacks, Barry. *The Company of Children*. Garden City, NY: Doubleday and Company, 1969.

Spinosa, Charles, Hubert L. Dreyfus, and Fernando Flores. *Disclosing New Worlds: Entrepreneurship, Democratic Action, and the Cultivation of Solidarity*. Cambridge, MA: MIT Press, 1997.

Stahl, Franklin W. *The Mechanics of Inheritance*. Englewood Cliffs, NJ: Prentice-Hall, 1964, 1969.

Stein, Dorothy. *Ada: A Life and a Legacy*. Cambridge, MA: MIT Press, 1985.

Sternberg, Robert J., ed. *Handbook of Human Intelligence*. Cambridge: Cambridge University Press, 1982.

Sternberg, Robert J. and Douglas K. Detterman, eds. *What Is Intelligence? Contemporary Viewpoints on its Nature and Definition*. Norwood, NJ: Ablex Publishing Corporation, 1986.

Stewart, Ian. *Does God Play Dice?* New York: Basil Blackwell, 1989.

Stock, Gregory. *Metaman: The Merging of Humans and Machines into a Global Super-organism*. New York: Simon and Schuster, 1993.

Stork, David G. *HAL's Legacy: 2001's Computer as Dream and Reality*. Cambridge, MA: MIT Press, 1996.

Strassmann, Paul A. *Information Payoff: The Transformation of Work in the Electronic Age*. New York: The Free Press, 1985.

Talbot, Michael. *The Holographic Universe*. New York: HarperCollins, 1991.

Tanimoto, Steven L. *The Elements of Artificial Intelligence: An Introduction Using LISP*. Rockville, MD: Computer Science Press, 1987.

Taylor, F. Sherwood. *A Short History of Science and Scientific Thought*. New York: W. W. Norton and Company, 1949.

Taylor, Philip A., ed. *The Industrial Revolution in Britain: Triumph or Disaster?* Lexington, MA: Heath, 1970.

Thearling, Kurt. "How We Will Build a Machine That Thinks." A Workshop at Thinking Machines Corporation, August 24–26, 1992.

Thomas, Abraham. *The Intuitive Algorithm*. New Delhi: Affiliated East-West PVT, 1991.

Thomis, Malcolm I. *The Luddites: Machine Breaking in Regency England*. Hamden, CT: Archon Books, 1970.

Thorpe, Charles E. *Vision and Navigation: The Carnegie Mellon Navlab*. Norwell, MA: Kluwer Academic, 1990.

Thurow, Lester C. *The Future of Capitalism: How Today's Economic Forces Shape Tomorrow's World*. New York: William Morrow, 1996.

Time-Life Books. *Computer Images*. Alexandria, VA: Time-Life Books, 1986.

Tjepkema, Sandra L. *A Bibliography of Computer Music: A Reference for Composers*. Iowa City: University of Iowa Press, 1981.

Toepperwein, L. L., et al. *Robotics Applications for Industry: A Practical Guide*. Park Ridge: Noyes Data Corporation, 1983.

Toffler, Alvin. *Powershift*. New York: Bantam Books, 1990.

———. *The Third Wave: The Classic Study of Tomorrow*. New York: Bantam Books, 1980.

Toffoli, Tommaso and Norman Margolis. *Cellular Automata Machines: A New Environment for Modeling*. Cambridge, MA: MIT Press, 1987.

Torrance, Stephen B., ed. *The Mind and the Machine: Philosophical Aspects of Artificial Intelligence*. Chichester, UK: Ellis Horwood, 1986.

Traub, Joseph F., ed. *Cohabiting with Computers*. Los Altos, CA: William Kaufmann, 1985.

Truesdell, L. E. *The Development of Punch Card Tabulation in the Bureau of the Census, 1890–1940*. Washington, D.C.: Government Printing Office, 1965.

Tufte, Edward R. *The Visual Display of Quantitative Information*. Cheshire, CT: Graphics Press, 1983.

———. *Visual Explanations: Images and Quantities, Evidence and Narrative*. Cheshire, CT: Graphics Press, 1997.

Turing, Alan. "Computing Machinery and Intelligence." Reprinted in *Minds and Machines*, edited by Alan Ross Anderson. Englewood Cliffs, NJ: Prentice-Hall, 1964.

———. "On Computable Numbers, with an Application to the *Entscheidungsproblem*" Proceedings, London Mathematical Society, 2, no. 42 (1936).

Turkle, Sherry. *The Second Self: Computers and the Human Spirit*. New York: Simon and Schuster, 1984.

Tye, Michael. *Ten Problems of Consciousness: A Representational Theory of the Phenomenal Mind*. Cambridge, MA: MIT Press, 1995.

Ullman, Shimon. *The Interpretation of Visual Motion*. Cambridge, MA: MIT Press, 1982.

Usher, A. P. *A History of Mechanical Inventions*. Second ed. Cambridge, MA: Harvard University Press, 1958.

Vaina, Lucia and Jaakko Hintikka, eds. *Cognitive Constraints on Communication*. Dordrecht, Netherlands: Reidel, 1985.

Van Heijenoort, Jean, ed. *From Frege to Gödel*. Cambridge, MA: Harvard University Press, 1967.

Varela, Francisco J., Evan Thompson, and Eleanor Rosch. *The Embodied Mind: Cognitive Science and Human Experience*. Cambridge, MA: MIT Press, 1991.

Vigne, V. "Technological Singularity." *Whole Earth Review*, Winter 1993.

von Neumann, John. *The Computer and the Brain*. New Haven, CT: Yale University Press, 1958.

Waddington, C. H. *The Strategy of the Genes*. London: George Allen and Unwin, 1957.

Waldrop, M. Mitchell. *Complexity: The Emerging Science at the Edge of Order and Chaos*. New York: Simon and Schuster, 1992.

————. *Man-Made Minds: The Promise of Artificial Intelligence*. New York: Walker and Company, 1987.

Waltz, D. "Massively Parallel AI." Paper presented at the American Association of Artificial Intelligence (AAAI) conference, August 1990.

Waltz, David. *Connectionist Models and Their Implications: Readings from Cognitive Science*. Norwood, NJ: Ablex, 1987.

Wang, Dr. An. *Lessons: An Autobiography*. Reading, MA: Addison-Wesley, 1986.

Wang, Hao. *A Logical Journey: From Gödel to Philosophy*. Cambridge, MA: MIT Press, 1996.

Warrick, Patricia S. *The Cybernetic Imagination in Science Fiction*. Cambridge, MA: MIT Press, 1980.

Watanabe, Satoshi. *Pattern Recognition: Human and Mechanical*. New York: John Wiley and Sons, 1985.

Waterman, D. A. and F. Hayes-Roth, eds. *Pattern-Directed Inference Systems*. Out of print.

Watson, J. B. *Behaviorism*. New York: Norton, 1925.

Watson, J. D. *The Double Helix*. New York: Atheneum, 1968.

Watt, Roger. *Understanding Vision*. London: Academic Press, 1991.

Webber, Bonnie Lynn and Nils J. Nilsson, eds. *Readings in Artificial Intelligence*. Los Altos, CA: Morgan Kaufmann, 1981.

Weinberg, Steven. *Dreams of a Final Theory*. New York: Pantheon Books, 1992.

————. *The First Three Minutes: A Modern View of the Origin of the Universe*. New York: Pantheon Books, 1977.

Weiner, Jonathan. *The Next One Hundred Years*. New York: Bantam Books, 1990.

Weinstock, Neal. *Computer Animation*. Reading, MA: Addison-Wesley, 1986.

Weiss, Sholom M. and Casimir A. Kulikowski. *A Practical Guide to Designing Expert Systems*. Totowa, NJ: Rowman and Allanheld, 1984.

Weizenbaum, Joseph. *Computer Power and Human Reason*. San Francisco: W. H. Freeman, 1976.

Werner, Gerhard. "Cognition as Self-Organizing Process." *Behavioral and Brain Sciences* 10, 2:183.

Westfall, Richard. *Never at Rest: A Biography of Isaac Newton*. Cambridge: Cambridge University Press, 1980.

White, K. D. *Greek and Roman Technology*. London: Thames and Hudson, 1984.

Whitehead, Alfred N. and Bertrand Russell. *Principia Mathematica*. 3 vols. Second ed. Cambridge: Cambridge University Press, 1925–1927.

Wick, David. *The Infamous Boundary: Seven Decades of Heresy in Quantum Physics*. Boston: Birkhäuser, 1995.

Wiener, Norbert. *Cybernetics: or Control and Communication in the Animal and the Machine*. Cambridge, MA: MIT Press, 1965.

————. *God and Golem, Inc.: A Comment on Certain Points Where Cybernetics Impinges on Religion*. Cambridge, MA: MIT Press, 1985.

Wills, Christopher. *The Runaway Brain: The Evolution of Human Uniqueness*. New York: Basic Books, 1993.

Winkless, Nels and Iben Browning. *Robots on Your Doorstep: A Book About Thinking Machines*. Portland, OR: Robotics Press, 1978.

Winner, Langdon. *Autonomous Technology: Technics-Out-of-Control as a Theme in Political Thought*. Cambridge, MA: MIT Press, 1977.

Winograd, Terry. *Understanding Computers and Cognition*. Norwood, NJ: Ablex, 1986.

————. *Understanding Natural Language*. New York: Academic Press, 1972.

Winston, Patrick Henry. *Artificial Intelligence*. Reading, MA: Addison-Wesley, 1984.

————. *The Psychology of Computer Vision*. New York: McGraw-Hill, 1975.

Winston, Patrick Henry and Richard Henry Brown, eds. *Artificial Intelligence: An MIT Perspective*. Vol. 1. Cambridge, MA: MIT Press, 1979.

————, eds. *Artificial Intelligence: An MIT Perspective*. Vol. 2. Cambridge, MA: MIT Press, 1979.

Winston, Patrick Henry. and Karen A. Prendergast. *The AI Business: Commercial Uses of Artificial Intelligence*. Cambridge, MA: MIT Press, 1984.

Wittgenstein, Ludwig. *Philosophical Investigations*. Oxford: Blackwell, 1953.

————. *Tractatus Logico-Philosophicus*. London: Routledge and Kegan Paul, 1961.

Yavelow, Christopher. *MacWorld Music and Sound Bible*. San Mateo, CA: IDG Books Worldwide, 1992.

Yazdani, M. and A. Narayanan, eds. *Artificial Intelligence: Human Effects*. Chichester, UK: Ellis Horwood, 1984.

Yovits, M. C. and S. Cameron, eds. *Self-Organizing Systems*. New York: Pergamon Press, 1960.

Zadeh, Lofti. *Information and Control*. Vol 8. New York: Academic Press, 1974.

Zeller, Eduard. *Plato and the Older Academy*. Reprint ed. New York: Russell and Russell, 1962.

Zue, Victor W., Francine R. Chen, and Lori Lamel. *Speech Spectrogram Reading: An Acoustic Study of English Words and Sentences*. Cambridge, MA: MIT Press. Lecture Notes and Spectrograms, July 26–30, 1982.

WEB LINKS

The following is a catalog organized by subject of World Wide Web sites relevant to topics in the book. Remember that compared to books listed in a bibliography, web sites are not nearly as long lasting. These sites were all verified when the book went to press, but inevitably some will become inactive. The Web, unfortunately, is littered with nonfunctioning sites.

SITES RELEVANT TO THE BOOK

Web site for the book *The Age of Spiritual Machines: When Computers Exceed Human Intelligence* by Ray Kurzweil:
<http://www.penguinputnam.com/kurzweil>
To e-mail the author:
raymond@kurzweiltech.com
To download a copy of Ray Kurzweil's Cybernetic Poet:
<http://www.kurzweiltech.com>
This book's publisher, Viking:
<http://www.penguinputnam.com>
For publications of Ray Kurzweil:
Go to <http://www.kurzweiltech.com> or <http://www.kurzweiledu.com> and then select "Publications"

WEB SITES FOR COMPANIES FOUNDED BY RAY KURZWEIL

Kurzweil Educational Systems, Inc. (creator of print-to-speech reading systems for persons with reading disabilities and visual impairment):
<http://www.kurzweiledu.com>
Kurzweil Technologies, Inc. (creator of Ray Kurzweil's Cybernetic Poet and other software projects):
<http://www.kurzweiltech.com>
The dictation division of Lernout & Hauspie Speech Products (formerly Kurzweil Applied Intelligence, Inc.), creator of speech recognition and natural language software systems:

<http://www.lhs.com/dictation/>
The overall Lernout & Hauspie web site:
<http://www.lhs.com/>
Kurzweil Music Systems, Inc., creator of computer-based music synthesizers, sold to
Young Chang in 1990:
<http://www.youngchang.com/kurzweil/index.html>
TextBridge Optical Character Recognition (OCR). Formerly Kurzweil OCR from
Kurzweil Computer Products, Inc. (sold to Xerox Corp. in 1980):
<http://www.xerox.com/scansoft/textbridge/>

ARTIFICIAL LIFE AND ARTIFICIAL INTELLIGENCE RESEARCH

The Artificial Intelligence Laboratory at Massachusetts Institute of Technology (MIT):
<http://www.ai.mit.edu/>
Artificial Life Online:
<http://alife.santafe.edu>
Contemporary Philosophy of Mind: An Annotated Bibliography:
<http://ling.ucsc.edu/~chalmers/biblio.html>
Machine Learning Laboratory, the University of Massachusetts, Amherst:
<http://www-ml.cs.umass.edu/>
The MIT Media Lab:
<http://www.media.mit.edu/>
SSIE 580B: Evolutionary Systems and Artificial Life, by Luis M. Rocha, Los Alamos
National Laboratory:
<http://www.c3.lanl.gov/~rocha/ss504_02.html>
Stewart Dean's Guide to Artificial Life:
<http://www.webslave.dircon.co.uk/alife/intro.html>

ASTRONOMY/PHYSICS

American Institute of Physics:
<http://www.aip.org/history/einstein/>
International Astronomical Union (IAU):
<http://www.intastun.org/>
Introduction to the Big Bang Theory:
<http://www.bowdoin.edu/dept/physics/astro.1997/astro4/bigbang.html>

BIOLOGY AND EVOLUTION

American Scientist Article: Reward Deficiency Syndrome:
<http://www.amsci.org/amsci/Articles/96Articles/Blum-full.html>
Animal Diversity Web Site, the Museum of Zoology at the University of Michigan:
<http://www.oit.itd.umich.edu/projects/ADW/>
Charles Darwin's *Origin of Species*:
<http://www.literature.org/Works/Charles-Darwin/origin/>
Evolution and Behavior:
<http://ccp.uchicago.edu/~jyin/evolution.html>
The Human Genome Project:
<http://www.nhgri.nih.gov/HGP/>
Information Processing in the Human Body:
<http://vadim.www.media.mit.edu/MAS862/Project.html>

Thomas Ray/Tierra:
<http://www.hip.atr.co.jp/~ray/>
The Visible Human Project:
<http://www.nlm.nih.gov/research/visible/visible_human.html>

BRAIN IMAGING RESEARCH

Brain Research Web Page, Jeffrey H. Lake Research:
<http://www.brainresearch.com/>
Applications of brain research:
<http://www.brainresearch.com/apps.html>
Amiram Grinvald's web site: Imaging the Brain in Action:
<http://www.weizmann.ac.il/brain/grinvald/grinvald.htm>
The Harvard Brain Tissue Resource Center:
<http://www.brainbank.mclean.org:8080>
The McLean Hospital Brain Imaging Center:
<http://www.mclean.org:8080/>
Optical Imaging, Inc., Home Page:
<http://opt-imaging.com/>
Research Imaging Center: Solving the Mysteries of the Mind, University of Texas
Health Science Center at San Antonio:
<http://biad63.uthscsa.edu/>
Visualization and Analysis of 3D Functional Brain Images, by Finn Å rup Nielsen, In-
stitute of Mathematical Modeling, Section for Digital Signal Processing, former Elec-
tronics Institute, Technical University of Denmark:
<http://hendrix.ei.dtu.dk/staff/students/fnielsen/thesis/finn/finn.html>
Weizmann Institute of Science:
<http://www.weizmann.ac.il/>
The Whole Brain Atlas:
<http://www.med.harvard.edu/AANLIB/home.html>

COMPUTER BUSINESS/MEDICAL APPLICATIONS

Automated Highway System DEMO; National AHS Consortium Home Page:
<http://monolith-mis.com/ahs/default.htm>
Biometric (The Face Recognition Home Page):
<http://cherry.kist.re.kr/center/html/sites.html>
Face Recognition Homepage:
<http://www.cs.rug.nl/~peterkr/FACE/face.html>
The Intelligent Vehicle Initiative: Advancing "Human-Centered" Smart Vehicles:
<http://www.tfhrc.gov/pubrds/pr97-10/p18.htm>
Kurzweil Educational Systems, Inc.:
<http://www.kurzweiledu.com/>
Kurzweil Music (Welcome to Kurzweil Music Systems):
<http://www.youngchang.com/kurzweil/index.html>
Laboratory for Financial Engineering at MIT:
<http://web.mit.edu/lfe/www/>
Lernout & Hauspie Speech Products:
<http://www.lhs.com/>
Medical Symptoms Matching Software:
<http://www.ozemail.com.au/~lisadev/sftdocpu.htm>

Miros Company Information:
<http://www.miros.com/About_Miros.htm>
Synaptics, Inc.:
<http://www.synaptics.com/>
Systran:
<http://www.systransoft.com/>

COMPUTERS AND ART/CREATIVITY

Arachnaut's Lair - Electronic Music Links:
<http://www.arachnaut.org/music/links.html>
ArtSpace: Computer Generated Art:
<http://www.uni.uiuc.edu/~artspace/compgen.html>
BRUTUS.1 Story Generator:
<http://www.rpi.edu/dept/ppcs/BRUTUS/brutus.html>
But Is It Computer Art?:
<http://www.cs.swarthmore.edu/~binde/art/index.html>
Computer Artworks, Ltd.:
<http://www.artworks.co.uk/welcome.htm>
Computer Generated Writing:
<http://www.notam.uio.no/~mariusw/c-g.writing/>
Northwest Cyberartists: Time Warp of Past Events:
<http://www.nwlink.com/cyberartists/timewarp.html>
Music Software:
<http://www.yahoo.com/Entertainment/Music/Software/>
An OBS Cyberspace Extension of *Being Digital*, by Nicholas Negroponte:
<http://www.obs-us.com/obs/english/books/nn/bdintro.htm>
Ray Kurzweil's Cybernetic Poet:
<http://www.kurzweiltech.com>
Recommended Reading, Computer Art:
<http://ananke.advanced.org/3543/resourcessites.html>
Virtual Muse: Experiments in Computer Poetry:
<http://camel.conncoll.edu/ccother/cohar/programs/index.html>

COMPUTERS AND CONSCIOUSNESS/SPIRITUALITY

Considerations on the Human Consciousness:
<http://www.mediacom.it/~v.colaciuri/consc.htm>
Extropy Online, Arterati on Ideas, by Natasha Vita More; Vinge's View of the Singularity:
<http://www.extropy.com/~exi/eo/articles/vinge.htm>
God and Computers:
<http://web.mit.edu/bpadams/www/gac/>
Kasparov vs. Deep Blue: The Rematch:
<http://www.nytimes.com/partners/microsites/chess/archive8.html>
Online papers on consciousness, compiled by David Chalmers:
<http://ling.ucsc.edu/~chalmers/mind.html>
Toward a Science of Consciousness 1998 "Tucson III," Conference, The University of Arizona, Tucson, Arizona. Support provided by the Fetzer Institute and the Institute of Noetic Sciences:
<http://www.zynet.co.uk/imprint/Tucson/>

COMPUTING SCIENCE RESEARCH

Defining Virtual Reality, Industry Consortium in the Institute for Communication Research, Department of Communication, Stanford University:
<http://www.cyborganic.com/people/jonathan/Academia/Papers/Web/defining-vr.html>
Computer Games: Past, Present, Future:
<http://www.bluetongue.com/~pang/DRAFT.html>
The Haptics Community Web Page:
<http://haptic.mech.nwu.edu>
Modeling and Simulation: Linking Entertainment and Defense:
<http://www.nap.edu/readingroom/books/modeling/index.html>
Physics News Update Number 219—The Density of Data. A link to Lambertus Hesselink's research on crystal computing:
<http://www.aip.org/enews/physnews/1995/split/pnu219-2.htm>
Student cracks encryption code. A link to an article in *USA Today* on how Ian Goldberg, the graduate student from the University of California, cracked the 40-bit encryption code:
<http://www.usatoday.com/life/cyber/tech/ct718.htm>

Autonomous Agents
Agent Web Links:
<http://www.cs.bham.ac.uk/~amw/agents/links/index.html>

Computer Vision
Computer Vision Research Groups:
<http://www.cs.cmu.edu/~cil/v-groups.html>

DNA Computing
"DNA-based computers could race past supercomputers, researchers predict." A link to an article in the *Chronicle of Higher Education* on DNA computing, by Vincent Kiernan:
<http://chronicle.com/data/articles.dir/art-44.dir/issue-14.dir/14a02301.htm>
Explanation of Molecular Computing with DNA, by Fred Hapgood, Moderator of the Nanosystems Interest Group at MIT:
<http://www.mitre.org/research/nanotech/hapgood_on_dna.html>
The University of Wisconsin: DNA Computing:
<http://corninfo.chem.wisc.edu/writings/DNAcomputing.html>

Expert Systems/Knowledge Engineering
Knowledge Engineering, Engineering Management Graduate Program at Christian Brothers University: Online Resources to a Variety of Links:
<http://www.cbu.edu/~pong/engm624.htm>

Genetic Algorithms/Evolutionary Computation
The Genetic Algorithms Archive at the Navy Center for Applied Research in Artificial Intelligence:
<http://www.aic.nrl.navy.mil/galist/>
The Hitchhiker's Guide to Evolutionary Computation, Issue 6.2: A List of Frequently Asked Questions (FAQ), edited by Jörg Heitkötter and David Beasley:
<ftp://ftp.cs.wayne.edu/pub/EC/FAQ/www/top.htm>
The Santa Fe Institute:
<http://www.santafe.edu>

Knowledge Management

ATM Links (Asynchronous Transfer Mode):
<http://www.ee.cityu.edu.hk/~splam/html/atmlinks.html>
Knowledge Management Network:
<http://kmn.cibit.hvu.nl/index.html>
Some Ongoing KBS/Ontology Projects and Groups:
<http://www.cs.utexas.edu/users/mfkb/related.html>

Nanotechnology

Eric Drexler's web site at the Foresight Institute (includes the complete text of *Engines of Creation*):
<http://www.foresight.org/EOC/index.html>
Richard Feynman's talk, "There's Plenty of Room at the Bottom":
<http://nano.xerox.com/nanotech/feynman.html>
Nanotechnology: Ralph Merkle's web site at the Xerox Palo Alto Research Center:
<http://sandbox.xerox.com/nano>
MicroElectroMechanical Systems and Fluid Dynamics Research Group Professor Chih-Ming Ho's Laboratory, University of California at Los Angeles:
<http://ho.seas.ucla.edu/new/main.htm>
Nanolink: Key Nanotechnology Sites on the Web:
<http://sunsite.nus.sg/MEMEX/nanolink.html>
Nanothinc:
<http://www.nanothinc.com/>
NEC Research and Development Letter: A summary of Dr. Sumio Iijima's research on nanotubes:
<http://www.labs.nec.co.jp/rdletter/letter01/index1.html>
An Overview of the Performance Envelope of Digital Micromirror Device (DMD) Based Projection Display System by Dr. Jeffrey Sampsell of Texas Instruments. A link to a paper describing the creation of micromirrors in a tiny, high-resolution projector:
<http://www.ti.com/dlp/docs/it/resources/white/overview/over.shtml>
Small Is Beautiful: A Collection of Nanotechnology Links:
<http://science.nas.nasa.gov/Groups/Nanotechnology/nanotech.html>
Center for Nanoscale Science and Technology at Rice University:
<http://cnst.rice.edu/>
The Smart Matter Research Group, Xerox Palo Alto Research Center:
<http://www.parc.xerox.com/spl/projects/smart-matter/>
Richard Smalley's home page:
<http://cnst.rice.edu/reshome.html>

Neural Implants/Neural Prosthetics

Membrane and Neurophysics Department, the Max Planck Institute for Biochemistry:
<http://mnphys.biochem.mpg.de/>
"Neural Prosthetics Come of Age as Research Continues," by Robert Finn, in the *Scientist*. A link to an article on the use of neural prosthetics in helping patients with neurological disorders:
<http://www.the-scientist.library.upenn.edu/yr1997/sept/research_970929.html>
Physics of Computation–Carver Mead's Group:
<http://www.pcmp.caltech.edu/>

Neural Nets

Brainmaker/California Scientific's home page:
<http://www.calsci.com/>
Hugo de Garis's web site on Brain Builder Group:
<http://www.hip.atr.co.jp/~degaris>
IEEE Neural Network Council Home Page:
<http://www.ewh.ieee.org/tc/nnc/>
Neural Network Frequently Asked Questions:
<ftp://ftp.sas.com/pub/neural/FAQ.html>
PROFIT Initiative at MIT's Sloan School of Management:
<http://scanner-group.mit.edu/>

Quantum Computing

The Information Mechanics Group/Lab for Computer Science at MIT:
<http://www-im.lcs.mit.edu/>
Quantum computation/cryptography at Los Alamos National Laboratory:
<http://qso.lanl.gov/qc/>
Physics and Media Group at the MIT Media Lab:
<http://physics.www.media.mit.edu/home.html>
Quantum Computation at IBM:
<http://www.research.ibm.com/quantuminfo/>

Supercomputers

Accelerated Strategic Computing Initiative:
<http://www.llnl.gov/asci>
Lawrence Livermore National Laboratory/University of California for the U.S. Department of Energy:
<http://www.llnl.gov/>
NEC Begins Designing World's Fastest Computer:
<http://www.nb-pacifica.com/headline/necbeginsdesigningwo_1208.shtml>

FUTURE VISIONS

ACM 97 "The Next 50 Years" (Association for Computing Machinery):
<http://research.microsoft.com/acm97/>
The Extropy Site (a web site and on-line magazine covering a wide range of advanced and future technologies)
<http://www.extropy.org>
SETI Institute web site:
<http://www.seti.org>
WTA: The World Transhumanist Association:
<http://www.transhumanism.com/>

HISTORY OF COMPUTERS

Advances of the 1960s:
<http://www.inwap.com/reboot/alliance/1960s.txt>
BYTE Magazine–December 1996/Cover Story/Progress and Pitfalls:
<http://www.byte.com/art/9612/sec6/art3.htm>

History of Computing: IEEE Computer Society:
<http://www.computer.org/50/>
The Historical Collection, the Computer Museum History Center:
<http://www.tcm.org/html/history/index.html>
Intel Museum Home Page: What is Moore's Law?:
<http://www.pentium.com/intel/museum/25anniv/hof/moore.htm>
SPACEWAR: Fanatic Life and Symbolic Death Among the Computer Bums, by Stewart
Brand:
<http://www.baumgart.com/rolling-stone/spacewar.html>
Timeline of Events in Computer History, from the Virtual History Museum Group:
<http://video.cs.vt.edu:90/cgi-bin/ShowMap>
Chronology of Events in the History of Computers:
<http://www3.islandnet.com/~kpolsson/comphist.htm>
Unisys History Newsletter:
<http//www.cc.gatech.edu/services/unisys-folklore/>

INDUSTRIAL REVOLUTION AND LUDDITES/NEOLUDDITE MOVEMENT

Anarcho-Primitivist, anticivilization, and neo-Luddite articles:
<http://elaine.teleport.com/~jaheriot/anarprim.htm>
What's a Luddite?:
<http://www.bigeastern.com/ludd/nl_whats.htm>
Luddites On-Line:
<http://www.luddites.com/index2.html>
The Unabomber Manifesto by Ted Kaczynski:
<http://www.soci.niu.edu/~critcrim/uni/uni.txt>

INDEX